Two week loan

Benthyciad pythefnos

Marine Science Frontiers for Europe

Springer
Berlin
Heidelberg
New York
Hong Kong
London
Milan
Paris
Tokyo

Gerold Wefer
Frank Lamy
Fauzi Mantoura
Editors

Marine Science Frontiers for Europe

With 120 Figures, 20 in Colour, and 8 Tables

 Springer

Editors:

Professor Dr. Gerold Wefer
Dr. Frank Lamy

Universität Bremen
Fachbereich Geowissenschaften
Klagenfurter Strasse
28359 Bremen
Germany

Profesor Dr. Fauzi Mantoura
IAEA, Marine Environment Laboratory
4, Quai Antoine 1er, B.P. 800
MC 98000 Monaco
Monaco

Photo Credits for Cover: V. Diekamp,
V. Frenzel, F.Garcia Pichel, G. Meinecke

ISBN 3-540-40168-7 Springer-Verlag Berlin Heidelberg New York

Library of Congress Cataloging-in-Publication Data applied for

Bibliograhic information published by Die Deutsche Bibliothek
Die Deutsche Bibliothek lists this publication in the Deutsche Nationalbibliografie; detailed
bibliographic data is available in the Internet at <http://dnb.ddb.de>.

Springer-Verlag Berlin Heidelberg New York
a member of BertelsmannSpringer Science+Business Media GmbH

http://www.springer.de

© Springer-Verlag Berlin Heidelberg 2003
Printed in Germany

Typesetting: Camera-ready
Cover design: E. Kirchner, Heidelberg

Printed on acid-free paper 32/3141/as 5 4 3 2 1 0

Preface

Europe is the continent with a high coast to surface ratio. About 50% of Europe's population lives permanently within 50 km of the coastal zone. European seas encompass a large range of settings and regimes, from estuaries, deltas to the coastal zone, through diverse shelf systems, and down the shelf edge into the deep ocean. The sustainable development of living and non-living marine resources, the protection of the marine environment and the provision of marine-based services are critical to economic prosperity and to the quality of life of European citizens. In recent times, human activity has spread farther out into the oceans as the ocean margins gain increasing attention as potential centers for hydrocarbon exploration and industrial fisheries. As a result of expected population growth over the coming decades, the economic importance of European seas is likely to increase. This generates many conflicts between competing uses, like hydrocarbon exploitation, fisheries, transport, tourism, etc., which has detrimental consequences to the marine environment.

Important tasks for future marine research will be the development of scientific foundations for a sustainable use of the living and non-living resources of the oceans, focusing on the threatened coastal regions and continental slopes. These foundations, however, must be based on thorough scientific understanding of the response of the natural marine systems to human activities, such as global climate change and the local or regional impacts from large-scale construction, resource extraction, etc.

A great challenge for marine and climate research in the future is to understand and model the connections between global climate development and its regional effects on ocean-margin systems, especially in shelf seas and coastal waters. The goal is to increase our ability to predict the evolution of shelf seas and their margins in response to natural or anthropogenic climate changes. This task is accompanied by a need to understand the direct human impacts on the natural system. The lack of knowledge about basic processes significantly hampers the prediction of the immediate impact, and even more its mid- to long-term consequences. Examples of this include coastal construction such as the building of dikes, ports, and pipelines, river deepening and correction, and extraction of mineral resources, steadily altering the coastal environment by changing the current patterns, local sea level, and the erosion and deposition of sediments.

A better understanding of these processes and relationships must form the foundation for the scientific assessment of future threats posed to the European Seas by human activities. Consequently, a quantitative risk assessment and development of strategies for precautionary activities, warning systems, and sustained ocean protection measures will be included among the important tasks for marine scientists in the coming decades, as will the assessment of the sedimentological and ecological consequences of the exploitation of living and non-living resources and the use of the marine environment for waste disposal. Furthermore despite major scientific advances in recent years, the ocean still offers exciting perspectives for frontier research on such fundamental issues as, for example, the origin and evolution of life.

To discuss some of the above outlined questions we brought together experts from various disciplines of the marine sciences with a strong interest in these systems, to promote discussion between workers in different fields by focussing on a common topic of great interest to society.

The meeting which took place in Bremen in Germany from February 18-21, 2001 was arranged in the framework of a "Hanse Conference" within the interdisciplinary program of the Hanse-Wissenschaftskolleg Delmenhorst, a foundation set up to promote interdisciplinary studies in collaboration between the universities of Bremen and Oldenburg. The aim of the Hanse Conferences in general is to provide opportunities for experts from different fields of the sciences and humanities to come together and explore the large framework of topics of common interest. What unites the participants is their desire to look over the fence to neighbouring disciplines. Young colleagues who wish to build an interdisciplinary career are particularly welcome.

In conducting the conference, we have attempted to avoid the disadvantages common to many large scientific meetings characterized by information-overload and lack of time for discussion. Instead, we have loosely followed the model of the "Dahlem Konferenzen". An advisory committee from different disciplines formulated the overall goal and the themes of four discussion groups. This committee was also responsible for producing an initial list of invited participants, a list subsequently expanded through the recommendations of invitees. We aimed for about 40 scientists, complemented by selected postdoctoral researchers. The conference was set for four days. Within each of the four theme sections, several participants were asked to provide background papers in their fields, as a basis for discussion. The aim is to have these papers sent as drafts to all participants before the conference, to stimulate the formulation of questions and critical comments.

The focus of activity within the Hanse Conferences is discussion, and not presentation of talks. The participants come with the background knowledge acquired through the study of the overview papers prepared for the conference. On the first day of the conference each of the four discussion groups agree on an agenda of topics derived from the questions and comments that arise from the study of the background papers. The following two days are dedicated to debating these topics, within the four discussion groups. On the fourth day, each group reviews a summary prepared by its rapporteur, who presents the most important results of the discussions. Suggestions for modifications to the summary are incorporated into the final summary, which is presented at the end of the conference in a plenary session by each of the rapporteurs. At this meeting, comments are invited by all participants on any of the points raised.

The final proceedings, which are published in this book cover important aspects of science frontiers in European marine research and are structured into the four thematic sections discussed during the conference. These include "Ocean-Climate Coupling, Variability and Change", "Coupled Biogeochemical Cycling and Controlling Factors", "Coastal and Shelf Processes, Science for Integrated Management", and "Ecosystem Functioning and Biodiversity". The final report from each group follows each of the thematic sections, which contain the revised background papers. All papers benefited from peer review. It is hoped that they will be useful in informing the ongoing discussions on preservation, exploration, exploitations, and risk assessment of the ocean wherever such debate may take place. We especially hope that high-school and college teachers find much material in these proceedings to enrich their courses in environmental sciences. In the educational realm, a marriage between physical understanding of the Earth's life support systems and an appreciation of history leading to responsibility will be necessary to provide the basis for political action which can deal with the challenge of the sustainable use of delicate marine ecosystems.

The Hanse Conference on "Marine Science Frontiers for Europe" was planned as part of the development of a Marine Scienc Plan for Europe, to be drafted by the Marine Board of the European Science Foundation (ESF) (Integrated Marine Science in Europe, ESF Marine Board Position Paper 5, November 2002). The final group reports formulate recommendations for future ocean margin research within an European Research Area under a joint European Marine Science Plan.

The papers benefited from detailed reviews by Avan Antia, Wolfgang Balzer, Wolfgang Berger, Keith Brander, Angelika Brandt, Peter Burbridge, Hein de Baar, Jan Willem de Leeuw, Jean-Pierre Gattuso, Gerd Graf, Carlo Heip, Venugopalan Ittekkot, Michel Kaiser, Remi Laane, Han Lindeboom, Fauzi Mantoura, Jochem Marotzke, John Patching, David Prandle, Karsten Reise, Will de Ruijter, Ulrich Saint-Paul, Jun She, Victor Smetacek, Doug Wallace, Paul Wassmann, Matthias Wolff.

The Hanse Conference was supported by the Hanse Wissenschaftskolleg Delmenhorst, the DFG Research Center for Ocean Margins (RCOM) as well as by the University of Bremen. Technical support was given by Barbara Donner, Adelheid Grimm-Geils, Christina Hayn, Ingeborg Mehser, Sarah Middendorf, and Lauraine Panaye. To each and all of those involved, our sincere thanks.

Gerold Wefer, Bremen
Frank Lamy, Bremen
Fauzi Mantoura, Plymouth

Bremen, December 2002

Contents

Ecosystem Functioning and Biodiversity

Tropical Pacific Influences on the North Atlantic
Thermohaline Circulation

M. Latif

Max-Planck-Institut für Meteorologie, Bundesstr. 55, D-20146 Hamburg, Germany
corresponding author (e-mail): latif@dkrz.de

Abstract: Most global climate models simulate a weakening of the North Atlantic thermohaline circulation (THC) in response to enhanced greenhouse warming. Both surface warming and freshening in high latitudes, the so-called sinking region, contribute to the weakening of the THC. Some models simulate even a complete breakdown of the THC at sufficiently strong forcing. Here results from a state-of-the-art global climate model are presented that does not simulate a weakening of the THC in response to greenhouse warming. Large-scale air-sea interactions in the tropics, similar to those operating during present-day El Niños, lead to anomalously high salinities in the tropical Atlantic. These are advected into the sinking region, thereby increasing the surface density and compensating the effects of the local warming and freshening. The results of the model study are corroborated by the analysis of observations.

Introduction

The Atlantic thermohaline circulation (THC) is an important component of the global climate system (Broecker 1991). It transports about 1 PW of heat poleward in the North Atlantic, thereby warming western Europe. The THC is forced partly by convection at high latitudes, which causes dense surface waters to sink to deeper ocean layers, forming the so-called North Atlantic Deep Water (NADW). Strong and rapid changes in the intensity of the NADW formation have been reported from paleoclimatic records (Broecker et al. 1985), and it is well established that such changes exert a strong impact on the climate over large land areas (e.g. Manabe and Stouffer 1995 and 1999; Schiller et al. 1997). Several papers have suggested that the thermohaline circulation may weaken in response to greenhouse warming (e.g. Mikolajewicz et al. 1990; Manabe et al. 1991; Stocker and Wright 1991; Cubasch et al. 1992; Manabe and Stouffer 1994; Rahmstorf 1999; Wood et al. 1999). Here, we investigate the sensitivity of the THC to greenhouse warming using a global climate model that has been applied in various climate variability and response studies (Roeckner et al. 1996; Bacher et al. 1997; Oberhuber et al. 1998; Timmermann et al. 1999; Roeckner et al. 1999; Christoph et al. 1998; Ulbrich and Christoph 1999). It is shown that the tropical feedbacks associated with an increased El Niño frequency can stabilise the THC (Latif et al. 2000).

Model

The model (ECHAM4/OPYC) is flux-corrected (using annual mean corrections) with respect to heat and freshwater. The application of flux correction reflects serious errors in the individual model components. The effects of flux correction on the response characteristics of a model are largely unknown, but it has been shown that flux correction may cause misleading results (e.g. Neelin and Dijkstra 1995). It should be noted, however, that many studies addressing the stability of the THC have been conducted with flux-corrected models (e.g. Manabe and Stouffer 1994).

Our model employs a horizontal resolution over most of the globe of 2.8°x2.8°. A special feature of the model is the higher meridional resolution of 0.5°

From WEFER G, LAMY F, MANTOURA F (eds), 2003, *Marine Science Frontiers for Europe.* Springer-Verlag Berlin Heidelberg New York Tokyo, pp 1-9

used in the tropical oceans, which enables a realistic simulation of the El Niño/Southern Oscillation (ENSO) phenomenon (Roeckner et al. 1996; Bacher et al. 1998; Oberhuber et al. 1998), the strongest natural interannual climate fluctuation. We performed two integrations. The first experiment is a 240-year long control integration with fixed present-day concentrations of greenhouse gases. In the second experiment the model was forced by increased greenhouse gas concentrations, giving approximately the historical increase in radiative forcing from 1860 to 1990 (Roeckner et al. 1999), and subsequently (up to 2100) by increases according to IPCC scenario IS92a (IPCC 1992). The effects of anthropogenic sulfate emissions are not included in the simulation.

Results

The control run simulates the THC in the North Atlantic reasonably well (Zhang et al. 1998), with a maximum overturning of about 23 Sv at a depth of about 2000 m and an NADW outflow at 30°S of about 15 Sv ($1Sv=10^6m^3/s$) at a depth of about 1500 m. However, the model fails to simulate the inflow of the very dense Antarctic Bottom Water (AABW) into the North Atlantic. The THC is relatively stable during the control integration, but exhibits some superimposed interdecadal variability (Fig. 1). The range of this internal THC variability is consistent with that simulated by other global climate models (Delworth et al. 1993; Timmermann et al. 1998). In the transient greenhouse warming simulation the THC remains also stable, with a slight decrease during the first half and a slight increase during the second half. Thus, our model predicts a stable THC rather than a weakening THC as simulated by most other global models (Rahmstorf 1999).

The physics responsible for the stabilisation of the THC in our model is related to tropical air-sea interactions. The regional distribution of the simulated sea surface temperature (SST) trend is characterised by warming in most ocean areas, with some cooling tendency in certain regions around Antarctica (not shown, see also Roeckner et al. 1999). Of particular importance is the El Niño-like warming trend in the eastern equatorial Pacific. As has been described in Timmermann et al. (1999),

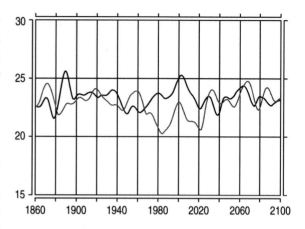

Fig. 1. Time series of the maximum overturning [Sv] in the North Atlantic for the control integration with present-day greenhouse gas concentrations (black line) and the transient greenhouse warming simulation (red line).

this warming results from air-sea interactions similar to those producing present-day El Niños and has strong impacts on the entire tropical climate system. The model simulates strongly enhanced precipitation over the central equatorial Pacific (Fig. 2), a feature observed during present-day El Niños (Philander 1990; Ropelewski et al. 1987). This leads to adiabatic warming and drying through anomalous subsidence over Northeast Brazil and the tropical Atlantic, thereby reducing precipitation and river runoff from the Amazon catchment area and increasing evaporation from the tropical Atlantic. The model simulates an additional freshwater export from the Atlantic to the Pacific of about 0.015 Sv/decade, in comparison to only 0.005 Sv/decade in another Max-Planck-Institute coupled model (ECHAM-3/LSG) simulating a weakening of the THC (Mikolajewicz and Voss 2000) as discussed by Rahmstorf (1999) (his Fig. 2). Such an enhanced freshwater export is also observed during present-day El Niños (Schmittner et al. 2000).

The reduced freshwater flux induces anomalously high salinities in the tropical Atlantic Ocean (Fig. 3b). We averaged the freshwater fluxes over two regions (30°S-45°N and 45°N-90°N) and display their time evolutions in Fig. 4a. The freshwater input to the Atlantic towards the end of the integration is strongly reduced by about 0.3 Sv in the

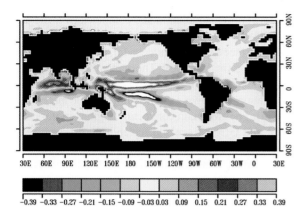

Fig. 2. Centennial linear trend in the freshwater flux (P-E+R, precipitation minus evaporation plus river runoff) [mm/(d*decade)] obtained from the transient greenhouse warming simulation and computed over the period 2000-2100. Note the freshening in the equatorial Pacific associated with the increased El Niño frequency and the strongly reduced freshwater influx over the tropical Atlantic.

tropical region, while it is enhanced by about 0.1 Sv in the middle and high latitudes. Thus, the changes in the freshwater flux will tend to enhance the salinity in the tropics and to reduce the salinity in the middle and higher latitudes, with changes in the tropics being considerably stronger in the tropics. We time-integrated the anomalous freshwater flux forcing into the Atlantic and Arctic poleward of 30°S. This quantity has been compared to the anomalous salt content of the upper 375m at the end of the integration. The actual salt gain of the upper Atlantic amounts to about two third of the total implied freshwater induced salt gain. We do not expect a perfect match, since the integration domain is not closed. However, this computation shows that the salinity anomalies can easily be explained by the changes in atmospheric moisture transport.

The poleward transport of the salt anomaly and the mixing within the subtropical gyre result in increased surface salinity over the entire North Atlantic Ocean. The poleward propagation of the anomalously high salinities is visualised by a Hovmoeller diagram showing the temporal evolution of the salinity anomalies averaged over the upper 375m in a meridional section along 45°W

(Fig. 4b). The salinity transport outweighs the counteracting local freshwater influx through enhanced precipitation (Fig. 2, Fig. 4a) and warming in the high latitudes of the North Atlantic (Fig. 3a), which are the principal feedbacks in models predicting a reduced THC.

We computed separately the salt and temperature contributions in the change of the surface density, which is the crucial quantity for deep water formation. The linear expansion coefficients for tempertaure and salinity for a temperature of 5°C and salinity of 35 psu, which are characteristic values for the sinking region, amount to -0.12 (kg/m³)/K and 0.79 (kg/m³)/psu, respectively. Given an SST change of about 0.3°C/decade and a salinity change of about 0.05 psu/decade (Figs. 3a,b) yields comparable density changes of about 0.4 kg/(m³*decade). A calculation using the complete equation of state yields a similar result, with almost no density change in the regions of strongest deep water formation (Fig. 3c). Thus, in our transient greenhouse warming simulation the poleward salt transport is an important feedback process, in addition to the freshening through increased precipitation and warming in the high latitudes, and the net outcome is a stable rather than a weakening THC.

Are the changes in the hydrological cycle large enough to have a significant effect on the THC? Typical threshold values for a complete collapse of the present-day THC are in the range of 0.1-0.4 Sv for anomalous freshwater input into the North Atlantic (e.g. Stocker and Wright 1991; Rahmstorf 1995; Schiller et al. 1997). Given a change of the order of -0.3 Sv in our simulation, it is not surprising that this effect is large enough to balance the locally induced weakening of the THC.

Further modeling and observational evidence

Next, simulations with an atmospheric general circulation model (AGCM) and some observations are analysed (Latif 2001). Unfortunately, long time series of the freshwater flux over the Atlantic are not available from observations. Instead AGCM simulations with observed SSTs prescribed globally for several decades were used to study further the connections between the tropical Pacific and

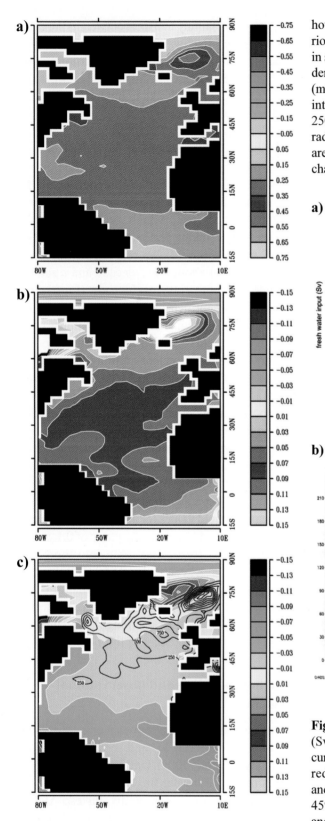

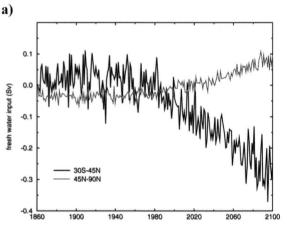

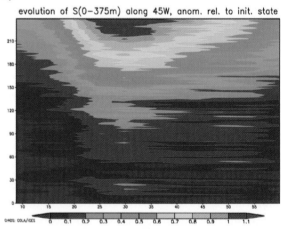

house warming simulation and computed over the period 2000-2100. **a)** Trend in SST [°C/decade], **b)** trend in sea surface salinity [psu/decade], **c)** trend in surface density [kg/ (m³*decade)]. The depth of the convection (m) simulated for the month of February in the control integration is shown as contours (contour interval is 250 m). The two main convection regions in the Labrador and GIN (Greenland-Iceland-Norwegian) Seas are simulated realistically by the model. No significant change in the density is simulated in these regions.

a)

b)

evolution of S(0–375m) along 45W, anom. rel. to init. state

Fig. 4. a) Time series of anomalous freshwater input (Sv) into the Atlantic in the region 30°S-45°N (black curve) and into the mid- and high latitudes (45°N-90°N, red curve). **b)** Temporal evolution of the salinity anomalies [psu] averaged over the upper 375m along 45°W. The northward propagation of the salinity anomalies from the subtropics to higher latitudes can be clearly seen by the tilt of the contours. The anomalies are calculated by subtracting the initial values.

Fig. 3. Centennial linear trends of selected quantities in the North Atlantic obtained from the transient green-

tropical Atlantic Oceans. The tropical atmosphere is highly predictable (e.g. Lau 1985), and it is believed that such simulations provide useful insights about the nature of the response of the tropical atmosphere to tropical Pacific SST anomalies associated with ENSO-type multi-decadal variability. The AGCM used here is ECHAM4 (Roeckner et al. 1996) with T42 resolution (2.8° x 2.8°), a model that has been used in many climate applications (see e.g. Roeckner et al. 1999). In particular, ECHAM4 served as the atmospheric component of the coupled model ECHAM4/OPYC described above. An ensemble of three AGCM integrations was performed for the period 1903-1994, and the ensemble mean is shown here. The SSTs used to drive the model are the GISST SSTs (Parker et al. 1995). The Kaplan (1997) dataset is used in the investigation of the SSTs themselves shown below, since it is the longer dataset. However, since the response characteristics of the model do not depend on the forcing SSTs, the use of the different SST datasets in this study is justified.

The atmosphere model simulates an out-of-phase relationship between decadal fluctuations in eastern tropical Pacific (Niño-3) SST and the freshwater flux over the tropical Atlantic (10°S-30°N) as shown in Fig. 5, confirming the observational results of Schmittner et al. (2000). The correlation between the two time series amounts to -0.75. Furthermore, both the tropical Pacific SST and the simulated freshwater flux exhibit rather strong trends during the last 50 years: While the tropical Pacific SST is slowly increasing by about 0.5°C, the freshwater flux over the tropical Atlantic is slowly decreasing by about 0.06 Sv (10^6m³/s) (Fig. 5). This behaviour is consistent with that found in the greenhouse warming simulation described above and supports the picture that there exists a potential "atmospheric bridge", by which the tropical Pacific and Atlantic Oceans can interact with each other. Further evidence for the existence of the atmospheric bridge comes from salinity observations taken at Bermuda (33°N, 65°W). A strong increase in the surface salinity is observed during the last few decades (Fig. 6), which is expected given the strong increase in tropical Pacific SST and the decrease of tropical Atlantic (model) freshwater flux during the recent

decades. These trends may continue in response to greenhouse warming as hypothesised above.

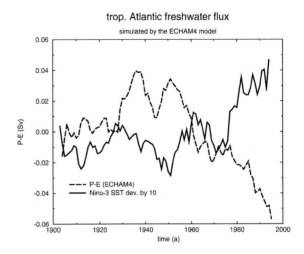

Fig. 5. Time series of the Niño-3 SST anomaly and the anomalous tropical Atlantic freshwater flux as simulated by the ECHAM4 model averaged over the region 10°S-30°N, which is the region with the most consistent response over the Atlantic. Both time series vary out of phase and exhibit strong trends during the last 50 years. Both time series were filtered with a 11-year running mean.

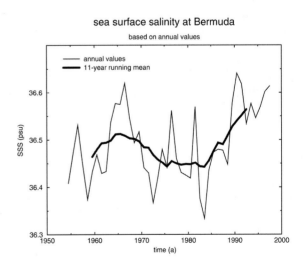

Fig. 6. Time series of the salinity observed at Bermuda (33°N, 65°W). Please note the strong increase of the salinity during the most recent decades.

In the next step, it is investigated whether the THC in the Atlantic responds to the variations in the freshwater flux. This can be done only indirectly by using SSTs, since the required direct observations of the freshwater flux, surface salinity, and the THC do not exist. Ocean model and coupled model studies have shown that variations in the THC are associated with changes in the poleward ocean heat transport and an interhemispheric SST dipole (see e.g. Manabe and Stouffer 1999 and references therein). Thus, this characteristic SST anomaly pattern can be used as a "fingerprint" to identify variations in the real THC. A correlation analysis was conducted using the Kaplan SST dataset. First, the zero-lag correlation of the low-pass filtered tropical Pacific (Niño-3) SST anomaly time series with the global SST anomaly field has been computed. The resulting correlation pattern

(Fig. 7a) shows the well known picture in the tropical Pacific, with an El Niño-like positive signal in the eastern and central equatorial Pacific and negative anomalies in the mid-latitudes of both hemispheres. A teleconnection to the Indian Ocean can be seen also, a feature known from present-day El Niño events. Strong negative correlations are found in the North Atlantic near Greenland, which is consistent with the recent increase of the North Atlantic Oscillation.

Based on the arguments described above, the time it will take to develop sufficiently strong salinity anomalies and to transport those poleward into the so called "sinking region" in the high latitudes is of the order of a few decades. Therefore, a time lag of 30 years has been introduced in the correlation analysis. The results of this lagged correlation analysis (Fig. 7b) exhibit strongest signals in the

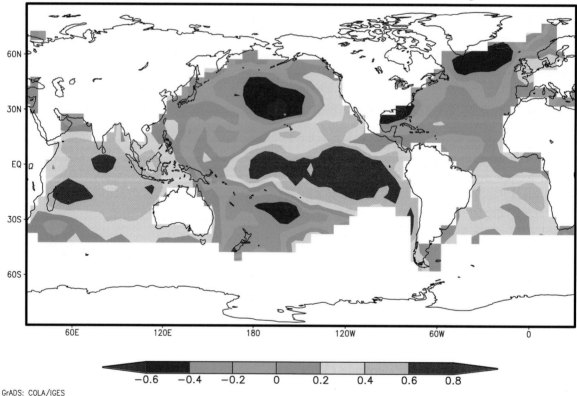

GrADS: COLA/IGES

Fig. 7. a) Spatial distribution of correlation coefficients between the Niño-3 SST anomaly time series and the global SST anomalies at zero lag.

Atlantic Ocean, although a tropical Pacific SST index has been used as reference. The SST anomaly pattern is the interhemispheric dipole identified in model studies to go along with changes in the THC. Anomalously warm temperatures are found in the northern part and anomalously cold temperatures in the southern part of the Atlantic Ocean. Thus, the results indicate that interdecadal changes in tropical Pacific SST are followed by basin-wide changes in Atlantic SST, and that periods of high (low) SSTs in the eastern tropical Pacific are followed by a strong (weak) THC in the Atlantic. The statistical significance of the results was tested using a t-test and assuming 10 degrees of freedom, and correlations above about 0.5 are significant at the

95% level. More important than the level of the significance, however, is the fact that physically motivated SST anomaly patterns result from the two correlation analyses.

Concluding remarks

The tropical air-sea interactions that stabilise the thermohaline circulation in our model simulation are not adequately represented in most global climate models applied hitherto to the problem of anthropogenic climate change, mainly because of their too coarse resolution in the tropics. This feedback needs to be studied more carefully in relation to the destabilising feedbacks considered in previous in-

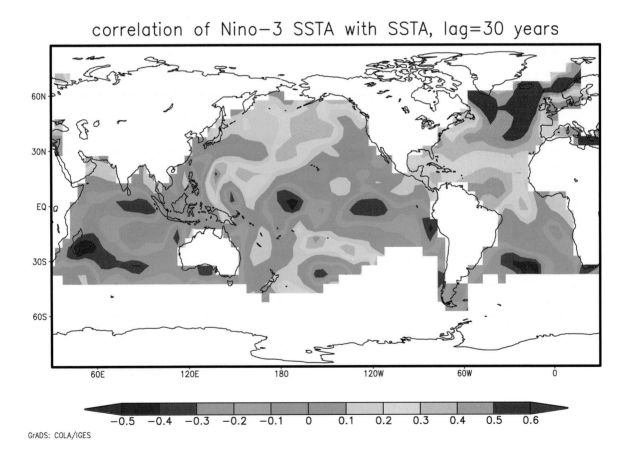

Fig. 7. b) Spatial distribution of correlation coefficients between the Niño-3 SST anomaly time series and the global SST anomalies at lag 30 years. The anomaly structure in **b)** is reminiscent of variations in the THC, indicating that variations in the THC follow variations in tropical Pacific SST with a time lag of 30 years. The data were low-pass filtered with a 11-year running mean prior to the correlation analyses.

vestigations. At this stage of analysis we can conclude only that the response of the THC to enhanced greenhouse warming is still an open question. The uncertainty arises from model shortcomings in the representation of key physical processes, such as the hydrological cycle. Our results may be important also in view of multi-decadal and paleoclimatic variations, because they indicate that variations in the tropics may synchronise changes globally. The hypothesis of the tropical influence on the THC is very difficult to verify by observations. There is, however, some observational evidence that multi-decadal variations in the tropical Pacific SST feed back on the North Atlantic THC.

References

Bacher A, Oberhuber JM, Roeckner E (1997) ENSO dynamics and seasonal cycle in the tropical Pacific as simulated by the ECHAM4/OPYC3 coupled general circulation model. Climate Dyn 14:431-450

Broecker WS (1991) The great ocean conveyor. Oceanography 4: 79-89

Broecker WS, Peteet DM, Rind D (1985) Does the ocean-atmosphere system have more than one stable mode of operation? Nature 315:21-26

Christoph M, Barnett TP, Roeckner E (1998) The Antarctic Circumpolar Wave in a Coupled Ocean-Atmosphere GCM. J Climate 11:1659-72

Cubasch U, Hasselmann K, Höck H, Maier-Reimer E, Mikolajewicz U, Santer BD, Sausen R (1992) Time-dependent greenhouse warming computations with a coupled ocean-atmosphere model. Climate Dyn 8:55-69

Delworth T, Manabe S, Stouffer RJ (1993) Interdecadal variations of the thermohaline circulation in a coupled ocean-atmosphere model. J Climate 6: 1993-2011

IPCC, Climate Change (1992) The Supplementary Report to the IPCC Scientific Assessment. Edited by JT Houghton, BA Callander and SKV Varney. Cambridge University Press 200 pp

Kaplan A, Cane MA, Kushnir Y, Clement AC, Blumenthal MB, Rajagopalan B (1997) Analyses of global sea surface temperature 1856-1991. J Geophys Res 102:27835-27860

Lau N-C (1985) Modeling the seasonal dependence of the atmospheric responses to observed El Niños 1962-1976. Mon Wea Rev 113:1970-1996

Latif M (2001) Tropical Pacific/Atlantic Ocean Interactions at multi-decadal time scale. Geophys Res Lett 28:539-542

Latif M, Roeckner E, Mikolajewicz U, Voss R (2000) Tropical stabilisation of the thermohaline circulation in a greenhouse warming simulation. J Climate 13:1809-13

Manabe S, Stouffer RJ (1994) Multiple-century response of a coupled ocean-atmosphere model to an increase of atmospheric carbon dioxide. J Climate 7:5-23

Manabe S, Stouffer RJ (1995) Simulation of abrupt climate change induced by freshwater input to the North Atlantic Ocean. Nature 378:165-167.

Manabe S, Stouffer RJ (1999) The role of thermohaline circulation in climate. Tellus 51:91-109

Manabe S, Stouffer R, Spelman M, Bryan K (1991) Transient responses of a coupled ocean-atmosphere model to gradual changes of atmospheric CO_2. Part I: Annual mean response. J Climate 4:785-818

Mikolajewicz U, Voss R (2000) The role of the individual air-sea flux components in CO_2-induced changes of the ocean's circulation and climate. Climate Dyn 16: 627-642

Mikolajewicz U, Santer BD, Maier-Reimer E (1990) Ocean response to greenhouse warming. Nature 345:589-593.

Neelin JD, Dijkstra HA (1995) Ocean-atmosphere interaction and the tropical climatology. Part I: The dangers of flux correction. J Climate 8:1325-1342

Oberhuber JM, Roeckner E, Christoph M, Esch M, Latif M (1998) Predicting the '97 El Niño event with a global climate model. Geophys Res Lett 25:2273-2276

Parker DE, Jackson M, Horton EB (1995) The GISST 2.2 sea surface temperature and sea ice climatology. Climate Research Technical Note 63, Hadley Centre, Meteorological Office, Bracknell, UK, 35 pp

Philander SGH (1990) El Niño, La Niña, and the Southern Oscillation. Academic Press, San Diego, 293 pp

Rahmstorf S (1995) Bifurcations of the Atlantic thermohaline circulation in response to changes in the hydrological cycle. Nature 378:145-149

Rahmstorf S (1997) Risk of sea-change in the Atlantic. Nature 388:825-826

Rahmstorf S (1999) Shifting seas in the greenhouse? Nature 399:523-524

Roeckner E, Oberhuber JM, Bacher A, Christoph M, Kirchner I (1996) ENSO variability and atmospheric response in a global atmosphere-ocean GCM. Climate Dyn 12:737-754

Roeckner E, Bengtsson L, Feichter J, Lelieveld J, Rodhe H (1999) Transient climate change simulations with a coupled atmosphere-ocean GCM including the troposheric sulfur cycle. J Climate 12:3004-3032

Ropelewski CF, Halpert M (1987) Global and regional

scale precipitation patterns associated with the El Niño/Southern Oscillation. Mon Wea Rev 115:1606-1627

Schiller A, Mikolajewicz U, Voss R (1997) The stability of the thermohaline circulation in a coupled ocean-atmosphere model. Climate Dyn 13:325-348

Schmittner A, Appenzeller C, Stocker TF (2000) Enhanced Atlantic freshwater export during El Niño. Geophys Res Lett 27:1163-1166

Stocker TF, Wright DG (1991) Rapid transitions of the ocean's deep circulation induced by changes in surface water fluxes. Nature 351:729-732

Timmermann A, Latif M, Voss R, Groetzner A (1998) Northern Hemisphere interdecadal variability: A coupled air-sea mode. J Climate 11:1906-1931

Timmermann A, Oberhuber J, Bacher A, Esch M, Latif M, Roeckner E (1999) Increased El Niño frequency in a climate model forced by future greenhouse warming. Nature 398:694-697

Ulbrich U, Christoph M (1999) A shift of the NAO and increasing storm track activity over Europe due to anthropogenic greenhouse gas forcing. Climate Dyn 15, 7:551-559

Wood RA, Keen AB, Mitchell JF, Gregory JM (1999) Changing spatial structure of the thermohaline circulation in response to atmospheric CO_2 forcing in a climate model. Nature 399:572-575

Zhang X-H, Oberhuber J, Bacher A, Roeckner E (1998) Interpretation of interbasin exchange in an isopycnal ocean model. Climate Dyn 14:725-740

Climate Records from Corals

T. Felis[*] and J. Pätzold

*Universität Bremen, Fachbereich Geowissenschaften, Klagenfurter Straße,
28359 Bremen, Germany*
* corresponding author (e-mail): tfelis@uni-bremen.de

Abstract: In many regions instrumental climate records are too short to resolve the full range of decadal- to multidecadal-scale natural climate variability. Massive annually banded corals from the tropical and subtropical oceans provide a paleoclimatic archive with a seasonal resolution, documenting past variations in water temperature, hydrologic balance, and ocean circulation. Recent coral-based paleoclimatic research has focused mainly on the tropics, providing important implications on the past variability of the El Niño-Southern Oscillation (ENSO) phenomenon and decadal tropical climate variability. However, new records from some of the rare subtropical/mid-latitude locations of coral growth were shown to reflect aspects of dominant modes of Northern Hemisphere climate variability, e.g. the North Atlantic Oscillation (NAO). This natural mode has important socio-economic impacts owing to its large-scale modulation of droughts, floods, storms, snowfall, and fish stocks at timescales relevant to society. Coral records extending over several centuries from key locations (e.g. northern Red Sea, Bermuda) provide the opportunity to assess recent shifts in the NAO with respect to the natural variability of the pre-instrumental period. Providing a better understanding of NAO dynamics, such paleoclimatic records, together with those derived from other paleoclimatic archives, are essential for the predictability of future European climate.

Introduction

Instrumental climate records are too short to resolve the full range of decadal- to multidecadal-scale natural climate variability. Banded corals, tree rings, ice cores, and varved sediments provide paleoclimatic archives which can be used to reconstruct past climate variability in the pre-instrumental period in annual resolution. These proxy climate indicators provide paleoclimatic records which are important for the assessment of perturbations to the natural climate variability by anthropogenic forcing, for climate predictability and for a better understanding of the dominant modes of the global climate system, e.g. the El Niño-Southern Oscillation (ENSO) phenomenon of tropical Pacific origin, the Asian and African monsoon, the North Atlantic Oscillation (NAO), the Pacific Decadal Oscillation (PDO), and the mechanisms of decadal climate variability. These natural modes have important socio-economic effects owing to their large-scale modulation of droughts, floods, storms, snowfall, or fish stocks at timescales relevant to society.

Massive "stony" (scleractinian) corals from the modern and fossil reefs of the tropical and subtropical oceans provide an important archive of past climate and ocean variability. These corals build skeletons of aragonite ($CaCO_3$) and grow at rates of millimeters to centimeters per year. During growth, annual density bands are produced in the skeleton that can be used for the development of chronologies. As corals grow they incorporate isotopic and elemental tracers reflecting the environmental conditions in the ambient seawater during skeleton secretion, e.g. water temperature, hydrologic balance (evaporation, precipitation, runoff), and ocean circulation. Compared to other paleoclimatic archives corals provide a clear seasonal resolution. Modern corals from living reefs provide continuous climate

From WEFER G, LAMY F, MANTOURA F (eds), 2003, *Marine Science Frontiers for Europe.* Springer-Verlag Berlin Heidelberg New York Tokyo, pp 11-27

records extending several centuries back from the present. Well-preserved fossil corals from emerged or submerged reef terraces provide information on climate variability during time-windows throughout the late Quaternary. The most commonly used corals in paleoclimatology are those of the genus Porites which are ideally suited for sub-annual sampling owing to their dense skeletons and rapid growth rates (about 1 cm per year).

Most reef-building (hermatypic) corals live in the upper ~40 m of the ocean where there is sufficient light for the photosynthetic activity of the coral's endosymbiotic algae (zooxanthellae). Furthermore the development of coral reefs is restricted to warm water temperatures. Most corals are located in regions where mean annual temperatures are around 24°C and/or mean winter minimum temperatures are not below 18°C, i.e. roughly between 23-24°N and S latitude. Therefore recent coral-based paleoclimatic research has focused mainly on the tropics providing important implications on past variability of the ENSO phenomenon and decadal tropical climate variability (e.g. Cole et al. 1993; Charles et al. 1997; Cole et al. 2000; Urban et al. 2000). However, some ocean currents transport warmer tropical waters to higher latitudes leading to coral growth also at some rare subtropical/mid-latitude locations. These locations have provided the opportunity for coral records from up to ~29°S in the southeastern Indian Ocean (Kuhnert et al. 1999); ~32°N in the North Atlantic (Pätzold and Wefer 1992; Pätzold et al. 1999; Berger et al. 2002), and ~28°N in the northern Red Sea (Felis et al. 2000). These paleoclimatic records were generated from living corals covering the past centuries. In contrast, fossil corals have revealed important aspects on climate variability during time-windows of up to decades length throughout the Holocene (Beck et al. 1997), especially the Mid-Holocene (Gagan et al. 1998; Corrège et al. 2000; Moustafa et al. 2000), and the last interglacial warm period (Hughen et al. 1999).

Coral-based paleoclimatic records from Bermuda and the northern Red Sea were shown to reflect aspects of dominant modes of Northern Hemisphere climate variability, in particular the oceanic (Bermuda) and atmospheric (northern Red Sea) signature of the NAO (Pätzold et al. 1999; Felis et al. 2000). The NAO has a strong influence on large-scale variations in the atmospheric circulation over the North Atlantic and its surrounding continents, controlling European winter climate (Hurrell 1995). Coral records of several centuries length from such key locations provide the opportunity to assess recent shifts in the NAO with respect to the natural variability of the pre-instrumental period. Providing a better understanding of NAO dynamics, such paleoclimatic records, together with those derived from tree rings, ice cores, and varved sediments, are essential for the predictability of future European climate.

An additional source of annual- to seasonal-resolution paleoclimatic records from the marine environment with relevance to European climate are the shells of the long-lived mollusc *Arctica islandica* (Weidman et al. 1994; Marchitto Jr. 2000). This annually-banded bivalves inhabit the continental shelves and slopes of the northern North Atlantic covering a geographic region where no warm-water corals grow. Individual *Arctica islandica* specimens are commonly living for about 100 years. Therefore the splicing of several accurately dated fossil shell records has the potential to provide continuous or near-continuous composite records covering the past 1000 years (Weidman and Goodfriend 1999).

In summary, coral records are successfully used since nearly a decade for seasonal-resolution reconstructions of tropical climate and ocean variability during the past centuries but also during periods of time with boundary conditions different from today. Future research should exploit the potential of corals from rare subtropical/mid-latitude locations (e.g. Bermuda, northern Red Sea) to better understand past mid-latitude climate variability which is important with respect to present and future European climate.

Annual density banding and age model

As corals grow, new skeleton is generated within the living tissue layer which always remains as a thin band (of several millimeters width) at the outermost surface of a colony. Centuries-old coral colonies can become several meters high considering typical growth rates for *Porites* of about 1 cm per year. Such large living corals are usually sampled by drilling a core vertically from the top to the bottom of a colony along the major axis of growth.

In general, coral skeletons reveal a density banding pattern of alternating bands of high- and low density, with each year being represented by a pair of such bands. The density variations result from changes in a coral's rate of calcification and/ or growth. The preliminary age model of a coral chronology is usually based on counting the annual density-band pairs. The counting starts at the top of a coral core within the tissue layer whose age is known from the date of collection of the core, provided a living colony was drilled. The preliminary age model based on banding is then refined using the seasonal cyclicity of isotopic or elemental tracers in the coral skeleton that reflects the seasonal cycle of temperature or light.

Well-preserved fossil corals can be dated either using radiocarbon or the ^{230}Th/^{234}U method. The ^{230}Th/^{234}U method is applied to date corals which are older than 30,000 to 45,000 years which is the limit of the ^{14}C method. Furthermore, ^{230}Th/^{234}U ages can be considered as absolute ages compared to ^{14}C ages, which are influenced by ocean reservoir effects and variations in the ^{14}C level of the atmosphere. The dating of a fossil coral provides a floating chronology for which top and bottom ages are known only approximately.

Climate tracers in corals

Annual growth rates

The annual growth rates of corals can be inferred from the annual density-band pattern and can provide a paleoclimatic tracer. Coral growth rates can reflect several environmental parameters such as temperature, nutrient or food availability, water transparency, and sediment input. An 800-year record of coral growth from Bermuda in the North Atlantic was shown to reflect temperature variability, probably as a result of enhanced coral growth during periods of increased wind-induced vertical mixing resulting in cooler but nutrient richer surface waters (Pätzold and Wefer 1992; Pätzold et al. 1999; Berger et al. 2002). However, due to the dependence on several environmental factors

growth rate variations in most corals are difficult to interpret in climatic terms.

Oxygen isotopes

The ratios of the isotopic species of oxygen (^{18}O/^{16}O) incorporated in coral skeletons during growth, reported as δ^{18}O, are primarily influenced both by the temperature and the δ^{18}O of the ambient seawater during skeleton precipitation. In localities where one of these two environmental factors dominates the other, coral δ^{18}O records can therefore provide information either on variations in water temperature or in δ^{18}O of the seawater with the latter being related to the hydrologic balance.

As temperature increases, there is a decrease in the δ^{18}O (depletion in ^{18}O) of the coral skeletal aragonite. Near-weekly resolution calibrations of Porites coral δ^{18}O variability suggest a temperature dependence of 0.18‰ per 1°C (Gagan et al. 1994). This ratio is supported by a regression of several long annually-averaged Indo-Pacific Porites coral δ^{18}O records against local SST anomaly which indicates a ratio of 0.19‰/°C (Evans et al. 2000).

Variations in the δ^{18}O of the seawater can result from evaporation (enrichment in ^{18}O), precipitation (enrichment in ^{16}O), or runoff (enrichment in ^{16}O), i.e. such variations reflect the hydrologic balance. Water mass transport can also play a role. High evaporation results in both an increase in the δ^{18}O of the seawater and a higher salinity; high precipitation or runoff has opposing effects. Because of this, salinity variations, if driven through evaporation, precipitation, or runoff, covary with variations in the δ^{18}O of the seawater.

The environmental interpretation of the ratios of the stable isotopic species of carbon (^{13}C/^{12}C) incorporated in coral skeletons, reported as coral δ^{13}C, is complicated because of interactions with physiological processes such as symbiont photosynthesis and respiration. The coral δ^{13}C signal is therefore difficult to apply in paleoclimatic research. In cases, a strong signal can emerge. For example, we found that a coral δ^{13}C record from the northern Red Sea documents interannual events of extraordinarily large plankton blooms caused by deep vertical water mass mixing in certain winters. We think this is a result of changes in the coral's food uptake, i.e. increased heterotrophic feeding on zooplankton during the periods of high plankton availability (Felis et al. 1998).

The strontium/calcium ratio

Because of the dependence of coral $\delta^{18}O$ on both temperature and seawater $\delta^{18}O$, with the latter covarying with salinity, there is the need for additional proxies either solely reflecting temperature or salinity, respectively. Concentrations of some of the various trace elements incorporated in coral skeletons during growth were shown to be dependent on the temperature of the ambient seawater during skeleton precipitation.

Sr/Ca ratios in corals were shown to provide a promising proxy for water temperature variability (Beck et al. 1992; McCulloch et al. 1994; Alibert and McCulloch 1997; Gagan et al. 1998). The slopes of the calibration equations show a strong similarity. However, there are still differences in the temperature calibrations between different studies. The average of several coral Sr/Ca calibrations suggests a temperature dependence of 0.062 mmol/mol per 1°C (Gagan et al. 2000), but it is not known whether this coral Sr/Ca-temperature relationship is generally valid for *Porites*.

Coral Sr/Ca ratios are influenced by the Sr/Ca ratio of the ambient seawater during skeleton precipitation. Due to the long residence times of Sr and Ca in the ocean the Sr/Ca ratios of seawater are supposed to be constant on glacial-interglacial timescales. However, some works indicate that seawater Sr/Ca ratios can vary significantly between sites (de Villiers et al. 1994) as well as at the same location over the annual cycle. Seawater Sr/Ca ratios were shown to be correlated with nutrient variability, e.g. at upwelling sites (de Villiers et al. 1994). Furthermore it was suggested that during glacial conditions the Sr/Ca ratio of seawater could have been significantly different due to the weathering and dissolution of Sr-enriched aragonitic carbonates exposed on the continental shelves (Stoll and Schrag 1998).

Other temperature-sensitive trace elements incorporated in coral skeletons are Mg and U. Recent studies revealed that U/Ca ratios in corals could provide a temperature proxy comparable in accuracy to Sr/Ca (Corrège et al. 2000) whereas Mg/Ca ratios probably do not (Schrag 1999).

Combined oxygen isotopes and Sr/Ca

Combined determinations of $\delta^{18}O$ and Sr/Ca in corals can provide information on $\delta^{18}O$ seawater as well as temperature variability, through removing the temperature component of the coral $\delta^{18}O$ variations which is derived from the coral Sr/Ca signal (McCulloch et al. 1994; Gagan et al. 1998). The residual coral $\delta^{18}O$ can be used to reconstruct variations in the hydrologic balance because variations in the $\delta^{18}O$ of seawater reflect evaporation, precipitation, and runoff, which in turn can be related to changes in atmospheric moisture transport. Water mass transport effects can also play a role. If the relationship between $\delta^{18}O$ of seawater and salinity is constant over time, the double-tracer technique of coupled $\delta^{18}O$ and Sr/Ca measurements in corals can be used to reconstruct past variations in ocean surface salinity.

Radiocarbon

Radiocarbon (^{14}C) incorporated in coral skeletons during growth reflects the ^{14}C content of the dissolved inorganic carbon (DIC) of the ambient seawater during skeleton precipitation and provides a useful tracer for ocean circulation and upwelling. The ^{14}C content of the surface ocean is controlled by the ^{14}C level of the atmosphere (equilibration time about a decade) and by mixing with waters which have a different ^{14}C signature. The latter can result from changes in the depth of the mixed layer or thermocline or changes in the rate of vertical mixing and upwelling which brings radiocarbon-depleted waters to the surface. Another factor is the horizontal advection of surface waters with a different ^{14}C signature from other oceanic source regions (e.g. Druffel and Griffin 1993). The ^{14}C content is reported as $\Delta^{14}C$ (‰), which is the $^{14}C/^{12}C$ ratio relative to 19th-century wood.

The atmospheric testing of nuclear weapons in the 1950s and early 1960s increased the $\Delta^{14}C$ of the surface ocean. During this time of rapidly increasing bomb ^{14}C in the atmosphere air-sea exchange was the primary controller on surface ocean

Δ^{14}C. This increased the natural Δ^{14}C gradient between surface and subsurface waters and makes coral Δ^{14}C records from the surface ocean very sensitive to changes in upwelling, but also to changes in the horizontal advection of water masses which upwelled elsewhere (e.g. Guilderson and Schrag 1998; Guilderson et al. 1998). Because the rate of biological processes on Δ^{14}C DIC as well as radioactive decay are negligible relative to surface water dynamics and the timeframe of interest, respectively, Δ^{14}C in surface waters is a quasi-conservative, passive advective tracer (Guilderson et al. 1998; Guilderson et al. 2000).

Climate records from corals

Past centuries

Current paleoclimatic records which are based on δ^{18}O, Sr/Ca, or Δ^{14}C determinations in modern corals do not extend back beyond the year 1600. This is due to the fact that most still growing massive corals which can be found in the modern reefs are not older than about 100 to 350 years. Furthermore, these centuries-old coral colonies are usually quite rare in most reef environments. Most long coral-based paleoclimatic records are the result of extensive surveys to discover the biggest colony of an area.

The generation of century-long coral stable isotope records in annual resolution started in the 1980s (Pätzold 1986). The longest coral stable isotope time series available is a 347-year record from Galápagos (Dunbar et al. 1994). More recent studies exploit the clear seasonal resolution which corals can provide. However, of the published coral records which are currently archived in the World Data Center-A for Paleoclimatology only 8 which have seasonal or higher resolution extend beyond 1850 (Fig. 1). These coral records are based on δ^{18}O with the exception of the Rarotonga record which is based on Sr/Ca (Linsley et al. 2000). The latter was generated by applying a newly developed method for rapid analysis of high-precision Sr/Ca ratios in corals (Schrag 1999). These coral records cover a latitudinal range of ~28°N to ~29°S in the Indian and Pacific Oceans (Fig. 2). Most of these records are correlated with local and regional climate variability but also reflect aspects of large-scale climate phenomena. Many of them indicate decadal-scale climate variability.

The Ras Umm Sidd coral δ^{18}O record from the northern Red Sea (Sinai, Egypt) (Felis et al. 2000) is correlated with instrumental observations of climate in the Middle East. The mean annual coral δ^{18}O signal apparently reflects varying proportions of both sea surface temperature (SST) and δ^{18}O seawater variability which in turn are related to large-scale aspects of Middle East climate variability on interannual and longer timescales. In conjunction with instrumental observations of climate the coral record suggests that colder periods are accompanied by more arid conditions in the northern Red Sea but increased rainfall in the southeastern Mediterranean, whereas warmer periods are accompanied by decreased rainfall in the latter and less arid conditions in the northern Red Sea.

The coral time series extending back to 1750 is dominated by a ~70-year oscillation (Fig. 3a). A comparable oscillation has been related to variations in the intensity of the thermohaline circulation in the North Atlantic (Delworth and Mann 2000). Interannual to interdecadal variability in the coral time series is correlated with instrumental indices of the NAO (Fig. 4), ENSO, and North Pacific climate variability. Apparently these modes consistently contributed to Middle East climate variability since at least 1750, preferentially at a period of ~5.7 years. These coral-based results suggest that atmospheric forcing by large-scale primarily Northern Hemisphere Pacific- and Atlantic-based climatic modes plays an important role in the northern Red Sea/Middle East region.

Another prominent mode in the coral time series is an oscillation with a period of 22-23 years. A similar oscillation has recently been identified as the most prominent mode in a tree-ring based reconstruction of the PDO (Biondi et al. 2001) which represents climate variability in the North Pacific region. The bidecadal oscillations in the Ras Umm Sidd coral record and in the PDO reconstruction are roughly in phase since about 1850 (Fig. 3b). This may suggest coherent bidecadal climate variability over large parts of the Northern Hemisphere extratropics for nearly 150 years. Recent results based on field correlations with sea level pressure

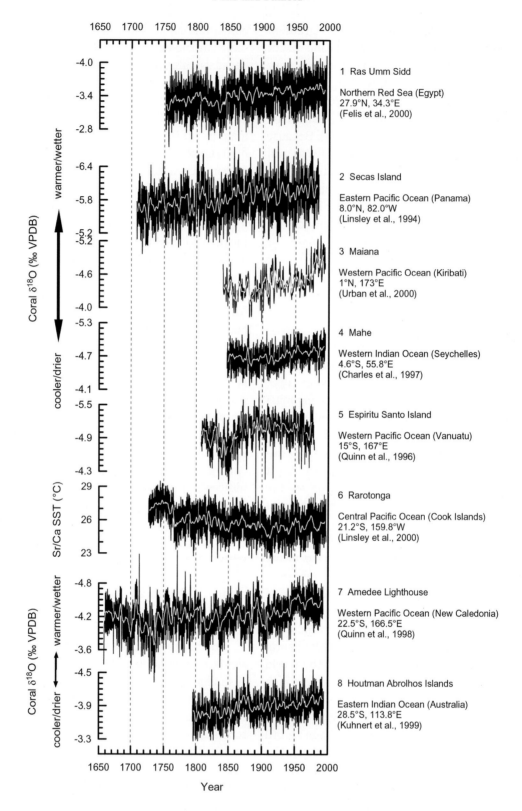

Fig. 1. Coral $\delta^{18}O$ and Sr/Ca records in seasonal or higher resolution extending back beyond 1850 which are archived at the World Data Center-A for Paleoclimatology (http://www.ngdc.noaa.gov/paleo/corals.html). Thick white lines represent 3-year running means.

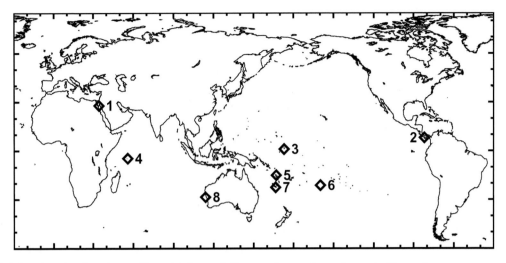

Fig. 2. Map showing locations of long high-resolution coral records as shown in Figure 1.

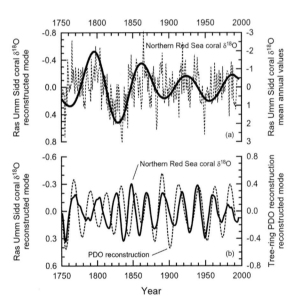

left: Fig. 3. a) The most prominent mode in the Ras Umm Sidd coral $\delta^{18}O$ record from the northern Red Sea has a period of ~70 years (solid line) (Felis et al. 2000) and is probably related to climate and ocean variability in the North Atlantic (Delworth and Mann 2000) Also shown is the detrended normalized mean annual coral $\delta^{18}O$ time series (dashed line). **b)** The second prominent mode in the coral record has a period of 22-23 years (solid line). The comparison with a tree-ring based reconstruction of the Pacific Decadal Oscillation (PDO) (dashed line) (Biondi et al. 2001) may suggest coherent bidecadal climate variability in the northern Red Sea/Middle East and the North Pacific during since about 1850.

suggest that the winter coral time series is linked to the Arctic Oscillation phenomenon, the Northern Hemisphere's dominant mode of atmospheric variability (Rimbu et al. 2001).

The Secas Island coral $\delta^{18}O$ record from the eastern Pacific Ocean (Panama) (Linsley et al. 1994) primarily reflects $\delta^{18}O$ seawater variability which is controlled by changes in precipitation. The precipitation pattern in this part of Central America is related to seasonal and interannual variability in

the latitudinal position of the Intertropical Convergence Zone. The coral time series extending back to 1707 is dominated by strong decadal variability.

Coral $\delta^{18}O$ records from the western equatorial Pacific primarily reflect $\delta^{18}O$ seawater variability mainly driven by changes in precipitation with minor contributions from relatively small changes in SST. During El Niño events increased precipitation associated with the eastward migration of the Indonesian Low and advection of fresher and

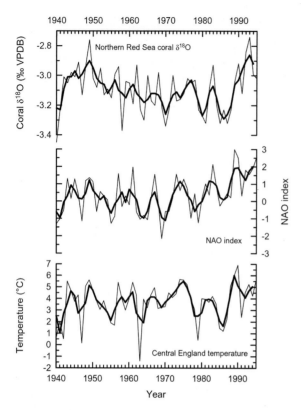

struction of ENSO extending back to 1840 provides evidence that variability in the tropical Pacific is linked to the region's mean climate. Cooler and drier background conditions during the mid to late 19th century when anthropogenic greenhouse forcing was absent were accompanied by prominent decadal variability and weak interannual variability. During a gradual transition towards warmer and wetter conditions in the early 20th century variability with a period of ~2.9 years intensified. Between 1920 and 1955 2-4 year variability was attenuated. With an abrupt shift towards warmer and wetter conditions in 1976 variability with a period of about 4 years becomes prominent. The results suggest that changes in the tropical Pacific mean climate and its variability have occurred during periods of natural as well as anthropogenic climate forcing.

The Mahe coral $\delta^{18}O$ record from the western equatorial Indian Ocean (Seychelles) (Charles et al. 1997) primarily reflects variations in SST. Interannual variability in the coral time series is correlated with Pacific climate records suggesting

Fig. 4. The winter Ras Umm Sidd coral $\delta^{18}O$ record from the northern Red Sea (Felis et al. 2000) and the winter index of the North Atlantic Oscillation (NAO) (Jones et al. 1997) show common variability. A similar variability is indicated by the winter time series of Central England temperatures (Parker et al. 1992) suggesting that both Central England temperatures and northern Red Sea coral $\delta^{18}O$ are influenced by the NAO. Coral $\delta^{18}O$ and Central England temperatures are for January-February; the NAO index is for December-March. Thick lines represent 3-year running means.

slightly warmer surface waters resulting from the eastward expansion of the western Pacific warm and fresh pool combine to generate strong coral $\delta^{18}O$ anomalies. Relatively cool and dry conditions during La Niña events produce coral $\delta^{18}O$ anomalies of opposite sign. Therefore corals from this region are excellent recorders of ENSO variability (Cole et al. 1993; Urban et al. 2000).

The Maiana coral $\delta^{18}O$ record from the western equatorial Pacific (Kiribati) (Urban et al. 2000) is strongly correlated with instrumental indices of ENSO variability (Fig. 5). This coral-based recon-

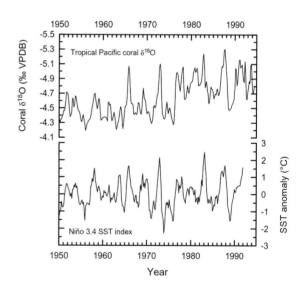

Fig. 5. The Maiana coral $\delta^{18}O$ record from the western equatorial Pacific and the Niño 3.4 sea surface temperature (SST) index, a commonly used measure of the state of the El Niño-Southern Oscillation, are strongly correlated (Urban et al. 2000). SSTs are from (Kaplan et al. 1998).

a consistent influence of ENSO on Indian Ocean SSTs for over a century. However, the coral time series extending back to 1846 is dominated by strong decadal variability which was suggested to be characteristic of the Asian monsoon system, implying important interactions between tropical and mid-latitude climate variability. This has been questioned recently by presenting evidence for a tropical Pacific forcing of decadal SST variability in the western equatorial Indian Ocean inferred from an annual-resolution coral $\delta^{18}O$ record from Malindi (Kenya) (Cole et al. 2000).

The Rarotonga coral Sr/Ca record from the central subtropical South Pacific (Cook Islands) (Linsley et al. 2000) provides an excellent proxy for SST variability. The coral time series extending back to 1726 is dominated by decadal variability. Several of the largest-scale SST variations at Rarotonga a coherent with SST regime shifts in the North Pacific, as indicated by the index of the PDO over the past 100 years. Because of this hemispheric symmetry it was suggested that tropical forcing may play an important role in at least some of the decadal variability which is observed in the Pacific Ocean.

The Amedee Lighthouse coral $\delta^{18}O$ record from the western South Pacific (Quinn et al. 1998) is correlated with variations in SST. The coral time series extending back to 1657 shows prominent decadal fluctuations, especially in the early 18th and early 19th century. Interannual-scale cooling events in the coral record coincide within 1 year with known volcanic eruptions (Crowley et al. 1997), e.g. 1808 (unknown source), 1813-1821 (several eruptions including Tambora 1815), 1835 (Coseguina), 1883 (Krakatau), and 1963 (Agung).

The Houtman Abrolhos Islands coral $\delta^{18}O$ record from the eastern subtropical Indian Ocean (Australia) (Kuhnert et al. 1999) is correlated to local SST variability. The location is influenced by the Leeuwin Current which is coupled to the Indonesian throughflow. The coral time series extending back to 1795 shows prominent pentadal and decadal variability.

The longest coral-based $\Delta^{14}C$ time series available is a 323-year record from the Great Barrier Reef which has a biannual resolution (Druffel and Griffin 1993). More recent studies on Pacific cor-

als provide seasonal to higher resolution records which give information on the seasonal to interannual variability in the surface circulation and thermocline structure in the Pacific basin (Fig. 6).

The Galápagos coral $\Delta^{14}C$ record from the eastern equatorial Pacific (Ecuador) (Guilderson and Schrag 1998) documents the seasonal upwelling of low $\Delta^{14}C$ subsurface waters. The interannual variability of the coral time series is dominated by ENSO. During El Niño events the depth of the thermocline increases and the upwelling of low $\Delta^{14}C$ water is reduced. The coral record shows that $\Delta^{14}C$ values during the upwelling season increased abruptly after the El Niño event of 1976. This suggests a reduction in the contribution of

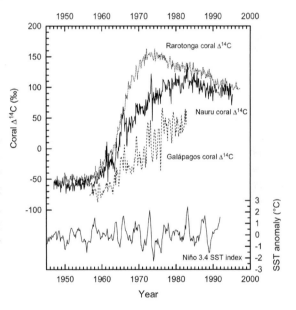

Fig. 6. Coral based $\Delta^{14}C$ records from different locations in the Pacific Ocean. Rarotonga, subtropical South Pacific (Guilderson et al. 2000); Nauru, western equatorial Pacific (Guilderson et al. 1998); Galápagos, eastern equatorial Pacific (Guilderson and Schrag 1998). The long-term trend in the coral $\Delta^{14}C$ records reflects the uptake of bomb ^{14}C in the ocean. Interannual variability in the equatorial records reflects changes in upwelling and surface circulation which in turn are related to the El Niño-Southern Oscillation (ENSO) phenomenon. The Niño 3.4 sea surface temperature index, a commonly used measure of the state of ENSO, is also shown. SSTs are from (Kaplan et al. 1998).

deeper, lower $\Delta^{14}C$ water to the upwelling since 1976, and together with a simultaneously occurring shift in upwelling season SSTs was interpreted as a shift in the vertical thermal structure of the eastern tropical Pacific towards a deepened thermo-cline.

The Nauru coral $\Delta^{14}C$ record (Guilderson et al. 1998) documents the interannual redistribution of surface waters in the western equatorial Pacific which is the result of mixing between waters of subtropical origin (higher $\Delta^{14}C$) and water upwelled in the eastern equatorial Pacific (lower $\Delta^{14}C$) then advected zonally by equatorial currents. The inter-annual variability in the coral time series is dominated by ENSO. During El Niño events coral $\Delta^{14}C$ values increase, reflecting the reduction of low $\Delta^{14}C$ water upwelling in the eastern Pacific and the invasion of high $\Delta^{14}C$ subtropical water into the western equatorial Pacific.

Holocene

Most modern corals are usually not growing for more than 100 to 350 years, but fossil corals provide an opportunity to reconstruct climate variability during time-windows throughout the Holocene. Well-preserved fossil corals can provide records of climate variability comparable in resolution and quality to those derived from modern corals. However, the longest Holocene time series available from a fossil coral is a 47-year record from Tasmaloum (Vanuatu) in the western South Pacific (Corrège et al. 2000). Most records based on fossil Holocene corals only cover periods of 5 to 20 years. This is due to the fact that finding a fossil coral that provides a century-long record is even more difficult than finding a comparable modern coral in the living reefs. Fossil corals are collected from uplifted or submerged reefs mostly by drilling. The chance to recover a long record along the major growth axis of a coral by drilling into a fossil reef is rather small. Direct collection of individual fossil colonies might provide a better solution but is also a matter of luck.

The Eilat coral $\delta^{18}O$ records from the northern Red Sea (Israel) show a higher seasonal amplitude for time-windows during the Mid-Holocene compared to modern corals (Moustafa et al. 2000). The

records of up to 18 years length were derived from corals which grew around 6000 to 4500 ^{14}C years before present (BP). The results suggest a higher seasonal cycle in SSTs and changes in the seasonal cycle of precipitation and/or evaporation during the mid-Holocene compared to now (Fig. 7). A possible explanation could be the occurrence of summer rains in this region of the Middle East during these times.

Coral Sr/Ca records from the southwest tropical Pacific provide information on the temperature variability during time-windows of up to 6 years during the early- to mid-Holocene (Beck et al. 1997). The corals grew around 10,300 to 4,200 calendar years BP and the Sr/Ca signal suggests that SSTs during the early Holocene in this part of the tropics (~16°S) were about 6.5°C cooler than today, but increased abruptly during the following 1500 years.

A coral from Tasmaloum (Vanuatu) which grew around 4150 calendar years BP provides a 47-year record of SST variability based on coral Sr/Ca and U/Ca (Corrège et al. 2000). Composite coral Sr/Ca and U/Ca derived temperatures suggests that SSTs in the southwest tropical Pacific during the mid-Holocene were comparable to modern SSTs with respect to the mean as well as typical ENSO variability. However, the variability in the seasonal amplitude as well as interannual SST variability during the mid-Holocene were considerably stronger than today. The coral time series shows several prominent interannual cooling events occurring at decadal-scale intervals as well as a decadal-scale modulation of the seasonal cycle (Fig. 8). The results could be interpreted in a way that phase shifts in the ENSO mode similar to today also occurred during the mid-Holocene but probably with stronger exchanges between the tropics and the extratropics.

Combined $\delta^{18}O$ and Sr/Ca records derived from Great Barrier Reef corals (Australia) provide important information on the temperature and surface-ocean hydrologic balance during the mid-Holocene (Gagan et al. 1998). Coral Sr/Ca ratios indicate that the tropical western Pacific was 1°C warmer ~5350 calendar years BP ago. The residual coral $\delta^{18}O$ signal as derived from the difference between the Sr/Ca and $\delta^{18}O$ records indicates that

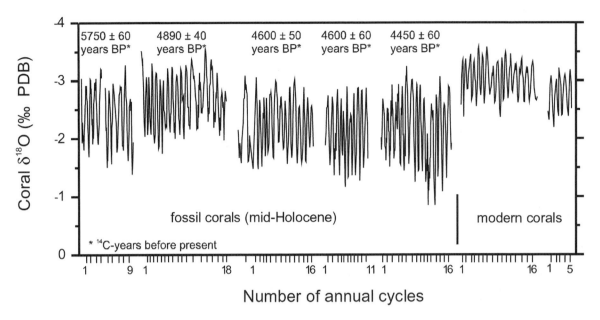

Fig. 7. Eilat coral δ^{18}O records from the northern Red Sea (Moustafa et al. 2000). Compared to modern corals the seasonal amplitude in corals which grew during the mid-Holocene is higher. This suggests a higher seasonal cycle in sea surface temperatures and changes in the precipitation and evaporation regime in this part of the Middle East compared to today.

the δ^{18}O of the surface water was 0.5‰ higher relative to modern seawater. The results suggest that the higher temperatures increased the evaporation resulting in higher δ^{18}O seawater. This δ^{18}O seawater anomaly may have been sustained by transport of part of the additional water vapor to extratropical latitudes.

Pleistocene

Because the growth of reef-building corals is restricted to the surface ocean the dating of fossil corals from uplifted or submerged Pleistocene reefs can provide a record of sea-level fluctuations (Bard et al. 1990). Well-preserved fossil corals can provide records of climate variability during time-windows of the Pleistocene comparable in resolution and quality to those derived from modern corals. Next to the difficulties in finding a long-lived fossil colony, corals from uplifted Pleistocene reef terraces are often affected by diagenesis making them not suitable for paleoclimatic reconstructions based on stable isotopes or trace element.

Coral δ^{18}O and Sr/Ca records derived from cores recovered from the submerged Barbados offshore reefs indicate that SSTs in the western equatorial Atlantic were 5°C colder than present values during the last glacial maximum (LGM) 18,000 to 19,000 years ago (Guilderson et al. 1994), a finding in conflict with the CLIMAP reconstructions (CLIMAP Project 1976, 1981).

Coral δ^{18}O and Sr/Ca records derived from a fossil colony collected on an uplifted reef terrace on Bunaken Island, North Sulawesi (Indonesia), reflect interannual variability in precipitation and SST in the western equatorial Pacific during the last interglacial period 124,000 years ago (Hughen et al. 1999). The 65-year long coral time series reveals ENSO variability similar to the modern instrumental record. This indicates that ENSO was robust during the last interglacial, a time when global climate was slightly warmer than Holocene. However, changes in ENSO magnitude and frequency after 1976 appear different with respect to the earlier instrumental and last interglacial records. The results were interpreted to support the hypothesis that ENSO behavior in recent decades is anomalous with respect to natural variability.

A fossil coral from the uplifted reef terraces of the Huon Peninsula (Papua New Guinea) provides

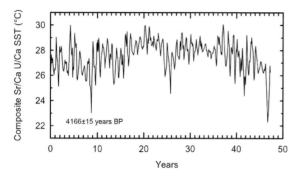

Fig. 8. The Tasmaloum composite sea surface temperature record derived from Sr/Ca and U/Ca analysis of a fossil coral from the southwest tropical Pacific (Corrège et al. 2000). Compared to today (not shown) the time series suggests a stronger SST variability with respect to seasonal amplitude and interannual variability as well as several prominent interannual cooling events during the mid-Holocene.

a record of SST from the western equatorial Pacific during the penultimate deglaciation around 130,000 years ago when sea level was 60 to 80 m lower than today (McCulloch et al. 1999). Coral $\delta^{18}O$ and Sr/Ca values indicate that SSTs were 6°C colder during the penultimate deglaciation than either last interglacial or present-day temperatures in this region of the tropics. The results again raise the question whether the tropics underwent significant cooling during glacial periods.

Deep-sea corals

In analogy to warm-water corals, solitary corals from the deep-sea can provide records of past climate and ocean variability. These non-reef-building (ahermatypic), non-photosynthetic corals usually live at depth of 500 to 2000 m. They are usually collected from the ocean floor by dredging. Therefore deep-sea corals provide one of the rare proxies to document past changes in intermediate and upper deep water masses which in turn can be related to changes in the surface ocean and the atmosphere (Smith et al. 1997; Adkins et al. 1998; Mangini et al. 1998).

Coupled ^{14}C and $^{230}Th/^{234}U$ dates from deep-sea corals can provide information about deep-water ventilation rates. Subtracting the $^{230}Th/^{234}U$-age (which represents the true age) from the ^{14}C-age

(which represents the coral's age plus the age of the deep water) gives the age of the deep water at the time of coral skeleton precipitation. This age represents the time since the deep water had last contact to the air at the ocean surface, i.e. the deep-water ventilation age (Mangini et al. 1998). Multiple ^{14}C dates throughout the life span of a deep-sea coral can provide information about variations in the deep-water ventilation rate (Adkins et al. 1998). This method revealed that the ventilation rate of the North Atlantic upper deep water varied greatly during the last deglaciation and provided evidence that the deep ocean changed on decadal-centennial time scales during rapid changes in the surface ocean and the atmosphere 15,400 years ago (Adkins et al. 1998). Another method suggests that paleotemperatures can be estimated from the stable isotopic composition of deep-sea corals (Smith et al. 1997).

Open problems

Oxygen isotope signals

Coral skeletons are depleted in ^{18}O and ^{13}C with respect to isotopic equilibrium with the ambient seawater, i.e. compared to inorganically precipitated aragonite. This isotopic disequilibrium or vital effect is most likely biologically mediated and attributed to so-called "kinetic" isotope effects which apparently result from discrimination against the heavy isotopes of oxygen and carbon during the hydration and hydroxylation of CO_2 (McConnaughey 1989a). The isotopic disequilibrium appears to be constant over time along the major growth axis of an individual coral colony, where growth and calcification rates are at their maximum (McConnaughey 1989a; Guilderson and Schrag 1999; Linsley et al. 1999). Samples for stable isotope analysis are therefore taken along the major axis of coral growth following single fans of corallites.

The mean isotopic disequilibrium offset from seawater can vary significantly for individual corals living in the same environment (Guilderson and Schrag 1999; Linsley et al. 1999), even within the same species of *Porites* (Linsley et al. 1999). Apparently it cannot solely be attributed to intercolony differences in average growth rates. The reported differences of 0.2 to 0.4‰ for coral $\delta^{18}O$ equal 1-2°C in terms of temperature interpretation. Therefore, caution is required when quantifications about mean climatic conditions are being made by comparing $\delta^{18}O$ records from individual corals which grew at different locations or during different periods of time. Coral $\delta^{18}O$ provides an excellent proxy for variability but should not be considered as an absolute proxy for SST or $\delta^{18}O$ of seawater.

It has been speculated that at least some of the low-frequency variability observed in individual coral $\delta^{18}O$ records may originate not directly from climatic influences but from variations in the mean disequilibrium offset through time as a result of complex biological or ecological processes (Evans et al. 2000). Differences between seasonal and mean annual SST-coral $\delta^{18}O$ calibrations have also been reported implying a larger interannual to decadal coral-derived SST variability than indicated by the instrumental SSTs. Changes in sea surface salinity on these timescales have been suggested as one possible explanation (Quinn et al. 1998; Felis et al. 2000). However, long salinity time series to test this assumption are usually rare. A recent study using 40 years of salinity observations from the Fiji area showed that on the seasonal timescale coral $\delta^{18}O$ in this region is driven by changes in SST whereas on the interannual timescale it is almost exclusively affected by sea surface salinity variations (Le Bec et al. 2000).

Sr/Ca signals

In analogy to stable isotopes, Sr/Ca ratios in coral skeletons show a vital effect that produces disequilibrium with the ambient seawater compared to inorganically precipitated aragonite. Similar to kinetic effects on coral stable isotopes this vital effect was shown to be partly biologically controlled, e.g. responding to growth rate variations (de Villiers et al. 1994). However, as for stable isotopes, it was shown that the effects of growth and calcification rates on coral Sr/Ca can be neglected when sampling follows a major growth axis along single fans of corallites (Alibert and McCulloch 1997).

Apparently the coral Sr/Ca-temperature calibrations do not vary significantly for individual corals living in the same environment, not even between different species of *Porites* and colonies with different growth and calcification rates, provided sampling follows optimal growth regions (Alibert and McCulloch 1997). This could imply an advantage in using Sr/Ca compared to using coral $\delta^{18}O$ to reconstruct changes in absolute mean temperatures at a given location.

There are notable differences in the temperature calibrations between studies from different sites which remain to be solved. The difference in the offsets between the regression lines (not the slopes!) from these studies can lead to differences of up to 3.5°C in absolute temperature estimates (Gagan et al. 2000). The difference in the Sr/Ca ratio of the seawater as measured at modern oligotrophic reef sites in the Pacific and Atlantic can represent an uncertainty of 1.2°C in the Sr/Ca thermometer (de Villiers et al. 1994), and seawater Sr/Ca variations over the annual cycle can correspond to temperature variations of 0.7 °C. Seawater Sr/Ca changes during glacial conditions were suggested to result in potential errors of up to 1.5°C (Stoll and Schrag 1998). All this is critical to the applicability of coral Sr/Ca ratios as an absolute proxy of past SST variations.

Future directions

More multicentury coral records in seasonal or higher resolution from tropical and subtropical locations are necessary to provide a more complete picture on the global patterns of decadal- and longer-scale climate variability during the past centuries. Such proxy records are the only marine archive to understand climate variations on these timescales. Apparently a bimonthly sampling resolution for corals seems to be a good compromise with respect to significant detection of interannual and decadal variability as well as laboratory expenditure (e.g. Felis et al. 2000; Urban et al. 2000). Applying the double tracer technique of coupled $\delta^{18}O$ and Sr/Ca determinations which until now has been only used for short periods to such bimonthly-resolution coral records will provide insights into the variability of temperature and hydrologic balance at key locations of the global climate system during the past centuries. Similar to other fields of paleoclimatology a multi-proxy approach is recommended for coral research. The combination of different proxies will reduce analytical errors and improve paleo estimates for environmental parameters.

A promising trend is the reconstruction of SST fields based on multicentury coral records. Work on the reconstruction of the SST field in the Pacific basin by using available coral $\delta^{18}O$ data sets has already started (Evans et al. submitted). In this first appoach only the leading modes of large-scale variability which are both evident in instrumental climate data and coral $\delta^{18}O$ records are being reconstructed.

Another trend is the splicing of accurately dated fossil coral records of several decades length to create longer records covering parts of the last millennium (Cobb et al. 1999). However, this is a difficult task considering vital effects and other difficulties, e.g. the lack of significant event years which could provide tie points as well as the precipitation of secondary aragonite leading to slight dating uncertainties.

Another promising trend is the generation of decades- to century-long records from fossil corals collected from uplifted or submerged reefs. Such records will provide information on climate variability during the last interglacial, the last glaciation, the last deglaciation, and the Holocene comparable in resolution and quality to those derived from modern corals. This will enable the assessment of natural climate variability during periods with boundary conditions different from today. Fossil corals from the Pacific basin will provide new insights into the ENSO phenomenon (e.g. Hughen et al. 1999). Fossil corals from the northern Red Sea where modern corals were shown to be related to prominent modes of Northern Hemisphere climate variability (Felis et al. 2000; Rimbu et al. 2001) will provide information on the existence of North Pacific- or North Atlantic-based atmospheric teleconnections on Middle East climate variability, e.g. the NAO, during these periods of time. It is an intriguing question whether the NAO was active during the entire Holocene or during the last interglacial. Seasonal resolution coral records from the northern Red Sea could provide some answers.

Future coral studies should proceed along the frontiers of coral reef distribution. These areas seem to be quite sensitive to environmental change, since different limiting factors affect coral growth. Ocean areas that have not been extensively covered include the tropical eastern and western Atlantic. Coral reefs develop along the western boundary of the Atlantic from the Caribbean down to about 20°S on the Abrolhos platform, Brazil. The reef distribution along the eastern Atlantic is rather limited but reaches up to the Cape Verde Islands. Tropical coral records from both sides of the Atlantic basin are potential sources to reconstruct the history of tropical Atlantic climate variability (Chang et al. 1997). Future research should also exploit the potential of corals from rare subtropical/mid-latitude locations (e.g. Bermuda, northern Red Sea) to better understand past mid-latitude climate variability which is important with respect to present and future European climate. In addition, further coral research in the Indian Ocean will help to detect modes of ocean climate variability on the decadal scale. Recently, a dipole mode linking the eastern and western tropical Indian ocean has been suggested (Saji et al. 1999; Webster et al. 1999). Current research in deep-sea coral reefs will provide informations about the potential of non-zooxanthellate corals in climate research (Freiwald et al. 1997), especially along the European margin of the North Atlantic.

Acknowledgments

We thank W. H. Berger and K. Brander for valuable comments on the manuscript.

References

Adkins JF, Cheng H, Boyle EA, Druffel ERM, Edwards RL (1998) Deep-sea coral evidence for rapid change in ventilation of the deep North Atlantic 15,400 years ago. Science 280:725-728

Alibert C, McCulloch MT (1997) Strontium/calcium ratios in modern Porites corals from the Great Barrier Reef as a proxy for sea surface temperature: Calibration of the thermometer and monitoring of ENSO. Paleoceanogr 12:345-363

Bard E, Hamelin B, Fairbanks RG (1990) U-Th ages

obtained by mass spectrometry in corals from Barbados: Sea level during the past 130,000 years. Nature 346:456-458

Beck JW, Edwards RL, Ito E, Taylor FW, Récy J, Rougerie F, Joannot P, Henin C (1992) Sea-surface temperature from coral skeletal strontium/calcium ratios. Science 257:644-647

Beck JW, Récy J, Taylor F, Edwards RL, Cabioch G (1997) Abrupt changes in early Holocene tropical sea surface temperature derived from coral records. Nature 385:705-707

Berger WH, Pätzold J, Wefer G (2002) Times of quiet, times of agitation: Sverdrup's conjecture and the Bermuda coral record. In: Wefer G, Berger WH, Behre K-H and Jansen E (eds) Climate Development and History of the North Atlantic Realm. Springer Berlin pp 89-99

Biondi F, Gershunov A, Cayan DR (2001) North Pacific decadal climate variability since AD 1661. J Clim 14:5-10

Chang P, Ji L, Li H (1997) A decadal climate variation in the tropical Atlantic Ocean from thermodynamic air-sea interactions. Nature 385:516-518

Charles CD, Hunter DE, Fairbanks RG (1997) Interaction between the ENSO and the Asian Monsoon in a coral record of tropical climate. Science 277:925-928

CLIMAP Project (1976) The surface of the ice-age Earth. Science 191:1131-1136

CLIMAP Project (1981) Seasonal reconstructions of the Earth's surface at the last glacial maximum. Geol Soc Ameri Tech Rep MC-36

Cobb KM, Charles CD, Kastner M (1999) Monthly-resolution windows on climate of the last millennium: Splicing fossil corals from the central tropical Pacific. Eos Trans AGU 80(46):F591

Cole JE, Fairbanks RG, Shen GT (1993) Recent variability in the Southern Oscillation: Isotopic results from a Tarawa Atoll coral. Science 260:1790-1793

Cole JE, Dunbar RB, McClanahan TR, Muthiga NA (2000) Tropical Pacific forcing of decadal SST variability in the western Indian Ocean over the past two centuries. Science 287:617-619

Corrège T, Delcroix T, Récy J, Beck W, Cabioch G, Le Cornec F (2000) Evidence for stronger El Niño-Southern Oscillation (ENSO) events in a mid-Holocene massive coral. Paleoceanogr 15:465-470

Crowley TJ, Quinn TM, Taylor FW, Henin C, Joannot P (1997) Evidence for a volcanic cooling signal in a 335-year coral record from New Caledonia. Paleoceanogr 12:633-639

de Villiers S, Shen GT, Nelson BK (1994) The Sr/Ca-temperature relationship in coralline aragonite: Influence of variability in (Sr/Ca)seawater and skeletal growth parameters. Geochim Cosmochim Acta 58:197-208

Delworth TL, Mann ME (2000) Observed and simulated multidecadal variability in the Northern Hemisphere. Clim Dyn 16:661-676

Druffel ERM, Griffin S (1993) Large variations of surface ocean radiocarbon: Evidence of circulation changes in the southwestern Pacific. J Geophys Res 98:20249-20259

Dunbar RB, Wellington GM, Colgan MW, Glynn PW (1994) Eastern Pacific sea surface temperature since 1600 A.D.: The $\delta^{18}O$ record of climate variability in Galápagos corals. Paleoceanogr 9:291-315

Evans MN, Kaplan A, Cane MA (2000) Intercomparison of coral oxygen isotope data and historical sea surface temperature (SST): Potential for coral-based SST field reconstructions. Paleoceanogr 15:551-563

Evans MN, Kaplan A, Cane MA (submitted) Sea surface temperature field reconstruction from coral $\delta^{18}O$ data using reduced space optimal interpolation. Paleoceanogr

Felis T, Pätzold J, Loya Y, Wefer G (1998) Vertical water mass mixing and plankton blooms recorded in skeletal stable carbon isotopes of a Red Sea coral. J Geophys Res 103:30731-30739

Felis T, Pätzold J, Loya Y, Fine M, Nawar AH, Wefer G (2000) A coral oxygen isotope record from the northern Red Sea documenting NAO, ENSO, and North Pacific teleconnections on Middle East climate variability since the year 1750. Paleoceanogr 15:679-694

Freiwald A, Henrich R, Pätzold J (1997) Anatomy of a deep-water coral reef mound from Stjernsund, west Finnmark, northern Norway. In: James NP, Clarke JAD (eds) Cool-Water Carbonates. Vol 56 SEPM, Tulsa, Oklahoma pp 141-162

Gagan MK, Chivas AR, Isdale PJ (1994) High-resolution isotopic records from corals using ocean temperature and mass-spawning chronometers. Earth Planet Sci Lett 121:549-558

Gagan MK, Ayliffe LK, Hopley D, Cali JA, Mortimer GE, Chappell J, McCulloch MT, Head MJ (1998) Temperature and surface-ocean water balance of the mid-Holocene tropical eastern Pacific. Science 279:1014-1018

Gagan MK, Ayliffe LK, Beck JW, Cole JE, Druffel ERM, Dunbar RB, Schrag DP (2000) New views of tropical paleoclimates from corals. Quat Sci Rev 19:45-64

Guilderson TP, Schrag DP (1998) Abrupt shift in subsurface temperatures in the tropical Pacific associated with changes in El Niño. Science 281:240-243

Guilderson TP, Schrag DP (1999) Reliability of coral isotope records from the western Pacific warm pool: A comparison using age-optimized records. Paleoceanogr 14:457-464

Guilderson TP, Fairbanks RG, Rubenstone JL (1994) Tropical temperature variations since 20,000 years ago: Modulating interhemispheric climate change. Science 263:663-665

Guilderson TP, Schrag DP, Kashgarian M, Southon J (1998) Radiocarbon variability in the western equatorial Pacific inferred from a high-resolution coral record from Nauru Island. J Geophys Res 103:24641-24650

Guilderson TP, Schrag DP, Goddard E, Kashgarian M, Wellington GM, Linsley BK (2000) Southwest subtropical Pacific surface water radiocarbon in a high-resolution coral record. Radiocarbon 42:249-256

Hughen KA, Schrag DP, Jacobsen SB (1999) El Niño during the last interglacial period recorded by a fossil coral from Indonesia. Geophys Res Lett 26:3129-3132

Hurrell JW (1995) Decadal trends in the North Atlantic Oscillation: Regional temperatures and precipitation. Science 269:676-679

Jones PD, Jonsson T, Wheeler D (1997) Extension to the North Atlantic Oscillation using early instrumental pressure observations from Gibraltar and south-west Iceland. Int J Clim 17:1433-1450

Kaplan A, Cane MA, Kushnir Y, Clement AC, Blumenthal MB, Rajagopalan B (1998) Analyses of global sea surface temperature 1856-1991. J Geophys Res 103:18567-18589

Kuhnert H, Pätzold J, Hatcher B, Wyrwoll K-H, Eisenhauer A, Collins LB, Zhu ZR, Wefer G (1999) A 200-year coral stable isotope record from a high-latitude reef off Western Australia. Coral Reefs 18:1-12

Le Bec N, Juillet-Leclerc A, Corrège T, Blamart D, Delcroix T (2000) A coral $\delta^{18}O$ record of ENSO driven sea surface salinity variability in Fiji (south-western tropical Pacific). Geophys Res Lett 27:3897-3900

Linsley BK, Dunbar RB, Wellington GM, Mucciarone DA (1994) A coral-based reconstruction of Intertropical Convergence Zone variability over Central America since 1707. J Geophys Res 99:9977-9994

Linsley BK, Messier RG, Dunbar RB (1999) Assessing between-colony oxygen isotope variability in the coral Porites lobata at Clipperton Atoll. Coral Reefs 18:13-27

Linsley BK, Wellington GM, Schrag DP (2000) Decadal sea surface temperature variability in the subtropical South Pacific from 1726-1997 AD. Science 290:1145-1148

Mangini A, Lomitschka M, Eichstädter R, Frank N,

Vogler S, Bonani G, Hajdas I, Pätzold J (1998) Coral provides way to age deep water. Nature 392:347-348

Marchitto Jr. TM, Jones GA, Goodfriend GA, Weidman CR (2000) Precise temporal correlation of Holocene mollusk shells using sclerochronology. Quat Res 53:236-246

McConnaughey T (1989a) ^{13}C and ^{18}O isotopic disequilibrium in biological carbonates: I. Patterns. Geochim Cosmochim Acta 53:151-162

McCulloch MT, Gagan MK, Mortimer GE, Chivas AR, Isdale PJ (1994) A high-resolution Sr/Ca and $\delta^{18}O$ coral record from the Great Barrier Reef, Australia, and the 1982-1983 El Niño. Geochim Cosmochim Acta 58:2747-2754

McCulloch MT, Tudhope AW, Esat TM, Mortimer GE, Chappell J, Pillans B, Chivas A, Omura A (1999) Coral record of equatorial sea-surface temperature during the penultimate deglaciation at Huon Peninsula. Science 283:202-204

Moustafa YA, Pätzold J, Loya Y, Wefer G (2000) Mid-Holocene stable isotope record of corals from the northern Red Sea. Int J Earth Sci 88:742-751

Parker DE, Legg TP, Folland CK (1992) A new daily Central England temperature series. Int J Clim 12:317-342

Pätzold J, Wefer G (1992) Bermuda coral reef record of the last 1000 years. 4th International Conference on Paleoceanography. Geol-Paläont Inst Univ Kiel, Kiel, Germany pp 224-225

Pätzold J (1986) Temperatur- und CO_2-Änderungen im tropischen Oberflächenwasser der Philippinen während der letzten 120 Jahre: Speicherung in stabilen Isotopen hermatyper Korallen. Berichte Geol-Paläont Inst Univ Kiel 12, Kiel, Germany pp 92

Pätzold J, Bickert T, Flemming B, Grobe H, Wefer G (1999) Holozänes Klima des Nordatlantiks rekonstruiert aus massiven Korallen von Bermuda. Natur und Museum 129:165-177

Quinn TM, Crowley TJ, Taylor FW, Henin C, Joannot P, Join Y (1998) A multicentury stable isotope record from a New Caledonia coral: Interannual and decadal sea surface temperature variability in the southwest Pacific since 1657 A.D. Paleoceanography 13:412-426

Rimbu N, Lohmann G, Felis T, Pätzold J (2001) Arctic Oscillation signature in a Red Sea coral. Geophys Res Lett 28:2959-2962

Saji NH, Goswami BN, Vinayachandran PN, Yamagata T (1999) A dipole mode in the tropical Indian Ocean. Nature 401:360-363

Schrag DP (1999) Rapid analysis of high-precision Sr/Ca ratios in corals and other marine carbonates. Paleoceanogr 14:97-102

Smith JE, Risk MJ, Schwarcz HP, McConnaughey TA

(1997) Rapid climate change in the North Atlantic during the Younger Dryas recorded by deep-sea corals. Nature 386:818-820

Stoll HM, Schrag DP (1998) Effects of Quaternary sea level cycles on strontium in seawater. Geochim Cosmochim Acta 62:1107-1118

Urban FE, Cole JE, Overpeck JT (2000) Influence of mean climate change on climate variability from a 155-year tropical Pacific coral record. Nature 407:989-993

Webster PJ, Moore AM, Loschnigg JP, Leben RR (1999) Coupled ocean-atmosphere dynamics in the Indian Ocean during 1997-98. Nature 401:356-360

Weidman CR, Goodfriend GA (1999) 1000-year paleotemperature record from fossil marine shells. Eos Trans AGU 80:F545

Weidman CR, Jones GA, Lohmann KC (1994) The long-lived mollusc Arctica islandica: A new paleoceano-graphic tool for the continental shelves of the northern North Atlantic. J Geophys Res 99:18305-18314

Fisheries and Climate

K. Brander

ICES, Palaegade 2-4, 1261 Copenhagen, Denmark
corresponding author (e-mail): keith@ices.dk

Abstract: The European Science Foundation report "Towards a European marine research area" identifies five major scientific challenges for priority action, of which the first is "Ocean-climate coupling" and the second is "Sustainable exploitation of living and non-living resources". This paper may be regarded as a link between these two, by examining the evidence that ocean climate affects fish stocks. The time series of fisheries and environmental information is relatively short to establish climatic effects on fish stocks, but evidence is presented from several areas of the northern North Atlantic during the 20th century. The warm period from 1930 to 1960 caused major changes in the distribution and abundance of many species, including cod, herring and capelin, which comprise a large proportion of the fisheries in this area. Recent research, which shows the influence of the North Atlantic Oscillation (NAO) on many environmental variables and biological systems, has helped to shape our understanding of climatic effects across the whole North Atlantic. Future research will benefit from well coordinated assembly, analysis and modeling of comprehensive and appropriate environmental and biological data and the results can be used to improve the assessment and management of fish stocks.

Introduction

"Il doit arriver aux poissons, comme aux animaux terrestres, que certaines années soient plus favorables que d'autres à leur multiplication et à leur accroissement, sans qu'on puisse en assigner précisément la cause" - H.-L. Duhamel du Monceau, 1769-1777. Traité général des pesches.

"The causes of fluctuations in the behaviour, occurrence and abundance of fish stocks have been attributed to widely different conditions in nature, and practically no progress has been made on some of these problems for a very long time." - G. Rollefsen, 1948. Introduction to the ICES Special Scientific Meeting on Climatic Changes.

It has long been known that fish stocks are affected by prevailing conditions (see quotes above), but we continue to struggle with establishing causes and processes. The effects of the changing environment on fisheries have become topical for a number of reasons. There is widespread interest in, and concern about, accelerating global change, of which declining fish stocks are perceived as a part. Fishing pressure is the major cause of the general decline and fisheries management has to date had limited success in halting or reversing it. Environmental changes add to the uncertainty in assessing and predicting the state of fish stocks.

This paper will present some of the evidence of the effects of climate change (long term environmental variability) on fisheries in the North Atlantic from West Greenland eastward and will consider how, in spite of incomplete understanding and uncertainties over their future states, environmental (including climate) information should be taken into account in assessing and managing fish stocks.

The effects of varying environment can be detected at all levels of biological organisation from ecosystems down to populations, individuals and even cells. The time and space scales at which environmental and biological entities are studied and linked range from climatic and basin scale to short term and local. At the larger scales the quantity of relevant data is limited and the scope for controlled experiment almost nil. In consequence many studies rely on correlations between time series which are often short and reliant on proxy data.

In many cases the functional form of the relationship between an environmental variable and a biological variable is non-linear. For example the growth rate of cod is high at an intermediate temperature value and declines at higher and lower temperatures (Björnsson et al. 2001). This particular relationship is direct, but many relationships between environment and biological processes are indirect or may involve several variables or interactions between biological components of the system being considered. Examples of direct, indirect and integrated relationships for European terrestrial and marine systems have been summarised and tabulated by Ottersen et al. (2001).

Given the multitude of possible direct and indirect processes by which environmental factors may influence biological systems, the task of selecting those which are actually responsible for observed variability is daunting. Even if this is successfully achieved, the processes which account for most of the observed variability during a particular time period may fail to do so subsequently, because other factors, which were not included, become more important. For example, during a period when spawning biomass is high, variability in recruitment may be due principally to environmental factors, but spawning biomass may subsequently decline to a level where it limits recruitment even if environmental factors are favourable. The history of investigations in this field is littered with correlations between environmental and biological variables which were statistically significant for a particular period of time, but which subsequently ceased to be so (Shepherd et al. 1984). The reason for this may be that the correlation was "by chance" in the first place, or it may be that the process was incomplete and that other factors became more important. Proposed mechanisms for environmental effects should be complete and consistent at all relevant scales. Given these difficulties, it is not surprising that progress in explaining phenomena, which were obvious to a scientific observer, like Duhamel du Monceau, in the eighteenth century, has been slow.

The role of the North Atlantic Oscillation (NAO) in research on climate and fish stocks

One of the most stimulating discoveries of the past ten years, for research on climate and fisheries, has been the extent to which biological systems, including fish stocks, respond to the many environmental factors, the variability of which can be substantially characterised by the state of the NAO (Ottersen et al. 2001). The NAO is a measure of the difference in atmospheric pressure between Iceland and Portugal. The winter (DJFM) index of the NAO (Hurrell 1995) provides a single annual value which, for the North Atlantic area, expresses a significant proportion of the variability in temperature, wind, cloud cover and rainfall and is also a useful proxy for marine environmental variables such as sea surface temperature (SST).

In relating particular biological variables, such as recruitment of fish stocks, to the NAO, one benefits from the background of associated climatic research and the expanding literature concerning NAO effects on other biological systems and other areas. The influence of the NAO is pervasive, but it is important to identify the local, proximate environmental variables which are producing a particular biological effect, since the NAO is not the direct driver. Even a strong statistical relationship between the NAO and a biological variable lacks credibility unless the underlying process and causes can be identified.

Two related aspects of the geography of the NAO should be kept in mind. Firstly, the action of the low and high pressure systems, whose gradient the NAO is measuring, is quite different over different parts of the North Atlantic. For example air temperatures over the NW and NE Atlantic tend to "sea-saw" in opposite directions and even at much smaller scales there are substantial differences in the response of local environmental variables (Ottersen et al. 2001). The NAO does not act uniformly. Secondly, the low and high pressure systems, which give rise to the NAO, are not completely geographically fixed, therefore local response to the NAO may also vary over time as the positions of low and high pressure areas shift. For example, the direction of winds over the Labrador Sea reversed over the last few years because of a shift in the pressure fields, even though the value of the NAO itself did not change substantially (Dickson et al. 2000).

A correlation between the NAO and a biological variable can be regarded as a two step process, the first being the response of one or more local environmental variables to the NAO and the second being the response of the biological variable to the local environmental variables. If the correlation with the NAO breaks down, then it is always worth considering whether this is because the local environmental variables are responding differently to the NAO.

Evidence that climate change affects fisheries

Few records of fish stocks go back more than 100 years, and the general decline in stocks due to the effects of fishing during this period overlays and obscures possible climatic effects. Nevertheless, in spite of the general decline, there have also been periods when stocks increased their range and

abundance and such changes are more likely to have environmental causes than to be caused by fishing pressure. One of the longest records, from outside the Atlantic, comes from counting scales of sardines and anchovies deposited in layered anoxic sediments (Fig. 1, Baumgartner et al. in press).

These stocks show large fluctuations over 1500 years prior to the advent of fishing. The variability is caused by changes in distribution and/or abundance which could be due to internal dynamics and interactions between the two species rather than climatic (environmental) factors. However additional evidence of synchronous large scale fluctuations in other fish species in different parts of the North Pacific (Fig. 2, Klyashtorin 1998) eliminates local dynamics and supports the existence of climate related causes, since there is no reason why local dynamics should cause synchronous variability across the whole Pacific.

Proxy records of fish stocks for the North Atlantic do not yet go back as far in time as the Californian series, but historic information on some fisheries goes back to the 10th century and provides semi-quantitative time series on periods when these flourished and declined. Detailed information on catches and matching environmental data is only

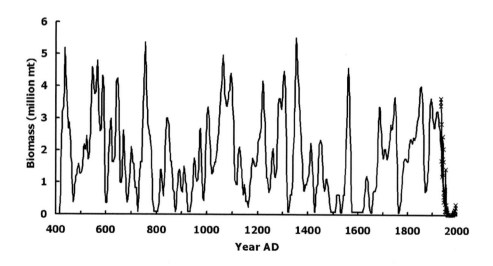

Fig. 1. Hindcast of Pacific sardine biomass, 15-year smoothing plotted with modern biomass estimates (1932-1995, marked x) based on natural log of scale deposition rates off Southern California. Data from Baumgartner et al. (in press)

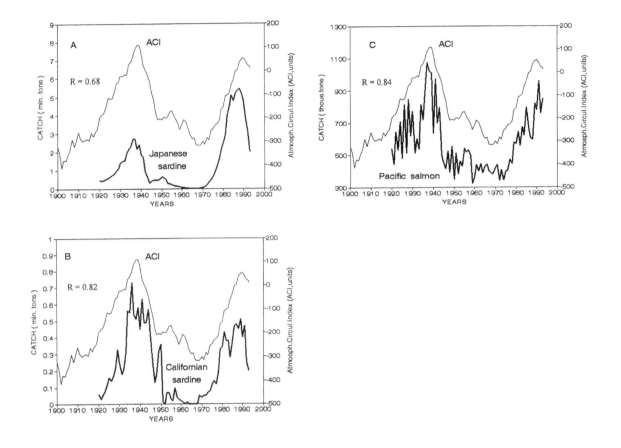

Fig. 2. Salmon and sardine catch in the Northern Pacific and Atmospheric Circulation Index (ACI) trend 1900-1993 after Klyashtorin (1998).

available for the past century or so; a relatively short period for establishing relationships between climate and fisheries empirically, particularly since the past century has also seen great increases in fishing pressure in all areas. Nevertheless, there have been large fluctuations in environmental conditions over much of the northern North Atlantic, and the effects of these on the distribution, abundance and growth of fish are evident. In the following brief accounts of observed effects of climate change the only environmental variable used is temperature. This is because the effects of temperature are well known and pervasive and because temperature records are more complete and available. Other environmental factors (e.g. wind, cloud, relative humidity) also have far-reaching consequence for marine productivity (by affecting vertical mixing, heat exchange, light penetration) and for the transport of early life stages of fish, but cannot be dealt with here.

Temperature of the northern North Atlantic during the 20th century

Time series of atmospheric temperature are longer and more complete than SST records, and in the area from Greenland to the Norwegian coast, the relationship between air temperature and SST is sufficiently close for the former to adequately represent long term variability in temperature (see Fig. 3). In this part of the North Atlantic the 20th century began with a cold period which lasted until 1920. Temperatures then rose by over one degree from 1920 to 1930 and remained above average until the 1960's (although with a substantial dip below the mean in the Norwegian Sea during the

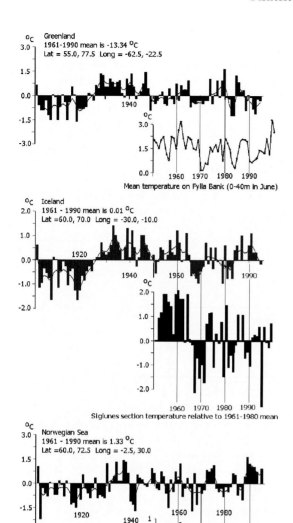

Fig. 3. Mean annual air temperature anomalies for Greenland, Iceland and the Norwegian Sea (1900-1995) from the IPPC Data Distribution Centre. SST data and anomalies are from the ICES Ocean Climate Status Summary.

1940s). Approximately decadal periods of cold and warm have alternated in all areas since the mid 60s. During the 1990s there was a general warming trend, but at Greenland temperatures even at the end of the decade were lower than they had been in the 1930s and 40s.

The warming period of the 1920s

The warming period of the 1920s caused poleward extension of the range of many fish and other marine and terrestrial species from Greenland to Iceland and further east as far as the Kara Sea. Records of changes in species distribution at this time provide some of the most convincing evidence of the pervasive effects of a change in climate on the marine ecosystem as a whole. Jensen (1939) published a paper "Concerning a change of climate during recent decades in the Arctic and Subarctic regions, from Greenland in the west to Eurasia in the east, and contemporary biological and physical changes", which presents much of this information. Some of the salient points concerning fish species are summarised below and in Tables 1 and 2.

The distribution of cod extended poleward by about 1000 km between 1920 and 1930 and can be followed in some detail, because fishing stations were established progressively further north as directed coastal fisheries were established by the Greenland Administration.

The international offshore fishery for cod subsequently reached a peak of over 400,000 tonnes in the early 1960s before collapsing. The decline was due to a combination of fishing pressure and adverse environmental change and the stock has not recovered since. The relationship between water temperature and recruitment levels is clear for this area (Brander 2000), with poor recruitment occurring at temperatures below about 1.5 °C (measured on Fylla Bank, 64°N, in June). The warming of the North Atlantic which has taken place over the last decade has also affected Greenland, but temperatures remained below 1.5°C until 1996, therefore it is too early to expect a recovery of the cod stock, which takes about seven years to reach maturity and which may be adversely affected by the trawl fishery for shrimps, which is now the mainstay of the Greenland fisheries.

One of the principal changes which occurred at Iceland during the 1920s warming period was in the major pelagic stocks, herring and capelin. Prior to 1920 the capelin spawned regularly on the south and south-west coasts, but from 1928 to 1935 very few were taken in these areas. In contrast, herring extended their spawning areas from the south

Changes in distribution and abundance	Fish species
Species previously absent, which appeared from 1920	*Melanogrammus aeglefinus, Brosme brosme, Molva molva*
Rare species which became more common and extended their ranges	*Pollachius virens* (new records of spawning fish), *Salmo salar, Squalus acanthias*
Species which became abundant and extended their ranges poleward	*Gadus morhua, Clupea harengus* (new records of spawning fish)
Arctic species which no longer occured in southern areas, but extended their northern limits	*Mallotus villosus, Gadus ogac, Reinhardtius hippoglossoides* (became much less common),

Table 1. West Greenland during the period of warming from 1920.

Changes in distribution and abundance	Fish species
Species previously absent, which appeared from 1920	*Notidanus griseus, Xiphias gladius, Trachurus trachurus*
Rare species which became more common and extended their ranges	*Glyptocephalus cynoglossus, Psetta maxima, Cetorhinus maximus, Thunnus thynnus, Scomber scombrus, Scomberesox saurus, Mola mola*
Species which became abundant and extended their ranges poleward	*Gadus morhua, Clupea harengus* (both extended their spawning distribution)
Arctic species which no longer occured in southern areas	*Mallotus villosus*

Table 2. Iceland during the period of warming from 1920.

and south-west coast to the east, north-west and north coasts over the same period (Saemundsson 1937).

Similar changes to those found at Iceland and Greenland during the warming period of the 1920's are also recoded from Jan Mayen, the Barents Sea, the Murman Coast, the White Sea, Novaya Zemlya and the Kara Sea, with cod and herring extending their range and becoming more abundant.

However, whereas Greenland and Iceland remained warm throughout the 1930s and 40s, the Norwegian Sea and Barents Sea cooled substantially in the late 1930s.

Climate effects on the Barents Sea

Our understanding of the processes underlying major fluctuations in the fish ecosystem of the Barents Sea is considerably more detailed than for most other areas of the northern North Atlantic (Rødseth 1998). The principal species involved in these changes are cod, capelin and herring, but many other components of the ecosystem have also been investigated. The early years of the 20th century, particularly 1902, were extremely cold in the Barents Sea, with extensive ice cover. This resulted in a crisis for the Norwegian fisheries, with low catches of small cod in very poor condition. Large numbers of seals moved down the Norwegian cost from the Barents Sea. A similar sequence of events occurred during the cold period in the 1980s, when the capelin stock collapsed, the cod were small and in poor condition and seals invaded the northern coast of Norway.

For cod in particular, the consequences of variability in temperature, transport and food during early life have been studied closely (Michalsen et al. 1998; Ottersen and Loeng 2000; Ottersen et al. 1998). Growth and survival rates of larvae and juveniles are higher in warm years and the large year classes of cod spread further east into the Barents Sea, where they encounter cooler water and their growth rate slows as a result.

Norwegian spring spawning herring

The stock biomass of Norwegian spring spawning herring increased almost ten-fold during the period from 1920 to 1930, when the Norwegian Sea and much of the North Atlantic went through a period of rapid warming (Toresen and Østvedt 2000). The herring stock declined rapidly from the late 1950s and by 1970 had decreased by over four orders of magnitude. From figure 4 one might conclude that the decline in the stock was due to the cooling which occurred from the late 1950's to the early 1970s, but although this may have been a contribu-

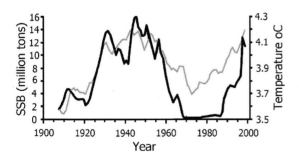

Fig. 4. Fluctuations of spawning stock biomass of Norwegian spring spawning herring (thick line) and mean annual temperature at the Kola section 1907-1998 (thin line, moving average over 19 years) after Toresen and Østvedt (2000).

tory factor, it is likely that heavy fishing pressure was mainly responsible for the collapse of the stock.

The collapse of the Norwegian spring spawning herring stock coincided with a retraction of the summer feeding distribution due to the southward and eastward shift in the location of the polar front which was located north of Iceland prior to 1965, but has remained west of Iceland since then. In spite of the rise in temperature in the Norwegian Sea and the complete recovery of the herring spawning stock, the temperature north of Iceland has remained cold and the fish have not re-occupied their earlier feeding areas.

In addition to the effect of the environment on the distribution of Norwegian spring spawning herring, the rapid cooling which took place during the mid-late '60's also resulted in reduced growth and recruitment (Toresen and Østvedt 2000) . The same is true for the Icelandic summer spawning herring (Jakobsson et al. 1993).

Bohuslän herring

One of the longest historic records of fisheries information is for the Bohuslän region of Sweden, in the eastern Skagerrak. The fishery depended on spent herring which migrated into the skerries and fjords during the autumn and overwintered there. The periods during which this occurred usually lasted for several decades and there have been about nine such periods since the 10th century. Although it has been claimed that "the Bohuslän periods are accompanied by negative NAO values" and that other European herring and sardine stocks also respond (positively or negatively) to environmental changes associated with the different phases of the NAO (Alheit and Hagen 1997), there is still some way to go in establishing this as a well-founded conclusion, based on a credible set of processes.

To begin with, the definition of a "Bohuslän period" is not rigorous because of the nature of the records and also because the definition includes both the migration of the fish (into the coastal area, where the fishery is located) and apparent changes in stock abundance. Secondly, no specific environmental factors and processes have been put forward and tested, to explain the switch between one period and the other. Thirdly, there are many ways of selecting and averaging NAO values and in the absence of a specific process it is difficult to justify the use of any particular one of these. Alheit and Hagen (1997) used the NAO index for January, which is shown together with the more widely used winter index (DJFM) in figure 5. There has only been one Bohuslän period (1877-1906) since 1821, when the instrumental record began, and it does not match very closely with the periods of negative NAO. The most prolonged negative NAO period, from 1952 to 1981, did not cause the characteristic shift in distribution at Bohuslän. While one may withold judgement on whether this constitutes falsifying evidence, because there may be different ways of defining both a "Bohuslän period" and a negative NAO value, it certainly is not supporting evidence. Recent reconstructions of the NAO from ice-cores, tree rings and documentary parameters go back to 1400 but are not consistent with each other (Jones et al. 2001), therefore the

reconstructed NAO cannot easily be compared with earlier Bohuslän periods in order to further test the relationship.

Evidence of recent distribution shifts

The mean temperature of the upper 300m of the North Atlantic warmed by just under 0.5°C between 1984 and 1999 (Levitus et al. 2000). In many areas the rate of warming was greater than this, but the underlying trend is overlain with substantial inter-annual and geographic variability. Records of fish catches from Portugal to Norway over this period show northward shifts in the distribution of many commercial and non-commercial species (Brander et al. in press). In the Ria Formosa lagoon in the Algarve region of Portugal, the change was largely due to the arrival of species from the Mediterranean and NW Africa, which had not previously been recorded in Portuguese waters. A variety of tropical species extended their ranges northward along the European continental slope since the early 1960s (Quero et al. 1998). In the sea areas around the British Isles red mullet (*Mullus surmuletus*) and bass (*Dicentrarchus labrax*) extended their ranges northward by several hundred kilometers since 1980 and have been recorded as far north as western Norway (Brander et al. in press).

Discussion

The information presented here on the effects of climate change during the 20th century on fish stocks is intended to give a broad overview of the major trends. Although the material concerns events which took place as long as 100 years ago, much of the joint analysis of environmental and fisheries data has been relatively recent. In particular the influence of the NAO and its use in characterising and giving a geographic framework to a range of environmental factors has stimulated renewed pan-Atlantic research (Ottersen et al. 2001).

The mechanisms and processes by which environmental factors influence biological systems have barely been mentioned here, but are essential in constructing credible explanations of climatic effects. They are also required if environmental

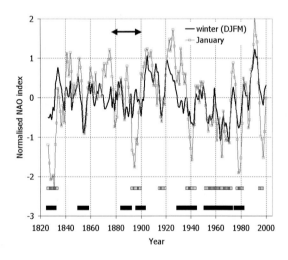

Fig. 5. Five-year running means of January and Winter (DJFM) standardised NAO indices (data from Climate Research Unit, UEA). The standardisation sets the mean for the entire time series (1824-2001) to zero. The double headed arrow indicates a Bohuslän herring period. The horizontal bars in the lower part of the graph indicate periods when the January (open squares) and Winter (solid bars) five year mean NAO indices were negative for at least five consecutive years.

information is to be applied in the detailed assessment and prediction of fish stocks which currently underpins fisheries management. Substantial research programmes have been underway for many years in areas such as the Barents Sea, to identify processes and to construct coupled bio-physical models which include environmental variability. Their application to fisheries and marine ecosystem management issues is in its infancy and a review of this would go beyond the scope of this paper. Nevertheless a few issue will be considered here, as they relate to the future research agenda for European Marine Science.

An impediment to progress in the past has been the difficulty of defining and obtaining appropriate marine environmental data, but rapid progress is taking place in collection, quality control and dissemination. When exploring possible mechanisms the range of scales of time, space and biological organisation with which we are concerned demands comparable ranges and resolution of environmental data. An example at the detailed end of the scale concerns the feeding, growth, migration and behaviour of cod in the Barents Sea, which are affected by the temperature field. Many of the processes have been quite thoroughly investigated and detailed seasonal information on the temperature field will help us to understand them and eventually use them for prediction, but in order to do so we also have to know the horizontal and vertical distribution of the fish in relation to the temperature field. New technologies, such as recording tags, which log the ambient temperature, depth and activity of individual fish make this possible. At the coarse end of the scale spectrum, it is obvious that the proxy temperature information presented in Figure 3 is less than perfect for the purpose for which it is used and that comprehensive, standardised data on sea temperature and other environmental parameters would be extremely valuable for detecting relationships and for driving models. Close collaboration between biologists, modelers and data scientists is needed in order to ensure that future data requirements are satisfied.

Climate variability causes changes in distribution, abundance, growth, maturity and behaviour of fish and therefore adds to the uncertainty in assessing and predicting the state of fish stocks.

Given that we know this and that we understand many of the processes (from fundamental physiology to large scale physical transport) which bring this about, it seems obvious that we should try to use environmental information when assessing and managing those stocks. To ignore such information is, in effect, to assume that the environment will remain unchanged. Although it may be difficult to predict future environmental states one can be confident that they will not remain the same as in the past.

One of the reasons given for ignoring environmental effects is that they can be used as an argument against curbing excessive fishing pressure, on the grounds that observed changes in stocks may be due to environmental change, not fishing. Since it is evident that most fish stocks are under serious threat from excessive fishing (which could in principle be reduced) whereas nothing can be done about environmental variability, it is argued that the message to reduce fishing pressure should not be obscured in any way. While it may be true that a simple message stands more chance of being heeded, the evidence that fish respond to environmental variability is well known - and has been for centuries - therefore it seems perverse to conceal it. Furthermore, under the precautionary principle there is a duty of care, which requires action in advance of scientific proof and which puts the burden of scientific proof on those who propose or cause change. The responsibility to reduce fishing pressure is plain, regardless of the changes brought about by climate.

References

Alheit J, Hagen E (1997) Long-term climate forcing of European herring and sardine populations. Fish Oceanogr 6:130-139

Baumgartner T, Ferreira V, Soutar A (in press) Fish Scale Deposition in the Santa Barbara Basin, southern California. Fish Bull US

Björnsson B, Steinarsson A, Oddgeirsson M (2001) Optimal temperature for growth and feed conversion of immature cod (*Gadus morhua* L). ICES J Mar Sci 58:29-38

Brander KM (2000) Effects of environmental variability on growth and recruitment in cod (*Gadus morhua*) using a comparative approach. Oceanol Acta

23:485-496

Brander KM, Blom G, Borges MF, Erzini K, Henderson GTD, Mackenzie B, Mendes H, Santos AMP, Toresen R (in press) Changes in fish distribution in the eastern North Atlantic; are we seeing a coherent response to changing temperature? ICES Mar Sci Symp

Dickson RR, Osborn TJ, Hurrell JW, Meincke J, Blindheim J, Adlandsvik B, Vijne T, Alekseev G, Maslowski W (2000) The Arctic Ocean response to the North Atlantic Oscillation. J Climate 13:2671-2696

Hurrell JW (1995) Decadal trends in the North Atlantic Oscillation; regional temperatures and precipitations. Science 269:676-679

Jakobsson J, Gudmundsdotir A, Stefansson G (1993) Stock-related changes in biological parameters of the Icelandic summer-spawning herring. Fish Oceanogr 2:260-277

Jensen AS (1939) Concerning a change of climate during recent decades in the Arctic and SubArctic regions, from Greenland in the west to Eurasia in the east, and contemporary biological and geophysical changes. Det Kgl Danske Videnskabernes Selskab Biologiske Medd XIV: pp 75

Jones PD, Osborn TJ, Briffa KR (2001) The Evolution of Climate Over the Last Millenium. Science 292:662-667

Klyashtorin LB (1998) Long term climate change and main commercial fish production in the Atlantic and Pacific. Fish Res 37:115-125

Levitus S, Antonov JI, Boyer TP, Stephens C (2000) Warming of the World Ocean. Science 287:2225-2229

Michalsen K, Ottersen G, Nakken O (1998) Growth of North-east Arctic cod (Gadus morhua L) in relation to ambient temperature. ICES J Mar Sci 55:863-877

Ottersen G, Loeng H (2000) Covariability in early growth and year-class strength of Barents Sea cod, haddock and herring: The environmental link. ICES J Mar Sci 57:339-348

Ottersen G, Michalsen K, Nakken O (1998) Ambient temperature and distribution of north-east Arctic cod. ICES J Mar Sci 55:67-85

Ottersen G, Planque B, Belgrano A, Post E, Reid PC, Stenseth NC (2001) Ecological effects of the North Atlantic Oscillation. Oecologia 128:1-14

Quero JC, Du Buit M-H, Vayne J-J (1998) Les observations de poissons tropicaux et le rechauffement des eaux dans l'Atlantique europeen. Oceanol Acta 21:345-351

Rødseth T (1998) Models for Multispecies Management. Physica Verlag, Heidelberg

Saemundsson B (1937) Andvari, 62 Ar Fiskirannsoknir. Reykjavik 35

Shepherd JG, Pope JG, Cousens RD (1984) Variations in fish stocks and hypotheses concerning links with climate. Rapp P-v Reun Cons int Explor Mer 185:255-267

Toresen R, Østvedt OJ (2000) Variation in abundance of Norwegian spring-spawning herring (*Clupea harengus*, Clupeidae) throughout the 20th century and the influence of climatic fluctuations. Fish Fish 1:231-256

Variability of the Thermohaline Circulation (THC)

J. Meincke[1*], D. Quadfasel[2], W.H. Berger[3], K. Brander[4], R.R. Dickson[5],
P.M. Haugan[6], M. Latif[7], J. Marotzke[8], J. Marshall[9], J.F. Minster[10], J. Pätzold[11],
G. Parrilla[12], W. de Ruijter[13] and F. Schott[14]

[1] University of Hamburg, Institute of Oceanography, Troplowitzstr. 7,
D-22529 Hamburg, Germany
[2] Niels Bohr Institute for Astronomy, Physics and Geophysics, Juliane Maries Vej 30
DK-2100 Copenhagen Ø, Denmark
[3] Scripps Institution of Oceanography, UCLA, San Diego, La Jolla, CA 92093-0215, USA
[4] ICES/GLOBEC Secretary, ICES, Palaegade 2-4, DK-1261 Copenhagen K, Denmark
[5] Centre of Environment, Fisheries and Aquaculture Science, The Laboratory,
Pakefield Road, Lowestoft, Suffolk NR 33 OHT, United Kingdom
[6] University of Bergen, Geophysical Institute, Allegaten 70, N-5007 Bergen, Norway
[7] Max-Planck-Institut für Meteorologie, Bundesstr. 55, D-20146 Hamburg, Germany
[8] University of Southampton, School of Ocean and Earth Science,
Southampton SO14 3 ZH, United Kingdom
[9] Massachusetts Institute of Technology, Dept. of Earth,
Atmospheric and Planetary Science, 77 Massachusetts Ave., Cambridge, MA 02139 /USA
[10] IFREMER, 155 rue Jean-Jaques Rousseau, 92138 Issy-les-Moulineaux Cedex, France
[11] University of Bremen, FB 5, Klagenfurter Straße, D-28359 Bremen, Germany
[12] Instituto Español de Oceanografia, Corazón de Maria 8, E-28002 Madrid, Spain
[13] Utrecht University, Institute for Marine and Atmospheric Research, PO Box 80005,
NL-3508 TA Utrecht, The Netherlands
[14] University of Kiel, Institute of Oceanography, Düsternbrooker Weg 24,
D-24105 Kiel, Germany
* corresponding author (e-mail): meincke@ifm.uni-hamburg.de

Abstract: Europe's relative warmth is maintained by the poleward surface branch of the Atlantic Ocean thermohaline circulation. There is paleoceanographic evidence for significant variability and even shifts between different modes of thermohaline circulation. Coupled ocean-atmosphere climate modelling allows first insight into the relative role of the various drivers of the Atlantic thermohaline circulation variability, i.e. the North Atlantic Oscillation, the tropical Atlantic variability, the ocean basin exchanges, small scale processes like high-latitude convection, overflows and mixing as well as effects of changes in the hydrological circle, the atmospheric CO_2-content and solar radiation. The strong need for continous model improvement requires concerted efforts in ocean time series observations and relevant process studies. New instrumentation and methods, both for *in situ* measurements and remote satellite sensing are becoming available to help on the way forward towards as improved understanding of North Atlantic climate varibility.

Rationale

About 90 % of the earth's population north of latitude 50 N live in Europe. This is a consequence of its climate which is 5 - 10 degrees warmer than the average for this latitude band, the largest such anomaly on earth (Fig. 1). A change or shift of our climate is thus likely to have a pronounced influence on human society. Europe's relative warmth is maintained by the poleward surface branch of the ocean thermohaline circulation (THC).

Most projections of greenhouse gas induced climate change anticipate a weakening of the THC in the North Atlantic in response to increased freshening and warming in the subpolar seas that lead to reduced convection (Rahmstorf and Ganapolski 1999; Manabe and Stouffer 1999). Since the overflow and descent of cold, dense waters across the Greenland-Scotland Ridge is a principal means by which the deep ocean is ventilated and renewed, the suggestion is that a reduction in upper-ocean density at high northern latitudes will weaken the THC (Doescher and Redler 1997; Rahmstorf 1996). However, the ocean fluxes at high northern latitudes are not the only constituents of this problem. Interocean fluxes of heat and salt in the southern hemisphere, wind-induced upwelling in the circum polar belt and atmospheric teleconnections also influence the strength and stability of the Atlantic overturning circulation (Latif 2000; Malanotte-Rizzoli et al. 2000).

Unfortunately, our models do not yet deal adequately with many of the mechanisms believed to control the THC, and our observations cannot yet supply many of the numbers they need. The problem lies in the wide range of spatial and temporal scales involved. Remote effects from other oceans and feedback loops via the atmosphere that affect the THC require a modelling on a global rather than a regional scale. Smaller scale processes such as exchanges through topographic gaps, eddy fluxes and turbulent mixing must either be resolved or properly parameterised. Our present instrumental observations are even insufficient to detect whether or not the THC is changing at all. Palaeoclimate records, however, show that massive and abrupt climate change has occurred in the northern hemisphere, especially during and just after the last ice age, with a THC change as the most plausible driver. Both paleo-climate records and models suggest that changes in the strength of the THC can occur very rapid, within a few decades (Rahmstorf 2002).

Data sets and model results allowing the recognition of long-term variability, including deep-ocean data, have only recently become available. The data base, however, is too incomplete to determine the underlying large-scale dispersal patterns adequately. An exception to this is the 'El Niño Southern Oscillation' phenomenon, which is fairly well understood with respect to the principal mechanisms involved (McPhaden et al. 1998). Because of the close atmosphere-ocean coupling, it can be predicted months in advance. For extra-tropical and higher latitude regions, this kind of prognosis is not yet possible. Significant research efforts are still required in this area. The northern Atlantic is of special importance in this connection, because of the direct effects of oceanic variations on the climate of Europe (Allen and Ingram 2002).

Science issues

Modes of variability and role of the ocean

Meridional overturning circulation

The earth receives most of its energy from the sun, in the form of short wave radiation. Some of this energy is reflected directly back into space, mainly by clouds, snow and ice, but about 70% is absorbed in the atmosphere, by the land surfaces and the oceans. Ultimately this energy is returned to space by the earth's long-wave temperature radiation. The earth is a sphere and its geometry leads to net gain of radiative energy near the equator and a net loss at high latitudes. The excess tropical heat is transported poleward by the atmosphere and the oceans. Each carries about half of this heat transport, the atmosphere by meso-scale eddy fluxes and the ocean by the overturning circulation as well as eddy fluxes. In the North Atlantic the Gulf Stream and the North Atlantic and Norwegian Currents transport warm water northward all the way to the Arctic Ocean. Along their path the currents continuously give their heat to the atmosphere and are so the main reason for the mild climate of Europe. The cooling of the water makes it so dense that it mixes to great depth and eventually returns as a deep and cold current flowing southward to the Atlantic. This

a)

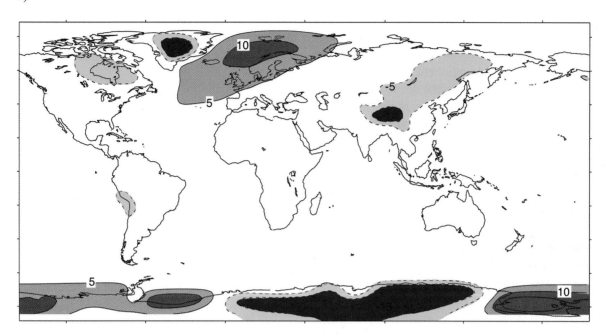

b)

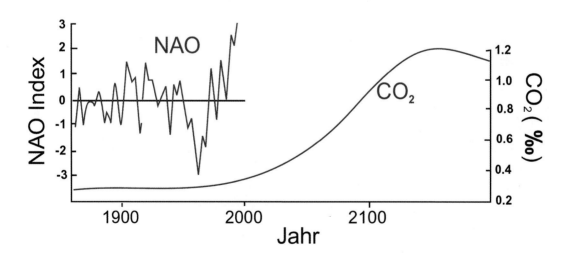

Fig. 1. a) Annual mean surface temperature anomalies form NCAR data, relative to zonal averages. There is a 5-10°C warm anomaly over NW Europe and the Nordic Seas, making this area highly sensitive to changes of the northward oceanic heat transport provided through the THC (Rahmstorf et al. 1999). **b)** Two examples for changes in atmospheric forcing: Observed changes of the NAO-index (3year running mean) and predicted increase of atmospheric CO_2 (after Rahmsdorf and Dickson pers. com.).

system is called "meridional overturning circulation" or "thermohaline circulation" which in this form is unique to the Atlantic Ocean (Schmitz and McCartney 1993) (Fig. 2).

The high northern latitudes and the ocean fluxes that connect them to adjacent seas provide the source of water masses for the deep branch of the THC. The circulation, however, is also driven by the sink terms of the deep water, i.e. the upwelling back into the upper layers of the ocean. This upwelling can either be directly driven by the winds, as in the case of the circumpolar belt, or induced by mixing across the base of the thermocline (Broecker 1991). The mixing is a complicated function of the circulation itself and its interaction with the topography and the stratification depends to a large extend on the local tidal and wind forcing (Munk and Wunsch 1998).

In addition, air-sea fluxes and interocean fluxes of heat and freshwater in the southern hemisphere are also controls on the strength and stability of the Atlantic overturning circulation. The characteristics of the northern sinking waters, in particular their salinity, is determined by the advective inflow into the southern Atlantic and by the integral surface buoyancy fluxes at the air-sea interface along its route to the Nordic Seas.

Long term variability and abrupt climate shifts

Ocean circulation models have shown that the thermohaline circulation can take on more than one stable equilibrium condition (Stommel 1961; Manabe and Stouffer 1988) (Fig.3). This results in the basic dynamic possibility of transitions between two conditions, or an abrupt breakdown of the thermohaline circulation within a period of only a few decades or even years. Model computations suggest that these processes have played a part in the observed rapid climate changes in the past (Marotzke 2000). The question whether such a breakdown can result from greenhouse warming

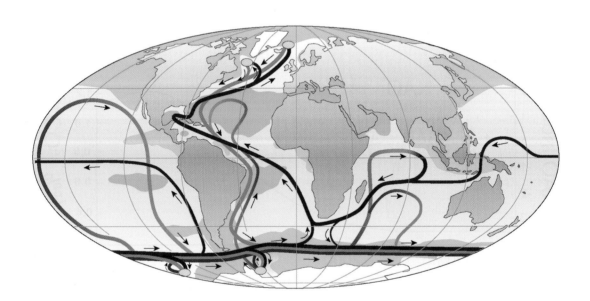

Fig. 2. Cartoon of the global thermohaline circulation (modified from Broecker (1987) by Rahmstorf 2002). Sinking of surface waters (red) in the yellow regions are the sources for deep currents (blue) and bottom currents (purple). Green shading indicates high surface salinities above 36/psu. Deep Water formation rates are estimated at 15+2 Sverdrup (1 Sverdrup= $10^6 m^3 s^{-1}$) in the North Atlantic and 21+6Sv in the Southern Ocean. The northward heat transport in the surface circulation warms the northern North Atlantic air temperatures by up to 10°C over the ocean (from Rahmstorf, reprinted with permission from Nature, Copyright (2002), Macmillan Magazines Limited).

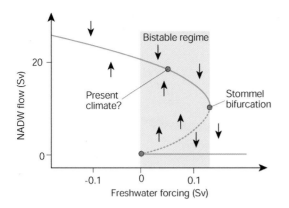

Fig. 3. The thermohaline circulation is nonlinear due to the combined effects of temperature and salinity on density. The plot illustrates the stability properties of the THC. Here the strength of the North Atlantic deep water transport is plotted against the freshwater input into the Atlantic (from Rahmstorf, reprinted with permission from Nature, Copyright (2002), Macmillan Magazines Limited).

of sediment horizons with increased amounts of ice-rafted debris (Heinrich Events) provide the most convincing sedimentological evidence of such short-term climate swings during the past ice ages (Heinrich 1988) (Fig. 4). These can be attributed to periodically recurring instabilities, for example, of the Laurentide ice sheet in northern North America, causing intensified glacial break-off and increased ice berg drifts in the northern North Atlantic. These massive calving episodes occur at the ends of middle-term cooling phases that extend over periods of seven to ten thousand years. Temperatures in the North Atlantic region decreased steadily during these times while the land ice sheets underwent significant growth. Sediment cores taken from various sites in the North Atlantic at depths of one to four thousand meters indicate that carbon isotope values and carbonate content are clearly decreased during periods of increased ice-berg drift. This points to a restriction or even suspension of the thermohaline circulation in the North Atlantic, which presumably would have contributed to a further cooling of the North Atlantic region (Keigwin et al. 1994).

is still not resolved and represents an important topic for further research.

In time series of paleoceanographic temperature estimates compiled from sediment cores in oceanographically sensitive regions of the North Atlantic, rapid climate changes have been demonstrated, especially in the form of changes in temperature-sensitive planktonic floral and faunal communities (Bianchi and McCave 1999; Sarntheim et al. 1994). The sporadic occurrences

The existing data sets indicate a global cooling of about one degree Celsius during the "Little Ice Age", the most significant climate event of the past 1,000 years for the northern hemisphere. This phenomenon lasted from the 15th to the beginning of the 19th century (Bradley and Jones 1995). The subsequent period of natural global warming over-

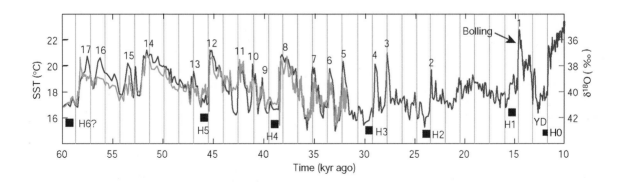

Fig. 4. Temperature reconstructions from subtropical Atlantic sediment cores (greenline, Sachs et al. 1999) and Greenland ices cores (blue-line, Grootes et al. 1993). Heinrich-events (H) are marked red, Dansgaard-Oeschger–events are numbered (from Rahmstorf, reprinted with permission from Nature, Copyright (2002), Macmillan Magazines Limited).

laps with the effects of increased industrial carbon dioxide emission and is used to study in detail the anthropogenic influence on the carbon cycle over the past 200 years. The natural and anthropogenic influences on the climate trend of the past 100 years are not easily distinguished. The "Little Ice Age" will be at the center of future Holocene climate research because it can be applied as a natural climate experiment, acting as a background upon which to interpret the anthropogenic influence on climate (Broecker 2000). Data profiles from 600-year-old sponge skeletons yield smaller temperature changes of less than one degree at the sea surface. From this it can be concluded that the post-Middle Ages cooling occurred primarily on the continents.

NAO/AO related variability

In mid-latitudes, the leading mode of atmospheric variability over the Atlantic region, the North Atlantic Oscillation (NAO), is profoundly linked to the leading mode of variability of the whole northern hemisphere circulation, the annular mode or Arctic Oscillation (AO) (Hurrell and v. Loon 1997; Deser 2000). This suggests that Atlantic effects are more far-reaching and significant than previously thought. The NAO exerts a dominant influence on temperatures, precipitation and storms, fisheries and ecosystems of the Atlantic sector and surrounding continents (see Hurrell 1995) (Fig. 5). It is the major factor controlling air-sea interaction over the Atlantic Ocean and modulates the site and intensity of the sinking branch of the ocean's overturning circulation. The NAO also seems to play a central role in real or perceived anthropogenic climate change. Understanding of the NAO and its time-dependence appear central to three of the main questions in the global change debate: has the climate warmed, and if so why and how? The THC in the North Atlantic accounts for most of the oceanic heat transport and is a major player in decreasing the pole-equator temperature gradient. The possibility of a significant weakening of the THC under global warming scenarios is a feature of coupled general circulation models (GCM) (Wood et al. 1999). This idea remains contentious. Yet, because of its large potential impact, the possibility

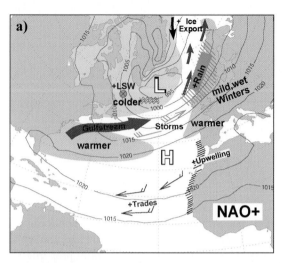

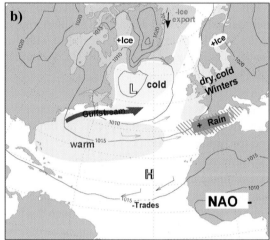

Fig. 5. a) Composite of effects of a NAO-extreme positive state. The enlarged pressure difference between the Azores High (H) and the Iceland Low (L) causes an intensified band of westerly winds stretching from southwest to northeast across the northern North Atlantic. Warm and humid air masses reach north western Europe and the Arctic, the North Atlantic Current and the ice flux from the Arctic are intensified. Massive outbreaks of cold arctic air affect the Labrador Sea and intensify convection. Enhanced trade winds cause strong upwelling off Northwest Africa. **b)** Composite of effects of a NAO-extreme negative state. A diminished pressure difference between the Azores High (H) and the Iceland Low (L) reduces the intensity of the band of westerly winds, which is more zonally oriented and reaches south eastern Europe. Northern Europe experiences dry and cold winters. The oceanic circulation is reduced, convection in the Labrador Sea is ceased.

that increased fresh-water input and atmospheric high-latitude temperature could suppress the THC must be taken seriously (Rahmstorf 2000).

Existing observations indicate indeed that decadal fluctuations of the northward water-mass and heat transport, as well as convection in the Labrador Sea, the European Nordic Seas, and the Arctic are dominated by the NAO. Phases of pronounced Azores highs and Iceland lows correspond with increased water-mass and heat transport of the North Atlantic Current and intensification of Labrador Sea convection. Interpretations of the still very incomplete data base indicate that less pronounced pressure differences weaken the North Atlantic Current and Labrador convection on the one hand and, on the other, strengthen convection in the Greenland Sea and over the Arctic shelf. Fluctuations in convection are synonymous with fluctuations in deep-water formation in the North Atlantic. This provides a significant mechanism for variations in the THC (Dickson et al. 2000).

The export of sea ice from the Arctic Ocean to the North Atlantic is also related to the oscillations observed there. The resulting meltwater fluctuations appear as salinity anomalies in the sub-polar gyre and can, as occurred during the so-called 'Great Salinity Anomaly' of the 70's, interrupt the primarily locally stimulated convection processes in the Labrador Sea and the Greenland Sea (Dickson et al. 1996). Initial investigations with an ocean-atmosphere model that also considers sea-ice processes has resulted in a process sequence: 'convection – intensity of the THC – intensity of the NAO – convection' that can be viewed as interdecadal variation in the North Atlantic (Morison et al. 2000). This result strongly suggests that low-frequency changes in the North Atlantic need to be considered with respect to a coupling concept, and that time series of relevant parameters can be expected to provide a certain degree of predictive potential.

This is interesting with respect to the correlation between decadal fluctuations of surface temperatures in the tropical Atlantic and deep-water production in the Labrador Sea, whereby the maximum temperature contrast is observed in the equatorial region about five years after the maximum convection depth in the Labrador Sea (Yang 2002).

To determine whether these relationships are the result of THC fluctuations and transport of Labrador Sea water to the tropical Atlantic, or if atmospheric coupling is also involved, will require further intensive investigations.

Tropical Atlantic variability

The equatorial zone is considered to be another key region for the Atlantic meridional circulation. Recent investigations with modern current-measurement arrays and high-resolution numerical models have shown that the circulation structure in the equatorial region is very complex (Schott et al. 1998). For example the warm water flowing northward in the western Atlantic makes large detours to the east in the equatorial zone before it continues into the Caribbean and the Florida Current or the Gulf Stream. This is caused by several under- and counter currents. The southward flowing deep water from the North Atlantic, part of the THC's cold branch, also seems to be subject to such detours and transformations (Stramma and Rhein 2001).

Shallow meridional circulation cells coupling the tropics with the mid-latitudes have been detected in the Pacific (Gu and Philander 1997; Latif and Barnett 1996) and are believed to exist also in the Atlantic Ocean. These advect decadal temperature anomalies from the subduction regions of the eastern subtropics towards the equator where they rise to the surface by upwelling and trigger unstable thermodynamic interactions between the ocean and atmosphere. A further mechanism is postulated for the coupling of the tropics with mid-latitudes. ENSO events could initiate meridionally travelling Kelvin Waves that, in turn, could induce slow, westward moving Rossby waves in the mid-latitudes. These then lead to decadal-scale shifts of the western boundary currents causing changes in the ocean's northward heat transport.

Variability with periods of a few years has been observed in the tropical Atlantic, with strong effects on the regional climates, specifically precipitation (Folland et al. 1986; Enfield 1996; Latif 2000). It appears to be an atmopheric teleconnection originating from the Pacific El Niño -Southern Oscillation (ENSO). An open research question so far is how this variability affects the cross-equatorial exchange of the large-scale THC and if it has consequences reaching far beyond the equatorial zone. There is also an independent Atantic equatorial variability of quasi-biennial period. It is the equatorial Atlantic counterpart to the Pacific ENSO, with strongest SST anomalies in the eastern equatorial Atlantic. It influences strongly the precipitation over western Africa and eastern South America. Again, its role in modifying cross-equatorial exchanges is unknown.

Indian Ocean modulation of the THC

The overturning circulation of the Atlantic is controlled not only by the heat and fresh water fluxes at its surface but also by the exchanges with the neighbouring oceans. Deep waters from the North Atlantic exported to the Indian and Pacific Oceans via the Antarctic Circumpolar Current have an upper layer return transport that enters both via the Drake Passage and around South Africa. In the latter case warm and salty Indian Ocean waters are injected into the Southeast Atlantic and propagate northward with the overturning circulation (De Ruijter et al. 1999). Modelling studies indicate that the strength of the overturning circulation responds significantly to variations in this interocean transport by its direct impact on the large scale density gradient (Weijer et al. in press). Thus an enhanced input of saline Indian Ocean waters strengthens and stabilizes the overturning circulation. A reduction would lead to a decreasing strength of the overturning and could bring it closer to a state of reduced stability and enhanced variability. In that case the destabilizing impact of fresh water inputs via the Arctic connections would become more dominant.

Effect of small-scale water mass transformation on the large-scale THC and fluxes

Shelf and open ocean convection

Deep-reaching subduction processes within water masses, such as convection in the open ocean or on shelf slopes, are of critical importance for the large-scale, three-dimensional thermohaline circulation process (Marshall and Schott 1999; Backhaus et al. 1997; Marotzke and Scott 1999). Deep convection can occur in the open ocean when stability of the water column is so low that driving atmospheric forces are sufficient to cause instability. The Greenland, Labrador, and Weddell Seas are key regions for convection events. Convection on the shelf slope is controlled by the accumulation of dense water on the shelves. Slope convection occurs widely in the marginal areas of the Arctic Ocean. Its persistent occurrence in the western Weddell Sea has also been documented. The water masses formed by convection do not enter directly into the large-scale circulation, rather they are subject to further mixing and transformation processes. So far, the representation of convection in circulation models has not been satisfactorily resolved. This will require an intensive investigation of those processes that are responsible for decadal variations in the water mass composition in convection regions, both by model studies and *in situ* measurements.

Overflows and entrainment

There is evidence that the flow of water over submarine channel and ridge systems, so-called overflow, has a basin wide influence on the circulation process. An especially important example of this is the overflow of dense water masses formed in the European Nordic Seas through the cross channels in the Greenland-Iceland-Scotland Ridge (Dickson et al. 2002; Hansen et al. 2001; Dickson and Brown 1994; Doescher and Redler 1997; Käse et al. submitted; Davies et al. 2001) . Although this flow represents only about a one-third share of the deep-water formation in the North Atlantic, it is apparently of basic importance with respect to the water mass composition and the dynamics of basin-wide circulation, as indicated by model simulations. Various aspects of overflow problematics need to be addressed in a more detailed manner than previously to provide an improved understanding of the climate system in the northern Atlantic. Significant points include to what extent the flow-through rates of the channels are hydraulically controlled, which mechanisms determine the dynamics and mixing of the intense, near-bottom slope currents south of the ridge, and how the effect of overflow can be incorporated into large-scale model simulations.

Mixing in the interiour

While decadal fluctuations of the thermohaline circulation cells may be primarily controlled by processes in the deep-water source regions, effects at longer-term, secular time scales can only be explained by considering those small-scale mixing processes that are responsible for the gradual warming of deep waters in the world ocean. Recent investigations indicate that these mixing processes depend on the roughness of the bottom topography and their interaction with near bottom currents, such as caused by tides and meso-scale eddies (Kunze and Toole 1997). Mixing is then achieved through breaking internal waves. Presently almost all model simulations assume an even-shaped distribution of mixing intensities. A more realistic representation of the mixing is required in order to quantitatively determine its influence on long-term changes of the thermohaline circulation.

Relevant components of the hydrological cycle: E-P, river run-off, ice flux

The ocean contain 97% of the earth' fresh water, the atmosphere holds about 0.001% of the total. The ocean plays a dominant role in the global hydrological cycle. Present estimates indicate that 86 % of global evaporation and 78% of global precipitation occur over the oceans (Baumgartner and Reichel 1975). Small changes in the ocean evaporation and precipitation patterns can influence the global hydrological cycle to a large extent, including the terrestrial one, which is so important for many human activities and industries: agriculture, hydro-electrical power production, floods, water resources in general, etc.

Change in the water cycle can also affect the THC of the ocean. Deep water convection could be stopped if surface salinities decrease because of enhanced freshwater input (Bryan 1986; Broecker 1987; Manabe and Stouffer 1995). The main factors controlling the surface salinity are the distributions of evaporation, precipitation, ice and continental run-off, which makes it fundamental to know the hydrologic cycle in the ocean (Schmitt 1995).

In spite of the importance of the ocean in the global hydrological cycle its role is still not well known and understood. Little is known about its average state and its variability. Some of the main causes of this ignorance stem from the difficulties of measuring in-situ some of the variables that play a main role in the water cycle (i.e. precipitation, sea surface salinity), the scarcity of long-term records and the availability of global climatologies.

Atlantic CO_2-storage and –fluxes

Oceanic storage of carbon (Gruber 1998) on seasonal to centennial time scales is determined by an interplay between biologically mediated transport and transformation processes, and physical transport. In particular, the ventilation of deep intermediate water by thermohaline circulation plays a cruical role in removing carbon from the surface mixed layer to the abyss, while the associated large-scale upwelling brings "old water" to the surface. Variability of deep mixing, overflows, ventilation of intermediate water and the compensating large scale upwelling will affect decadal to centennial scale atmospheric CO_2 for given emission scenarios. The first global carbon cycle ocean circulation model scenario runs with reduced North Atlantic overturning circulation (Sarmiento and LeQuéré 1996) showed that not only the strength but also the way, in which this circulation is represented in the model significantly affects the future evolution of the ocean carbon sink.

Currently available data indicate that the strongest total ocean carbon sinks appear to be near convection regions in the North Atlantic –Nordic Seas and possibly in the Southern Ocean. Most models also indicate a large North Atlantic uptake (Wallace 2001) of anthropogenic CO_2 i.e. the perturbation of the air-sea fluxes by increased CO_2 levels in the atmosphere. However, observation based estimates of carbon transports (Holfort et al. 1998; Lundberg and Haugan 1996) indicate that the anthropogenic CO_2 stored in the northern parts of the North Atlantic is mainly advected in from the south. This would be consistent with expectations of maximum uptake of anthropogenic CO_2 over regions where old water is upwelled and equilibrates with the increasing atmospheric CO_2 content. The interplay between ocean carbon uptake across the sea surface and the transport of this carbon to deep sequestration is intimately linked to the thermohaline circulation.

Interannual and interdecadal anomalies in sea ice cover, surface heat loss and circulation strength are expected to affect the associated total and anthropogenic CO_2 uptake. One may hypothesize that NAO related variations e.g. in the distributions of Atlantic and Arctic water in the Norwegian-Greenland Seas and relative contributions of Labrador Seas vs. Greeenland Seas convection (Dickson et al. 1996) would affect carbon transports. Data for quantifiying such effects on the carbon system have been largely lacking so far (Skeljvan et al. 1999, for evidence of strong interannual variability of CO_2 fluxes in the Norwegian Greenland Seas). Coordinated carbon measurements from voluntary observing ships and time series stations are now technologically possible. By combining these data with physical fields, carbon uptake in different parts of the North Atlantic may be estimated as a function of time.

Observed interannual variability in growth rate of atmospheric CO_2 is large compared to annual emissions. This has most often been ascribed to terrestrial biology, but oceanic uptake variations may be of comparable strenght. By quantifiying variability of air-sea exchange in the North Atlantic we have the possibility to better constrain the atmospheric budgets, in particular the strength of terrestrial sinks in Eurasia and North America.

Radiative changes

It is becoming increasingly apparent that solar forcing of the earth's climate is not constant (Cubasch et al. 1997). Total solar radiative output has varied by 0.1% over the last two solar cycles; this is thought to be too small to significantly influence climate, although it may have been larger back in time. Solar output of UV has varied by 10-50% over the last solar cycle; this has possibly affected stratospheric ozone, and thus stratospheric temperatures, with the potential to influence the large-scale dynamics of the troposphere. Recently it has been shown that cloud properties and solar modulated galactic cosmic rays (GCR) are correlated (Svensmark 1998); this introduces a quite different solar influence through a chain involving the solar wind, GCR, and clouds. The suggestion here is that atmospheric ionisation produced by GCR affects cloud microphysical properties and hence their radiative impact on climate. Although these processes are still largely uncertain, historical evidence suggests that solar variability has played a role in past climate change.

How stable is the THC in a Greenhouse world?

Basic description of strength as function of time

Figure 6 from the IPCC report 2001 (IPCC Climate Change 2001) shows the results of nine climate models, for changes in the strength of the Atlantic THC for the period 1850 to 2100. Most models simulate a considerable weakening of the THC, but some models show no trend at all. It is presently not understood at all whether the THC will indeed undergo a drastic weakening in a greenhouse world. Moreover, it is unclear whether such a weakening would be reversible, or rather characterised by the irreversible crossing of a threshold value, followed by an abrupt transition to a qualitatively different circulation state.

The most basic requirements for future research are: A continuous observational record of the strength of the Atlantic THC. Present observations of the THC are insufficient to detect whether it is changing. Long-term observations of the forcing factors of the THC are needed, however for a number of them no single observational estimate exists to this date. The development of climate models that faithfully incorporate all the feedbacks that are important for the stability of the THC are of high priority.

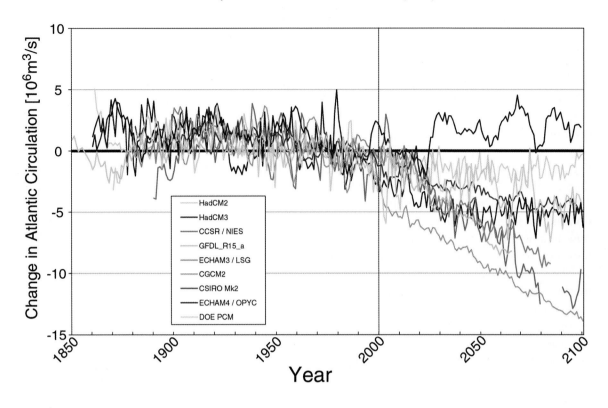

Fig. 6. Change of the volume transport of the Atlantic meridional overturning circulation, predicted from a variety of coupled ocean-atmosphere model. Atmospheric forcing according to the "business as usual scenario" with respect to atmospheric CO_2-increase. (IPCC 2001)

Governing forcing and feedback mechanisms

The mechanisms believed to control the strength of the THC include: the northward flux of warm and salty Atlantic surface water; the freshwater flux out of the Arctic; the speed and density of the deep overflows crossing the Greenland-Scotland Ridge; open-ocean convection; mixing near the ocean margins, including the sea surface; ice-ocean and atmosphere-ocean interactions; freshwater input from the atmosphere and rivers. These processes and transports are poorly observed and understood, and are only crudely represented in the present generation of climate models. Many of the forcing factors listed above are strongly influenced by the dominant modes of atmospheric variability, in particular the North Atlantic Oscillation and related phenomena. In addition, there are a number of potentially crucial tele-connections, such as the influence of prolonged El Niño periods (Schmittner et al. 2000; Latif 2000), the interactions between deep waters of northern and southern origins (Wood et al. 1999; Doescher et al 1997), and the inflow of Indian Ocean waters into the South Atlantic (de Ruijter 1999).

There is apparently a close relationship between climate forcing and deep-water formation, or thermohaline circulation, in the North Atlantic. As a result of various feedback processes, the latter exhibits a strongly dynamic nonlinear behavior. With respect to this, model studies have identified an especially important mechanism called salt-advection feedback, by which the weakening of thermohaline circulation leads to decreased salt transport to the high latitudes. This leads to a further weakening of the circulation, resulting in a positive feedback that is reflected quite well in present models (Rahmstorf 1997). The critical point is that under similar external forcing conditions, fundamentally different equilibrium states can exist.

In addition to salt advection, further various feedback mechanisms have been found, which have been represented primarily only in idealized or low-resolution models, including some only locally operating feedbacks. These include those in which fresh water input at the surface could lead to an interruption of convection and thereby affect thermohaline circulation (Manabe and Stouffer 1997). Feedbacks between thermohaline circulation and sea ice are also viewed as significant factors in some model studies. These local events, however, are not well reproduced in present models. The quantitative role of the various oceanic mechanisms is poorly known and it can only be clarified by inclusion of atmospheric transport information.

The described feedback mechanisms are driven in the models by heat flow and fresh-water flow, which operate at the sea surface. Changes in the driving forces, particularly increased fresh-water input at high latitudes, can initiate a transition of the oceanic system to a different equilibrium state (Fig. 7). The rate of this transformation depends on the respective feedback processes. They generally take place within a few centuries. Based on circulation models, however, the deep circulation can also change fundamentally, up to a complete breakdown, in less than ten years. The causes for this are local processes at the sea surface. Once it is shut down, hysteresis effects can prevent the resumption of circulation, even after the triggering anomaly is no longer in effect. Results of model simulations indicate that climatic states associated with overall cooling in the Atlantic of up to 6 degrees may be possible even in the absence of strong thermohaline circulation (Rahmstorf 2002) (Fig. 8). The sensitivity of thermohaline circulation is not well known as opposed to changes in atmospheric forces and it is also strongly model dependent. Climate model computations for greenhouse scenarios estimate a significant reduction of Atlantic deep convection in the next century. This prognosis, however, is somewhat uncertain, especially because of the low resolution of the ocean model. It is therefore unclear whether the conditions leading to a breakdown of the North

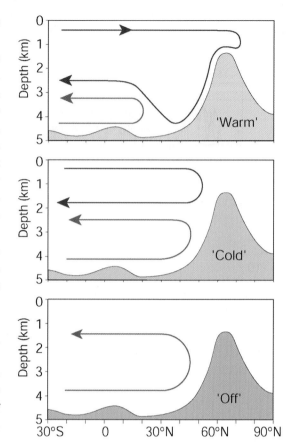

Fig. 7. Three different modes of Atlantic meridional circulation during the last glacial period. The ridge at 65°N symbolizes the Greenland-Scotland ridge. The red line represents the North Atlantic overturning circulation, the blue line represents Antarctic bottom water (from Rahmstorf, reprinted with permission from Nature, Copyright (2002), Macmillan Magazines Limited).

Atlantic Deep Water formation can be attained (Mikolajewicz and Voss 2000; Latif et al. 2000; Rahmstorf 2002; Stouffer and Manabe 1999) .

In principle, abrupt changes such as the breakdown or remobilization of the thermohaline circulation can also occur as a result of internal oceanic processes. In simplified ocean models, more-or-less regular swings of the deep-water circulation are observed, with alternating conditions of weaker and stronger circulation. Depending on the small-scale mixing of various water masses, a particular circulation pattern will remain stable for centuries or millennia, while transitions between the different states may last for only a few decades. However,

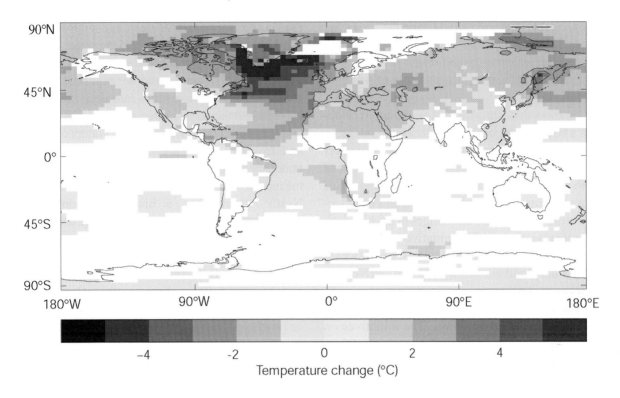

Fig. 8. Changes of surface air temperature caused by a shut down of the North Atlantic Deep Water formation in the coupled ocean-atmosphere circulation model of the UK-Hadley Center (Vellinga et al. 2002, with kind permission from Kluwer Academic Publishers).

only a vague suggestion of such long-term fluctuations has been found so far in coupled models. It is not clear to what extent the signature of these swings agrees with the signals found in paleo data such as the Dansgaard-Oeschger Oscillations.

Implementation issues

Observing the THC variability

Long-term committment for ocean-observations

Several current initiatives will together provide the system of critical measurements that are needed to understand the role of the oceans in decadal to centennial climate variability in the North Atlantic realm. These initiatives are aimed at detecting THC-changes in the North Atlantic and understand

them in the context of coupled variability to the Arctic-Subarctic system in the North and to the Tropical-Subtropical system in the South as well as to the processes internal to the North Atlantic system itself (Fig. 9).

Role of the Arctic and Subarctic seas-the ASOF Array: The basic intention of ASOF is to establish a long-term observational network that will investigate the role of Arctic and Subarctic seas in modulating the overturning circulation in the North Atlantic. The need for ASOF arises because present models do not deal adequately with many of the mechanisms believed to control the THC, and our observations cannot yet supply many of the numbers they need. E.g. we have only embryonic ideas as to the causes of long-term variability in the dense overflows which "drive" the THC, and no measurements of the freshwater flux between the Arctic Ocean and Atlantic by either of its two main pathways that are supposed to shut it down. Understandably then, we would take the view that these key mechanisms and processes are too crudely represented in the present generation of climate models, and it is the business of ASOF to provide them.

To meet that aim, ASOF would plan to establish a coordinated, circum-Arctic system of ocean flux measurements with decadal 'stamina' to cover all of the gateways that connect the Arctic Ocean with Subarctic seas (see Fig. 9). These time-space requirements are set by the decadal and pan-Arctic nature of the observed changes in the high latitude marine climate. The initiative includes a focus on the Labrador Sea as the site through which all the deep and bottom waters that "drive" the THC must pass.

The role of the tropical ocean – the PIRATA and western boundary arrays: The initiatives on time-series work in the tropical-subtropical Atlantic are aimed at linking the Atlantic upper oceanic and atmospheric tropical modes to the ENSO system of the tropical Pacific and to investigate the role that the equator has as a dynamical barrier for the cold and the warm limb of the THC, i.e.the southward flow of North Atlantic deep waters and the northward flow of subtropical/tropical surface waters of Indian Ocean and South Atlantic origin.The PIRATA-Array in the Tropical Atlantic (see Fig. 9 with PIRATA and western boundary arrays at 15N and 11S) is designed to serve as an equivalent of the Pacific TAO array for observing tropical Atlantic variability as well as identify any downstream effects of ENSO on the tropical Atlantic. It consists of Atlas buoys for meteorological observations, which carry T/S sensors for measuring upper-layer and thermocline stratification variability. Along the western boundary, near 11S off Brasil, the German CLIVAR group will maintain a boundary array to measure the variability of transports and water mass properties of the equatorward warm water transport that supplies the inflow into the equatorial zone and links up with the PIRATA observations. Likewise an array for measuring the fluxes of North Atlantic deep water into the equatorial zone will be kept in place near 15N off the Antilles.

Processes internal to the North Atlantic: Processes driving the THC in the North Atlantic are the convective formation of deep and intermediate waters, the overflows of intermediate waters across the submarine ridges separating the Nordic Seas from the North Atlantic, and the export of waters from the convective regimes into deep boundary currents and the ocean's interior. Advection pattern and transformation of upper layer waters of polar and tropical origin are closely linked to the underlying processes mentioned before and so are the modes of atmospheric variability. Although the North Atlantic has a history of continued observational activities there is a scarce data base on decadal variability in the open ocean. Notable exceptions are the time series observations on convection in the Labrador Sea, on water mass variability near Bermuda and on the inflow of Atlantic water to the Nordic Seas. Data are lacking on processes like the export of water from the convective regimes into the boundary currents.

There are several initiatives aiming at longterm measurements of significance for decadal changes (see Fig. 9). They are planning for repeats of trans-Atlantic sections, for a network of timeseries-stations in the open ocean as well as the maintenance of current meter arrays at key locations for boundary currents. Acoustic tomography is being applied to determine integral properties on the basin scale, to monitor changes in convection activity and measure the changes in stratification and heat storage. Together with the continous employment of profiling drifters and satellite-altimetry it will become possible to estimate inventories and changes in fluxes, that will be related to the changes imposed from the Arctic and the Tropics.

The use of paleoclimate data

In order to extend the historical observations of decadal climate variability numerous high resolution paleoclimatic records are available. Potential paleoarchives for oceanic parameters include high resolution sediment records and biogenic skeletal growth chronologies e.g. from corals and molluscs.

Ideally, laminated sediments should be studied to achieve the highest temporal resolution possible in ocean sediments. Recent investigations from the tropical Atlantic (Cariaco basin) revealed many abrupt sub-decadal to century-scale oscillations

ASOF: Arctic Subarctic Ocean Flux Array
RAPID CLIMATE CHANGE: (UK)
SFB-460: Subpolar North Atlantic Study (Germany)
OVIDE: Observation of the Interannual and Decadal
Variability of the Subpolar gyre in the N. Atlantic (France)
ANIMATE: Atlantic Network of Interdisciplinary Moorings
and time series for Europe (EU)
GYROSCOPE: ARGO funded by EU
MOVE: Meridional Overturning Variability Experiment
(Germany)
GAGE: Guyana Abyssal Gyre Experiment (USA)
PIRATA Extensions: SE, SW, NE
EGEE: Study of the Oceanic Circulation and its Variability in
the Gulf of Guinea (France)

Fig. 9. Map showing ongoing and planned oceanographic projects in the Atlantic Ocean for the decade around the year 2000 (Court. CLIVAR-IPO, Southampton).

that are synchronous with climate changes at high latitudes in the North Atlantic (Hughen et al. 1996). Sediment grain-size data from the Iceland basin were used to reconstruct past changes in the speed of deep-water flow. The study site is under the influence of Iceland-Scotland Overflow Water, which is an important component making up the THC (Bianchi and McCave 1999).

The oxygen isotopic composition of benthic foraminifera can provide quantitative reconstructions of upper ocean flows at key locations (Lynch-Stieglitz et al. 1999). These archives reveal a seasonal resolution and comprise a variety of proxy records of environmental and climatic conditions. Long growth chronologies cover several centuries and possibly reach up a 1000 years. Comparison of coral records from the different ocean basins will help to reveal and confirm the different modes and climatic teleconnections between the climate subsystems. Recent coral based research has mainly concentrated on climate modes of the tropical Pacific as well as the Indian Ocean. Future research will exploit the potential of deep-sea corals, which provide one of the rare proxies to document past changes in North Atlantic intermediate and deep water masses (Adkins et al. 1998).

Modelling the THC variability

Climate modelling with improved resolution

Though a range of climate models suggests that greenhouse warming can lead to THC weakening, these models all have relatively crude spatial resolution in their oceanic components. It has never been demonstrated that the THC can undergo dramatic weakening in ocean climate models of the resolution and sophistication that we believe are needed to reproduce quantitatively observed features of ocean circulation, such as the narrowness of fronts and boundary currents. Coupled ocean-atmosphere models with eddy permitting ocean models (1/3 deg) are currently developed world-wide to address the question of the stability of the thermo-haline circulation in more detail.

The great spectrum of current fluctuations characteristic of the global circulation, clearly observed in all measurement programs, represents one of the central problems of computer simulation. Great progress in the modeling of ocean circulation has been achieved in the past decade because of improvements in the measurement database by the WOCE program and by improvements in the capabilities of high-performance computers. High-resolution circulation models driven by realistic atmospheric conditions can now reproduce the fundamental aspects of oceanic eddy activity as well as the principally wind-driven variability at synoptic seasonal and interannual time scales (Fig. 10). A fundamental challenge for future model development is improvement in the representation of some critical, very small-scale oceanographic processes that control the thermohaline circulation

and thereby the reaction of the ocean-atmosphere system at decadal and longer time scales. Among them are the processes of convection, the export from convective regions into boundary currents, the overflows and mixing in the ocean's interior by the interaction of variable flows and topography.

Predictability

The requirements and starting point for any prediction include, first, a dynamic and model-oriented understanding of the relevant processes and, secondly, a quantitative determination of the physical condition of the ocean at a given time. Information from observations alone is not sufficient for either of these purposes, rather, a synthesis of observation data with models is necessary. Significant advances have been made in the development of methods for assimilating data into circulation models. The successful El Niño predictions, for example, would have been impossible without an operational system of data assimilation. Due to their spatial/temporal homogeneity, measurements from moored arrays and satellite observations are particularly well suited for assimilation. Because of the high variability of processes impacting climate and the chaotic nature of circulation caused by medium-scale variability, however, assimilation techniques applied to circulation in the middle and high latitudes are still in the developmental stages. The development of practicable methods for introducing various kinds of observation data into realistic models therefore presents a challenge for the coming years.

Predictions of processes in the high latitudes are not yet possible. It seems certain that decadal-scale oceanic variability in the middle and high latitudes are produced, to a considerable extent, by atmospheric fluctuations, particularly heat flow. These fluctuations, which have a bearing on the inherent nonlinearity of atmospheric circulation, are linked to large-scale patterns such as the North Atlantic Oscillation that are not predictable at present. Longer periods are reinforced by the great thermal capacity of the ocean yielding, in principle, the possibility of a certain degree of predictability for fluctuations in oceanic circulation.

Because of the short 'memory' of the atmosphere, processes in the global ocean-atmosphere

a)   **b)**

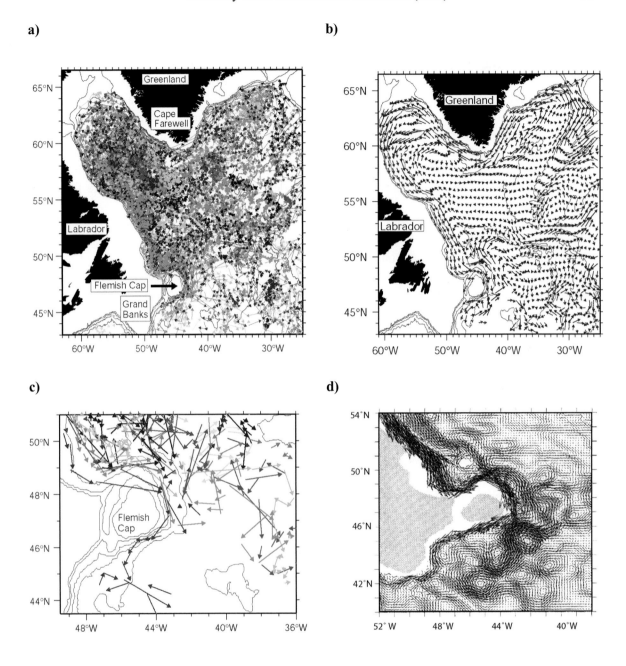

c)   **d)**

Fig. 10. a) Trajectories of floats in the Labrador and Irminger Seas at nominal depths of 400, 700 and 1500m. Each float is represented by one colour, data are from Nov. 1994 to April 1999 (from Lavender et al. 2000). **b)** Objectively mapped mean circulation at 700m depth. Blue arrows: Distances travelled over 30 days at speeds <5cm/s. Red arrows: Distances travelled over 8 days at speed >5cm/s. Note the narrow current filaments and their relation to the topography (from Lavender et al. 2000). **c)** Displacements of floats at 1500m depth at Flemish Cap. Note the incoherent southward flow along the boundary south of Flemish Cap (from Lavender et al. 2000). **d)** Result of a high-resolution (1/12 degr.) model run on the circulation at 1800m depth in the area of the Newfoundland Bank (C. Böning, pers. com). From Lavender et al., reprinted with permission from Nature, Copyright (2000), Macmillan Magazines Limited.

system and thereby the climate can, in principle, only be predicted to the extent that atmospheric variability is induced or reinforced by the ocean. So far little is known about the strength of this reaction. Model computations, however, provide evidence of fluctuations of the coupled system that are produced by feedback from the ocean. They appear to be linked to the time lag of oceanic wave and transport processes whose periods vary between 10 and 60 years depending on the region. It seems to be possible, therefore, to predict at least some portion of the large-scale spatial variability of the Atlantic surface temperature, which is linked to fluctuations of European climate, on a decadal time scale. The extent and limits of predictability are significant research topics for the future.

Linking the shelf seas into the global system

Shelf and coastal seas are components of the global climate system. Their coupling to it is achieved via atmospheric fluxes (wind, heat, freshwater), advective exchanges with the open ocean, and runoff from the continent, including chemical and sedimentological components.. For the North Atlantic the shelf areas of NW-Europe and Canada are especially important since they include the large freshwater reservoirs of the Baltic Sea and the Hudson Bay. Implementation efforts are needed to develop the means for coupling of the global climate models with the shelf sea models of different complexity. Assimilating the numerous long time series observations from shelf stations into a coupled ocean-shelf climate model will allow to discriminate between natural and anthropogenic causes of the observed variability and will lead to identifying regional effects of global climate change.

The ocean's euphotic layer

It is the conditioning of the ocean's very upper layers that determines its primary productivity. Among the controlling parameters are the downward penetration of light, the rates of gas exchange with the atmosphere and the availability of kinetic and potential energy. The scales inherent to these parameters are small, both in the space and the time domain, thus causing the well known patchiness of the plankton distribution. Presently any quantitative modelling of oceanic productivity lacks appropriate data on the actual physical structure of the oceanic euphotic layer. Progress is expected from the combined use of satellite remote sensing of the surface and direct determinations of the TS-structure of the upper layers by the mass deployment of profiling drifters. With these data assimilated into high resolution ocean circulation models a new basis will be available for ocean ecosystem modelling.

Measurement technology

The results of physical oceanographic research at sea described above were also made possible by technical advances in measurement systems. These include Acoustic Doppler Current Profilers that can measure the depth distributions of currents for hundreds of meters, either from a cruising ship or a moored underwater station, employing the back scatter of sound waves from suspended particles (Fig. 11). Another system consists of profiling deep drifters. They drift with the current at pre-determined depths and then ascend to the surface at programmed time intervals to record a profile of the water mass parameters. These data and their positions are sent to a station on land via satellite before they descend again to continue their trip (Fig. 10). A future version of the deep drifters -the gliders- will be able to return to a deep target position after surfacing and transmitting its data, enabling it to provide successive profiles from the same area. This prevents the deep drifter from being carried out of the region of study by current motion and significantly enhances input of data into model simulations. The value of the profiling drifters can be greatly enhanced by adding sensors for biogeochemical parameters to their payload. However care has to be taken to keep the costs for a drifter in adequate balance with the requirement for mass deployment.

Generally, the technology for long-term measurements at moored stations has improved to the extent that maintenance-free placements can be carried out over periods of several years. Considerable efforts are presently put into achieving two-way communication between long term moored

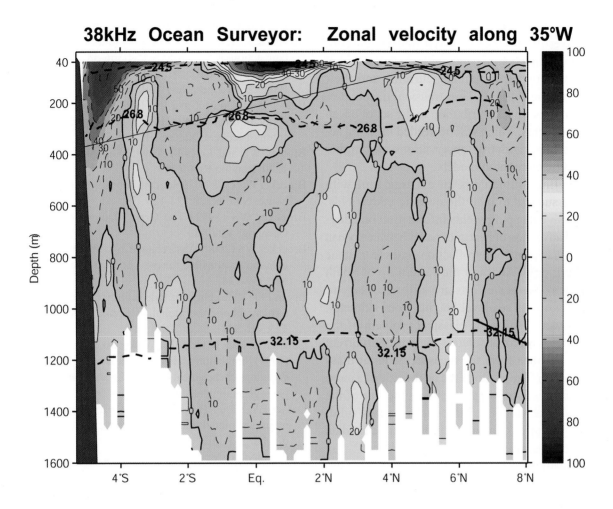

Fig. 11. Observation of zonal velocity in the equatorial region along 35°W. The data were aquired with a 38KHZ RDI-Ocean Surveyor during Meteor-cruise M53 in May 2002 (M. Dengler, F. Schott pers. com.). The figure demonstrates the most recent status in monitoring velocity profiles from underway-measurements.

sensor packages, which allows for data-transmission to shore and for the checking and re-programming of the sampling.

For future demands a measurement network of different autonomous systems needs to be planned that allows for the effective documentation of decadal oceanic variability for selected ocean areas. Present discussions include the installation of profiling CTD's and current meters in key-locations of ocean flux variability and moored measurement systems for acoustic tomography, by which the relatively small temperature changes of the climate signal can be determined on ocean basin scale. In

addition, the mass-deployment of deep drifters is planned. This will yield watermass inventories for the ocean basins and together with the flux-arrays the detection of THC-changes will be possible.

Special effort will be needed to measure the freshwater export from the Arctic system. Whereas first time series of ice export are building up from combined *in situ* and remote sensing measurements, the methods of obtaining the liquid freshwater transport in ice-covered waters are presently still under development.

Considerable improvements in the speed and in the precision of anthropogenic transient tracer measurements can provide data fields, which allow to resolve for the major dynamical features of the circulation components for deep and intermediate waters, e.g. boundary currents and their recirculation cells etc. Thus transient tracers provide an important contribution to the set of ocean measurements, which will be needed in future activities of assimilating observations into higher resolution circulation models.

Satellite altimetry is an existing tool with proven success to map the oceanic eddy fields. Successive mapping yields information on eddy-related advection of heat and matter between ocean basins via the major retroflection zones (e.g. at the southern tip of Africa or the eastern tip of South America). Larger efforts are still necessary to obtain absolute sea level heights from altimetry, which would solve the classical problem of the reference level for absolute 3-dimensional current determinations from the measurements of mass distribution. These efforts are on the side of a more accurate geoid-information and on the side of the *in situ* observations of the oceanic mass field, which has to be compatible to the altimetric data both in time and space.

Measurement techniques should be further developed in the future for small-scale processes, and these should be applied toward quantification of the key processes that drive or modify the large-scale water mass distribution and circulation. These include in particular fine scale measurements of stratification and currentshear in overflows and boundary currents near significant topographic features to obtain direct estimates of entrainment and mixing. The goal is a more realistic parameterisa-tion of the small-scale processes that are not resolved by models.

References

Adkins JF, Cheng H, Boyle EA, Druffel ERM and Edwards RL (1998) Deep-sea coral evidence for rapid change in ventilation of the deep north Atlantic 15.400 years ago. Science 280:725-728

Allen MR and Ingram WJ (2002) Constraints on future changes in climate and the hydrologic cycle. Nature 419:224-232

Backhaus JO, Fohrmann H, Kämpf J, Rubino A (1997) Formation and export of water masses produced in Arctic shelf polynijas – process studies of oceanic convection. ICES J Mar Sci 54:366-382

Baumgartner A and Reichel E (1975) The World Water Balance. Elsevier, New York 179 p

Bianchi GG and McCave IN (1999) Holocene periodicity in North Atlantic climate and deep-ocean flow south of Iceland. Nature 397:515-517

Bradley WS and Jones PD (1995) Climate since AD 1500. London Routledge

Broecker WS (1987) Unpleasant surprises in the greenhouse. Nature 328:123-126

Broecker WS (1991) The Great Ocean Conveyor. Oceanogr 42:79-89

Broecker WS (2000) Was a change in thermohaline circulation responsible for the Little Ice Age? PNAS 97:1339-1342

Bryan FO (1986) High latitude salinity effects and interhemispheric thermohaline circulations. Nature 323:301-304

Cubasch U, Voss R, Hegerl G, Waskewitz J, Crowley TJ (1997) Simulation of the influence of solar radiation variations on the global climate with an ocean-atmosphere general circulation model. Clim Dyn 13:757-767

Davies PA, Käse RH, Ahmed I (2001) Laboratory and numerical model studies of a negatively-buoyant jet discharged into a homogenous rotating fluid. Geophys Astrophys Fluid Dyn 95:127-183

De Ruijter WPM, Biastoch A, Drijfhout SS, Lutjeharms JRE, Matano RP, Pichevin T, van Leeuwen PJ, Weijer W (1999) Indian-Atlantic Inter-Ocean Exchange: Dynamics, estimation and impact. J Geophys Res 104:20.885-20.910

Deser C (2000) On the teleconnectivity of the "Arctic Oscillation". Geophys Res Lett 27:779-782

Dickson RR and Brown J (1994) The production of North Atlantic Deep Water: Sources, rates and pathways. J Geophys Res 99(C6):12319-12341

Dickson RR, Lazier JRN, Meincke J, Rhines PB, Swift J (1996) Long-term coordinated changes in the convective activity of the North Atlantic. Prog Oceanogr 38:241-295

Dickson RR, Osborn TJ, Hurrell JW, Meincke J, Blindheim J, Adlandsvik B, Vinje T, Alekseev G, Maslowski W (2000) The Arctic Ocean response to the North Atlantic Oscillation. J Climate 13:2671-2696

Dickson RR, Yashayaev I, Meincke J, Turrell B, Dye S, Holfort J (2002) Rapid freshening of the deep North Atlantic Ocean over the past four decades. Nature 416:832-836

Doescher R and Redler R (1997) The Relative Importance of Northern Overflow and Subpolar Deep

Convection for the North Atlantic Thermohaline Circulation. J Phys Oceanogr 27(9):1894–1902

Enfield D (1996) Relationship of the Inter-American Rainfall to tropical Atlantic and Pacific SST Variability. Geophys Res Lett 23:3305-3308

Folland CK, Palmer TN, Parker DE (1986) Sahel rainfall and worldwide sea temperatures. Nature 320:602-607

Grootes PM, Strüwer M, White JWC, Johnsen S, Jouzrel J (1993) Comparison of oxygen isotope records from the GISP2 and GRIP Greenland ice cores. Nature 366:552-554

Gruber N (1998) Anthropogenic CO_2 in the Atlantic Ocean. Glob Biogeochem Cycl 12(1):165-191

Gu D and Philander SGH (1997) Interdecadal climate fluctuations that depend on exchanges between the tropics and the subtropics. Science 275:805-807

Hansen B, Turrell WR, Østerhus S (2001) Decreasing overflow from the Nordic seas into the Atlantic Ocean through the Faroer-Shetland Channel since 1950. Nature 411:927-930

Heinrich H (1988) Origin and consequences of cyclic ice rafting in the northeast Atlantic Ocean during the past 130.000 years. Quat Res 29:142-152

Holfort J, Johnson KM, Siedler G, Wallace DWR (1998) Meridional Ocean. Glob Biogeochem Cycl 12:479-499

Hughen KA, Overpeck JT, Petersen LC, Trumbore S (1996) Rapid climatic changes in the tropical Atlantic region during the last deglaciation. Nature 380:51-54

Hurrell JW (1995) Decadal trends in the North Atlantic Oscillation regional temperatures and precipitation. Science 269:676-679

Hurrell JW and van Loon H (1997) Decadal variations in climate associated with the North Atlantic Oscillation. Clim Change 36:301-326

IPCC Climate Change (2001) The Scientific Basis Cambridge-Univ Press 98 p

Käse R, Girton JB, Sanford TB (submitted) Structure and Variability of the Denmark Strait Overflow. Model and Observations

Keigwin LD, Curry WB, Lehmann SJ, Johnsen S (1994) The role of the deep ocean in North Atlantic climate change between 70 and 130 kyr ago. Nature 371:323- 326

Kunze E, Toole JM (1997) Tidally-Driven Vorticity, Diurnal Shear and Turbulence Atop Fieberling Seamount. J Phys Oceanogr 27:2663-2693

Latif M (2000) Tropical Pacific/Atlantic ocean interactions at multi-decadal time scales. Geophys Res Lett 28:539–542

Latif M and Barnett TP (1996) Decadal climate variability over the North Pacific and North America:

Dynamics and Predictability. J Clim 9:2407-2423

Latif M, Roeckner E, Mikolajewicz U, Voss R (2000) Tropical stabilisation of the thermohaline circulation in a greenhouse warming simulation. J Clim 13:1809-1813

Lavender KL, Davis RE, Seeber L, Armbruster JG (2000) Mid-depth recirculation observed in the interior Labrador and Irminger seas by direct velocity measurements. Nature 407:66-71

Lundberg P and Haugan PM (1996) A Nordic-Sea – Arctic Ocean Carbon Budget from Volume Flows and Inorganic Carbon Data. Glo Biogeochem Cycl 10(3):439-510

Lynch-Stieglitz J, Curry WB, Slowey N (1999) A geostrophic transport estimate for the Florida Current form the oxygen isotope composition of benthic foraminifera. Paleoceanogr 14:360-373

Malanotte-Rizzoli P, Hedstrom K, Arango HG, Haidvogel DB (2000) Water mass pathways between the subtropical and tropical ocean in a climatological simulation of the North Atlantic Ocean circulation. Dyn Atmos Oceans 32:331-371

Manabe S and Stouffer RJ (1988) Two stable equilibria of a coupled ocean-atmosphere model. J Climate 1:841-866

Manabe S and Stouffer RJ (1995) Simulation of abrupt climate change induced by freshwater input of the North Atlantic Ocean. Nature 378:165-167

Manabe S and Stouffer RJ (1997) Coupled ocean-atmosphere response to freshwater input: comparison with younger dryas event . Paleoceanogr 12:2321-336

Manabe S and Stouffer RJ (1999) The role of thermohaline circulation in climate. Tellus 51A:91-109

Marotzke J (2000) Abrupt climate change and thermohaline circulation: Mechanisms and predictability. PNAS 97:1347-1350

Marotzke J and Scott JR (1999) Convective mixing and the thermohaline circulation. J Phys Oceanogr 29:2962-2970

Marshall J and Schott F (1999) Open-ocean convection: observations, theory and models. Rev Geophys 37(1):1-64

McPhaden MJ, Busalacchi AJ, Cheney R, Donguy J-R, Gage KS, Halpern D, Ji M, Julian P, Meyers G, Mitchum G, Niiler PP, Picaut J, Reynolds RW, Smith N, Takeuchi K (1998) The Tropical Ocean Global Atmosphere (TOGA) observing system: A decade of progress. J Geophys Res 103:14169-14240

Mikolajewicz U and Voss R (2000) The role of the individual air-sea flux components in CO_2 –induced changes of the ocean's circulation and climate. Clim Dyn 16:327-642

Morison JH, Aagaard K, Steele M (2000) Recent Environmental Changes in the Arctic: A Review. Arctic 53:4

Munk W and Wunsch C (1998) Abyssal recipes II: Energetics of tidal and wind mixing. Deep-Sea Res 45:1977-2010

Rahmstorf S (1996) On the freshwater forcing and transport of the Atlantic thermohaline circulation. Clim Dyn 12:799-811

Rahmstorf S (1997) Risk of sea-change in the Atlantic. Nature 388:825-826

Rahmstorf S (2000) The thermohaline ocean circulation - a system with dangerous thresholds? Clim Change 46:247-256

Rahmstorf S (2002) Ocean Circulation and climate during the past 120.000 years. Nature 419:207-214

Rahmstorf S and Ganapolski A (1999) Long-term global warming scenarios computed with an efficient coupled climate model. Clim Change 43:353-367

Sachs JP and Lehmann SJ (1999) Subtropical North Atlantic temperatures 60.000 to 30.000 years ago. Science 286:756-759

Sarmiento JL and LeQuéré C (1996) Oceanic carbon dioxide uptake in a model of century-scale global warming. Science 274:1346-1350

Sarntheim M, Winn K, Jung SJA, Duplessy J-C, Labeyrie L, Erlenkeuser H, Ganssen G (1994) Changes in East Atlantic Deep Water circulation over the last 30,000 years: Eight time slice reconstructions. Palaeoce-anogr 9(2):209-267

Schmitt RW (1995) The ocean component of the global water cycle (US National Report onto the IUGG, Rev of Geophysics, 33 (Supplement, Pt 2, 1395-1409)

Schmittner A, Appenzeller C, Stocker TF (2000) Enhanced Atlantic freshwater export during El Niño. Geophys Res Lett 27:1163-1166

Schmitz WJ and McCartney MS (1993) On the North Atlantic circulation. Rev Geophys 31(1):29-49

Schott F, Fischer J, Stramma L (1998) Transports and pathways of the upper-layer circulation in the western tropical Atlantic. J Phys Ocenaogr 28(10):1904-1928

Skjelvan I, Johannesson T, Miller L (1999) Interannual variability of CO_2 in the Greenland Norwegian Seas. Tellus 51b:477-489

Stommel H (1961) Thermohaline convection with two stable regimes of flow. Tellus 13:224-230

Stouffer RJ and Manabe S (1999) Response of a coupled ocean-atmosphere model to Increasing atmospheric carbon dioxide: Sensitivity to the rate of increase. J Clim Part 1 12(8):2224-2237

Stramma L and Rhein M (2001) Variability in the deep western boundary current in the equatorial Atlantic at 44°W. Geophys Res Lett 28:1623-1626

Svensmark H (1998) Influence of Cosmic Rays on Earth's Climate. Phys Rev Lett 81(22):5027-5030

Vellinga M and Wood RA (2002) global climatic impacts of a collapse of the Atlantic thermohaline circulation. Clim Change 54:251-267

Wallace DWR (2001) introduction to special section: Ocean measurements and models of carbon sources and sinks. Glob Biogeochem Cycl 15(1):3-11

Weijer W, De Ruijter WPM, Sterl A, Drijfhout SS (in press) Response of the Atlantic overturning circulation to inter-ocean leakages of buoyancy into the South Atlantic. Glob Planetary Change

Wood RA, Keen AB, Mitchell JFB, Gregory JM (1999) Changing spatial structure of the thermohaline circulation in response to atmospheric CO_2 forcing in climate model. Nature 399:572-575

Yang J (2002) A linkage for decadal climate variations in the Labrador Sea and the tropical Atlantic Ocean. (Abstract)

Biogeochemical Processes in Estuaries

R. Wollast

Université Libre de Bruxelles, Laboratory of Chemical Oceanography,
Campus de la Plaine, CP 208, Bd. du Triomphe, 1050 Brussels, Belgium
corresponding author (e-mail): rwollast@ulb.ac.be

Abstract: The hydrodynamic properties of estuaries are extremely variable and depend on the relative ratio of tidal mixing of freshwater and seawater. This ratio influences essential properties of the estuarine system such as the degree of vertical stratification of the water column, the residence time of freshwater, and the existence and intensity of a turbidity maximum. The estuaries, which are the obligated pathways for the transfer of freshwater to the ocean, act often as a filter for many constituents, whose efficiency increases with the residence time of freshwater in the system. This is especially the case for suspended particulate matter, which is often removed to a large extent from the water column by flocculation and sedimentation, generally occurring in the area of the turbidity maximum. Particulate organic matter and trace metals are therefore largely trapped in the estuarine sediments and never reach the sea. The turbidity maximum may, furthermore, generate anoxic conditions, which affect the behaviour of numerous components such as nutrients and redox-sensitive metals like iron and manganese. Finally, the intense biological activity, characteristic of the estuarine environment, is responsible for emissions of various biogases (CO_2, CH_4, N_2O) to the atmosphere. Due to the complexity of the hydrodynamic properties, the existence of strong gradients of chemical and biological parameters as well as the numerous processes occurring in estuaries, the evaluation of elemental fluxes requires the development of elaborated coupled hydrodynamic-biogeochemical models. Large errors could be generated by the use of simplified hypothesis (such as in the AZE method). In addition, estuaries are usually far from steady-state conditions and the estimation of fluxes requires intensive measurements at the boundaries of the systems.

Introduction

Being obligatory pathways between fresh water and marine systems, estuaries play an essential role in the transfer of material from the continents to the oceans. Both the hydrodynamics and the biogeochemical processes characteristic of this environment are however very complex and render the evaluation of the transfer fluxes of dissolved and solid components difficult. Most of the present-day evaluations of the river input to the coastal zone are still based on the concentration of the component of interest at the estuarine mouth multiplied by the fresh water discharge, neglecting the process which may occur within the estuary. In a region heavily populated like Western Europe (see Table 1), the fresh water entering the estuaries is strongly disturbed by human activities and transports excess concentrations of organic matter, nutrients and a large spectrum of organic or inorganic contaminants. In addition, European estuaries are usually areas of intense harbor activities linked to a large variety of industries. Navigation of increasingly large ships implies continuous dredging and other hydraulic works. For these reasons, estuaries are very disturbed ecosystems whose assimilative capacity of pollutants has often been overestimated.

Definition, classification and physical parameters

The most widely accepted definition of estuaries is that given by Pritchard (1967): "An estuary is a semi-enclosed coastal body of water which has a free connection with the open sea and within which sea water is measurably diluted with fresh water derived from land drainage".

However, this definition does not take into account the area of the river subjected to the oscillations of the tide but where only fresh water flows.

From WEFER G, LAMY F, MANTOURA F (eds), 2003, *Marine Science Frontiers for Europe.* Springer-Verlag Berlin Heidelberg New York Tokyo, pp 61-77

River	Mean water discharge ($m^3 sec^{-1}$)	Surface area (10^3 km^2)	Residence time (days)	Inhabitants per km^2	Inhabitants per m^3 sec^{-1}
Elbe	715	146	15-30	230	47000
Rhine	2200	224	2.5	190	19000
Scheldt	120	21.6	30-90	209	37600
Thames	350	14	30	775	31000
Seine	435	78.6	30	211	38000
Loire	830	115	30	135	18700
Gironde	1000	71	30-90	68	4800
Douro	1700	115	3 - 7	101	6800
Rhone	1670	97	2	103	6000
Po	1540	65	2	167	7000

Table 1. Characteristics of some important European estuaries including the surface area of the hydrographic basin, the mean residence time of freshwater in the estuarine zone and the density of inhabitants in the basin.

This represents often a large stretch of the river which cannot be ignored in estuarine studies, especially if sediment transport is considered. The limit of tidal influence on surface elevation can be traced as the point where the river bed rises above the sea level, in the absence of dams and sluices. This is the most commonly used limit of the hydrodynamic models. From a geomorphological standpoint, various types of subdivisions can be considered besides the classical coastal plain estuary (or drowned river valley), fjord type estuaries, bar-build estuaries and some lagoons are also included in the definition of Pritchard.

From a biogeochemical point of view, the characteristics of the water circulation induced by the mixing of fresh water and salt water are however more important. In a salt-wedge estuary, fresh water tends to spread out over the denser salt water, which flow landwards along the bottom. Under these stratified conditions, which mainly occur when the river discharge is large and the tidal amplitude low, vertical mixing is limited and the residence time of fresh water in the estuarine zone is very short. In European estuaries, this is typically the case for the Rhine, the Rhone and the Po where relatively little salt water mixes with fresh water. Nearly all fresh water is discharged into the adjacent coastal ocean within the surface water layer giving rise to a well-established river plume and an "external" estuarine circulation induced by density gradients.

The tidal effect becomes more important when the fresh water discharge diminishes and the salt intrusion increases. Friction with the bottom and at the pycnocline interface tend to mix vertically the water masses giving rise to a well-mixed estuary. Under these conditions, the volume of salt water entering the estuary per tide is much larger than the volume of fresh water discharged by the river during the same period of time. The ratio of these volumes is a convenient way to characterize the type of estuarine circulation. Except for the three stratified rivers mentioned above, many of the European estuaries may be considered as moderately to well-mixed estuaries. Due to its large dilution by seawater, the residence time of fresh water and river born material in the estuarine zone may reach several months (see Table 1) and this enables potential changes of its composition and properties.

The circulation pattern in fjords is more complicated because of their greater depth (typically 300-400 m) and often the existence of a sill at the entrance which isolates deeper water from the ocean. Fresh water entering fjords forms a low salinity surface water with little vertical mixing. In the Norwegian fjords, the sill depth is often so shallow that the deeper basin waters remain stagnant for prolonged periods.

In the bar-build estuaries and lagoons, the width of the sea-water inlet is small compared to the horizontal dimensions of the estuary. This inlet does not always allow free passage of sea-water flow at all stages of the tide. In these systems with reduced circulation, the influence of wind may become important. There are many lagoons bordering the coast of Italy, Spain and France in the Mediterranean Sea and those of Spain and Portugal along the North Atlantic. Due to the reduced circulation, lagoons are often highly polluted but they do not contribute significantly to the continental input.

Behaviour and transfer of material through estuaries

We will mainly discuss in this section the processes affecting the transfer of material through moderately to well-mixed estuaries where the residence time of fresh water is long enough to affect the transfer of many dissolved or particulate species from land to sea. In the framework of this book, emphasis will be put on european estuaries.

Particulate matter

The behaviour of particulate matter in estuaries is particularly complex. Most of the particulate matter, organic and inorganic, transported by rivers exhibits a negative surface charge. The increase of ionic strength during the mixing of fresh water and sea-water leads to the compression of the charged double layer. As a consequence, the river borne particles, often colloidal, are flocculated and more able to settle. This process occurs at relatively low

salinities (1 - 5 ‰) and favours sedimentation in this area. Furthermore the hydrodynamic properties of the estuarine circulation also favour the flocculated material to settle and accumulate within this zone (Fig. 1). The net effect of the density current induced by the horizontal salinity gradient, is that the residual bottom flow, integrated over a tidal period, is predominantly oriented landwards. These strong currents are responsible for an intense bottom transport and the river borne solid is trapped to a certain extent by this net non tidal current. In addition these currents are also able to carry upstream, by bedload transport, large amounts of coarse marine sediments. The conjunction of these currents and the flocculation process are favouring the occurrence of a turbidity maximum often located at the upper part of the estuary. This zone is also characterized by intensive shoaling where sediments of continental or marine origin tend to accumulate. This phenomenon is particularly spectacular in the Gironde estuary where the turbidity maximum may extend over more than 100 km and the concentration of suspended matter in the water column may exceed there 100 gl[-1] (Jouanneau et Latouche 1981).

The bottom currents generated by the tides are also responsible for a deposition-resuspension cycle of sediments. If the sediments are organic rich, the resuspension generates anoxic events in the

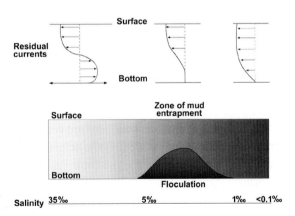

Fig. 1. Influence of density currents on the residual transport of suspended matter and on sediment accumulation in a partially well mixed estuary.

water column which may have deleterious effects on aquatic life. This fact has been well established in the Gironde (Abril et al. 1999), Loire (Thouvenin 1994) and in the Humber estuaries (Uncles et al. 1998). Besides this short term fluctuation, there are also large fluctuations of the concentration of suspended matter influenced by the river discharge. In coastal plain estuaries most of the sediment transport to the sea occurs during flood events, due to the resuspension of the material accumulated in the river and estuarine system during periods of normal discharge.

The evaluation of a budget for suspended material is therefore complex and requires more than discontinuous measurements of turbidity and water velocities at discrete stations. In addition to the longitudinal gradient of suspended matter, there is also a strong vertical gradient due to resuspension by bottom flow. The bed load transport, which is particularly difficult to measure directly, is of major importance in the sedimentary dynamics of an estuary. Finally, dredging complicates further attempts to evaluate budgets for the particulate material. On geological time scales, it is obvious that estuaries and deltas are very active systems characterized by intensive accumulation of sediment and particularly of organic rich muds (Berner 1982). Therefore physical models have been used in the past, to understand the behaviour of cohesive sediments in estuaries and to predict the effects of dredging activities. Today, several mathematical models of the sediment transport in estuaries have been developed and represent probably the best approach to quantify the fluxes of solids in these environments (Burchard and Baumert 1998; Guan et al. 1998; Brenon and Le Hir 1999). Some of these models are able to predict the occurrence and properties of the turbidity maximum.

Natural and artificial radionuclides are very useful markers which allow to identify the origin of the suspended matter and to estimate their residence time within the estuarine zone. For example, some specific radionuclides discharged by reprocessing plants such as those of La Hague and Sellafield, are good tracers for particles of marine origin which can be identified in the European estuarine sediments (Martin et al. 1993). The utilisation of short half-live radionuclides such as [234]Th and [7]Be looks as another

promising tool in order to gain a better understanding of the particle dynamics, especially in the turbidity maximum (Feng et al. 1999).

Dissolved and particulate carbon

Recent estimation of the amount of dissolved (DOC) and particulate organic carbon (POC) transported by rivers on a global basis are in rather good agreement and amount to about 0.4 - 0.5 GtC yr^{-1} (Meybeck 1993; Ludwig et al. 1996) of which 0.1 GtC yr^{-1} is of anthropogenic origin. The organic load is roughly equally distributed between the dissolved and the particulate phase. There are however still large uncertainties concerning the fate of the dissolved and particulate organic matter in estuaries and in the adjacent coastal zone.

It is generally considered that the organic matter entering the estuarine zone is relatively refractory and that only a small fraction can be respired before to be transferred to the shelf. Spitzy and Ittekkot (1991) have evaluated the speciation of the river borne organic matter and concluded that only 15% of both DOC and POC is degradable in the estuarine environment. This agrees well with the observation that DOC behaves quite conservatively in estuaries (Laane 1980; Mantoura and Woodwards 1983) or in other words that the concentration of DOC is simply proportional to salinity in these environments (Fig. 2). This is even true in heavily polluted estuaries such as the Scheldt and the Elbe which receive high loads of anthropogenic organic matter.

The behaviour of particulate organic carbon (POC) is more complex and diverse (Fig. 3). Planktonic and benthic primary production within the estuary may constitute a significant source of organic matter. The primary production is most often limited by the light penetration which is reduced due to the high concentration of suspended matter. The decrease of turbidity in the upper part of the estuary favours the development of phyto-plankton in the brackish water part or in the adjacent coastal zone. The classical pattern often found in estuaries is the production of large amounts of phytoplankton as soon as the turbidity has been reduced due to flocculation and sedimentation of suspended matter of continental origin. In this

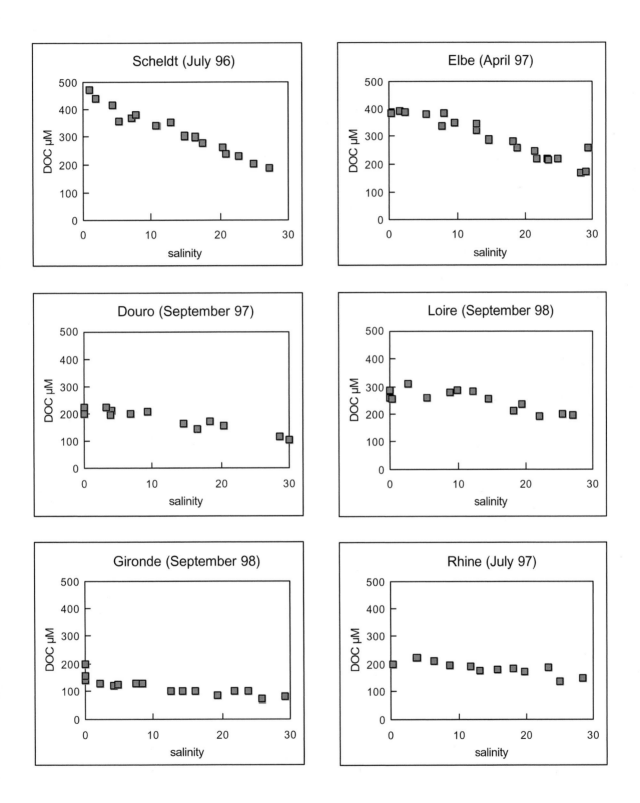

Fig. 2. Evolution of the concentration of dissolved organic carbon as a function of salinity in various european estuaries (BIOGEST Project, database: web site: http://www.ulg.ac.be/oceanbio/biogest/biogest.htm).

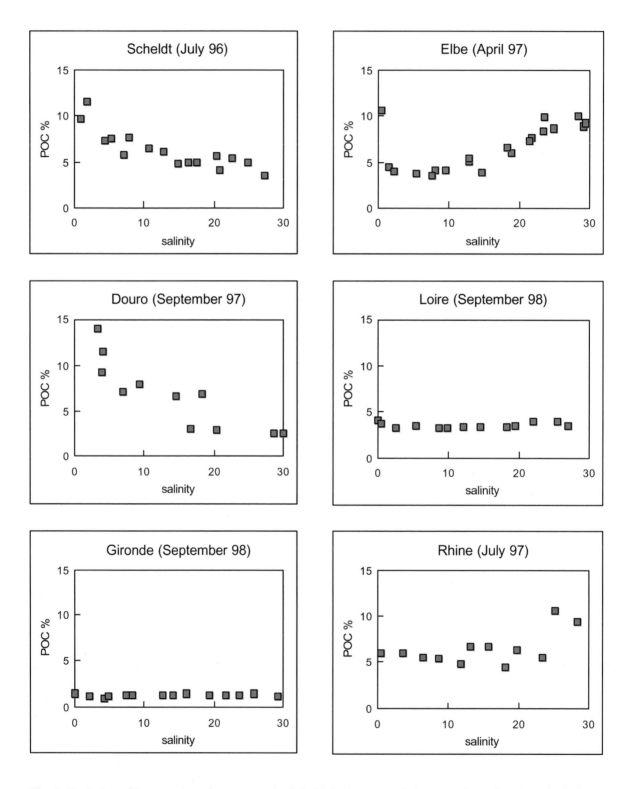

Fig. 3. Evolution of the organic carbon content (weight %) in the suspended matter of as a function of salinity in various european estuaries (BIOGEST Project, database: web site: http://www.ulg.ac.be/oceanbio/biogest/biogest.htm).

case,the percentage of POC in the suspended matter exhibits an increase with salinity. This situation can be seen in the Rhine and Elbe during spring and summer. In highly eutrophicated systems, the maximum of primary production is observed in the fresh water part and the suspended matter is characterized by a high POC content upstream with a progressive decrease with salinity. This is typically the case of the Scheldt and Douro. Note that both processes may coexist in the same estuary. Very high concentrations of chlorophyll-a (often exceeding 100 µg/l) can be observed in the fresh water part of estuaries (Meybeck et al. 1988; Rendel et al. 1997), but this biomass decays very rapidly when the salinity increases (Fig. 4). The decrease of chlorophyll-a is partly due to the mortality of fresh water phytoplankton when salinity increases but also to the flocculation and deposition of the particulate matter.In strong tidal estuaries, the contribution of benthic organisms living on the sand banks is often predominant, which complicates the evaluation of the primary production.

By comparing organic carbon distribution in sediments from the deltas of major world rivers with published data on riverine transport of organic carbon, Berner (1982) demonstrated that a major part of the riverine POC is deposited in estuaries and deltas. The amount of organic carbon trapped in these areas is intimately related to the rate of deposition and accumulation of suspended matter in the estuaries, which is known to be high but remains poorly quantified. It is however well established that the organic rich sediments deposited in estuaries are strongly anoxic and that their resuspension may generate deleterious effects in the water column. Due to the resuspension - deposition cycles of the sediments, turbulent mixing dominates in the benthic boundary layer and enhances the exchange of dissolved constituents across the water-sediment interface.

Respiration measurements as well as investigation of the composition of the pore water indicate that the organic matter is intensively degraded after deposition in the muddy sediments. Anaerobic conditions prevail and one can identify the successive use of nitrate, manganese, iron and sulfate as oxidants. Methanogenesis is also observed mostly in the low salinity zone of estuaries where the

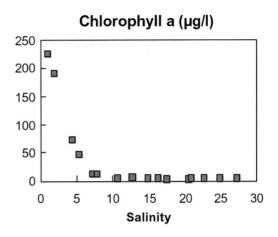

Fig. 4. Evolution of the concentration of chlorophyll-a with salinity in the Scheldt estuary in July 1996 (BIOGEST Project, database: web site: http://www.ulg.ac.be/oceanbio/biogest/biogest.htm)

amount of sulfate available is limited. The products of the respiration are partially transferred to the water column and dissolved Mn^{++}, Fe^{++} and H_2S can be detected in the water column of the muddy areas, even if this water column remains well oxygenated.

The balance between autotrophy and heterotrophy has been significantly modified by human activity but in an antagonistic way. The increase of nutrient load has lead to enhancement of net ecosystem production but on the other hand increased respiration of the organic carbon load has shifted the system to heterotrophy. A compilation of the data published in the literature (Gattuso et al. 1998) indicates that estuaries are most often net heterotrophic with a mean ratio of gross primary production to respiration equal to 0.8 (Fig. 5).

There are still considerable efforts required to understand the factors controlling the primary production, the behaviour and fate of phytoplankton in eutrophicated estuaries.The relative importance of production and consumption of biological particles has been reviewed thoroughly by Heip et al. (1995). It is likely that light is not the only controlling factor of photosynthesis in these systems. Furthermore, the intensive vertical mixing imposes to the phytoplankton short term fluctuation

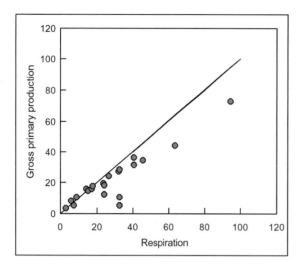

Fig. 5. Gross primary production (Pg) versus respiration (R) in world estuaries according to Gattuso et al. (1998). The straight line corresponds to a 1:1 relationship. The regression equation obtained gives $P_g = 1 + 0.76\ R$; r = 0.92 ; N = 21.

of the light intensity, but it is not clear how this modifies the photosynthetic efficiency of the cells. More specifically, the influence of changing salinity on the photosynthetic activity and mortality needs to be better documented. Finally, the importance of benthic productivity on tidal flats has not been studied carefully until now. According to Heip et al. (1995), it may be in the range of 25 to 50% of the total primary production. Long term respiration experiments of suspended matter or deposited sediments are also necessary in order to quantify the fate of organic matter. The stable isotopes ^{13}C and ^{15}N have been used with success to identify the origin and fate of organic matter in estuaries (Thornton and McManus 1994). Their use should be intensified in order to gain a better understanding of the carbon and nitrogen cycle in these complex environments.

Nutrients

Agriculture, industry and domestic activities, all contribute to the enhancement of the concentration of nutrients in estuaries. In developed and heavily populated countries as in Europe, the concentration of N and P nutrients have been increased almost by one order of magnitude with respect to the pristine levels. The only exception is the concentration of silica which did not increase or has even decreased in several rivers, most often where dams have been constructed, retaining the opal of fresh water diatoms in the sediments of the artificial lakes (Humborg et al. 2000). This is of considerable importance for the coastal zone where the reduced Si flux may contribute to the shift in species dominance from diatoms to flagellates (Billen et al. 1991; Rendel et al. 1997).

Due to the long residence time of fresh water in estuaries and to the intense biological activity, the nutrients exhibit a strong non conservative behaviour. First, because of their multiple source, there are strong seasonal fluctuations of the river input of dissolved N, P and Si species. Furthermore because of the changing physico-chemical conditions and of the strong biological activity, modifications of the nutrient species distribution may occur in the estuarine zone and modify their transport to the sea.

Ammonium originates mainly from domestic waste water, intensive stock farming and *in situ* degradation of organic matter. It is unstable in well oxygenated water and is transformed by nitrifying bacteria into nitrate and nitrate. This process has strong effects on the water quality of the estuary. It has been shown by model calculations (Vanderborght et al. 2002) that nitrification is predominantly responsible for the oxygen minimum observed in the Scheldt estuary (Fig. 6). Abril et al. (2000) came to the same conclusion in the Gironde estuary. Furthermore, the nitrification reaction produces protons, according to the reaction:

$$NH_4^+ + 2O_2 = NO_3^- + 2H^+ + H_2O$$

This reaction is also mainly responsible for the pH minimum often observed in estuaries, which in turn decreases the alkalinity, increases pCO_2 and may led to the dissolution of calcium carbonate.

Nitrate in turn may undergo denitrification, mainly in the area of the turbidity maximum where anoxia is favourable to the bacterial use of NO_3^- as an oxidant. Deposited sediments are also active denitrifying environments inducing the consumption of nitrate, which diffuses from the water column to the benthic boundary layer (Middelburg

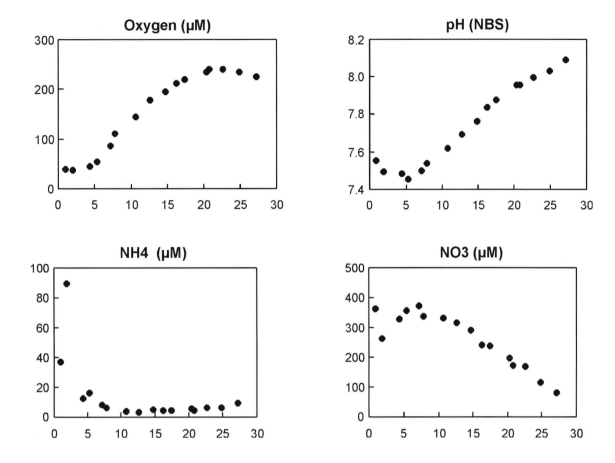

Fig. 6. Evolution of the concentration of oxygen, NH_4^+, NO_3^- and pH with salinity in the Scheldt estuary (July 1996) (after Vanderborght et al. 2002).

et al. 1995). N_2 and N_2O produced during this process are transferred to the atmosphere and lost for the aquatic system. Finally, both nitrate and ammonia are consumed during photosynthesis. The seasonal evolution of N-nutrients is often characterized by a marked decrease of their concentration during the summer, mainly du to the photosynthetic activity of phytoplankton and macrophytes.

All these processes have a strong influence on the transfer of nitrogen species from the continents to the oceans. They occur in all estuaries but to a variable extent, depending on the residence time of fresh water in the system, as shown in Figure 7 resulting from an overview of North Atlantic estuaries by Nixon et al. (1996). According to this study, estuarine process remove 30-65% of the total river born nitrogen and thus a significant fraction of the continental input does not reach the marine system.

The behaviour of phosphorus is not simpler. The freshwater discharge of this element has been essentially affected by the use of polyphosphates in washing powders, degradation of detrital organic matter and to a smaller extent by discharge of industrial waste water. In the estuarine zone it may be further released by sediments under anoxic conditions, in relation to the dissolution of iron oxyhydroxides which have a high affinity for phosphate. Conversely, precipitation of Fe^{2+} in the water column is sequestering phosphate ions which are removed from the water column by sedimentation. Dissolved phosphorus is also significantly consumed during photosynthesis when light is sufficient. The net effect of these processes is a removal of phosphorus from the water column. According to Nixon et al. (1996) 10-15% of the total phosphorus transported by rivers of the North

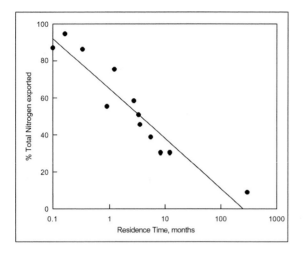

Fig. 7. Percent of total nitrogen input from land exported by estuaries as a function of mean water residence time in the system (after Nixon et al. 1996).

Atlantic area is entrapped in the delta and estuarine sediments and never reach the coastal ocean. This conclusion is however based on a restricted number of estuaries and requires more experimental data.

The behaviour of silica looks less complicated. Since the N:Si and P:Si ratio have been strongly decreased, Si is almost completely removed during the summer when the phytoplankton activity is at the highest. One question however remains : why is the diatom bloom occurring so late in the year compared to what happens in the marine system where the diatom bloom in our latitudes are typically observed in early spring. Light availability may be one explanation. There is a lack of recent studies on the transfer of silica in estuaries and especially on the possible role of eutrophication in the scavenging of this element in estuarine sediments.

The transfer of nutrients through estuaries is often influenced by processes occurring in the sediments. Furthermore, bottom turbulence enhances exchanges with the water column. The early diagenetic processes occurring in the upper sediment may thus significantly affect the composition of the water column in these shallow environments. There is at present a lack of studies of the early diagenesis in estuarine sediments and of the resulting coupling of benthic and pelagic processes.

Trace metals

The distribution of trace elements between the dissolved and the particulate phase is strongly dependent on the physico-chemical processes occurring in the aquatic system. Many trace elements are also involved in biological processes either actively, due to their metabolic role, or passively due to the adsorption capacity of organic particles. Their speciation may also be influenced by the changes of physico-chemical properties related to the biological activity. For these reasons, trace elements exhibit often a non conservative behaviour in estuarine system, where strong gradients of physico-chemical parameters (ionic strength, redox potential, pH, turbidity,...) are observed and where the biological activity is intense. Furthermore, in these shallow areas, the fluxes generated at the sediment-water interface related to early diagenesis may also affect significantly the concentration of trace metals in the water column. We will discuss below, a few examples of the well established behaviour of trace elements in estuaries. Evolution of the concentration of dissolved metals as a function of salinity observed in the Scheldt estuary will be given as a typical example to illustrate the processes which may affect their transfer to the sea.

The non-conservative behaviour of dissolved iron and manganese is a well established fact in many estuaries. The concentration of these elements exhibit high values in the fresh water part (Fig. 8) with frequently the occurrence of a peak corresponding to the turbidity maximum, where anoxic condition may be encountered. Iron and manganese are highly involved as electron acceptor in the bacterial respiration of organic matter in muddy sediments, inducing high concentrations of Fe^{2+} and Mn^{2+} in pore waters. The sediments act thus as a potential source of these metals for the water column. However, dissolved iron and manganese are rapidly precipitated in the overlying water column as oxy-hydroxides when oxic conditions are prevailing or restored. Removal of iron during estuarine mixing is always reported as being high. Martin and Windom (1991) have estimated that on the average 90% of dissolved Fe present in the river water is removed by various estuarine processes on a global scale. It is also

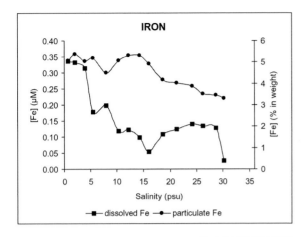

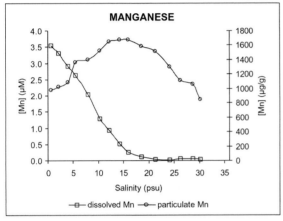

Fig. 8. Evolution with salinity of dissolved and particulate iron and manganese in the Scheldt estuary (November 1995; after Herzl 2000).

observed that the fresh water entering the estuary transport colloidal iron (often analysed as dissolved iron passing through a 0.45 μm porosity filter) which is flocculated and removed from the water column by sedimentation (Sholkovitz 1978). Experiments performed on natural water samples after addition of the radionuclides [54]Mn and [59]Fe (Herzl 2000) indicate that the transfer of iron is mainly due to the activity of bacterial and phyto-planktonic micro-organisms. For manganese, abiotic transfer due to adsorption or precipitation of Mn^{++} appears to be dominant. In addition, photo-reduction of newly formed amorphous Mn oxi-hydroxides occurs also in the lower Scheldt estuary, when turbidity becomes low enough to allow high light intensities. The removal of dissolved Mn in estuaries is less efficient that that of iron due to its

rapid remobilization from reduced sediments (Paucot and Wollast 1997; Martin and Windom 1991). Cobalt behaves very similarly to manganese.

Cadmium is also known for its strong non-conservative behaviour in estuaries (Fig. 9). The concentration of dissolved Cd exhibits first a marked minimum of concentration in the fresh water part of the estuaries and increases rapidly with increasing salinity. Like other heavy metals (e.a. Hg, Pb), Cd has a tendency to produce very stable complexes with the chloride ion. As the salinity increases, the solubility of particulate Cd is enhanced by the formation of the chloride complexes leading to the dissolution of the reactive solid Cd compound. This reaction is very rapid and has been identified even in estuaries with a very short residence time such as the Rhone (Elbaz-Poulichet et al. 1996). The concentration of dissolved Cd then decreases by dilution with seawater. Elbaz-Poulichet et al. (1987) have estimated that this mobilization process might increase the riverine flux of dissolved Cd by a factor of 2 to 30. The behaviour of zinc and copper in estuaries is more complex. Both release and uptake have been reported during mixing of fresh water and sea water (Martin and Windom 1991). Physico-chemical as well as biological processes have been evoked to explain their non-conservative properties. Considering their metablic role, one may expect a rapid uptake of these elements in the area of high productivity and a release when intensive respiration of the detrital organic matter occurs (e.g. Windom et al. 1999).

Furthermore, the solubility of many elements is decreased in the fresh water zone where muddy sediments prevail. This is due to the release of H_2S from anaerobic sediments to the water column. The distribution of dissolved copper and zinc are two examples which have been identified to be affected by the presence of H_2S in the Scheldt estuary (Paucot and Wollast 1997). Nickel and chromium exhibit generally a more conservative behaviour in estuaries.

A large fraction of the particulate metals are trapped in the fine sediments which are accumulating in the estuaries. These metals are either co-precipitated with or adsorbed onto Fe and Mn oxi-hydroxides and clay minerals. They are preferen-

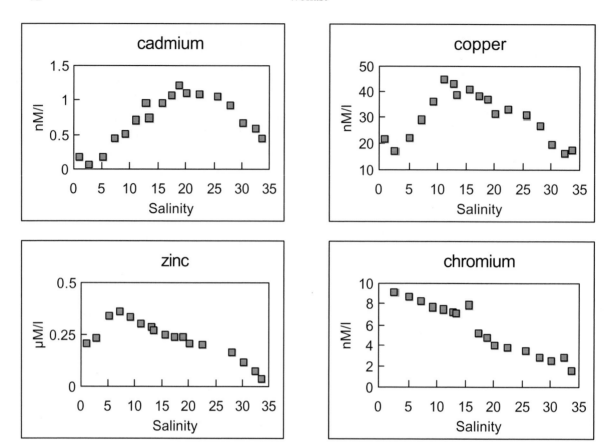

Fig. 9. Evolution of the concentration of dissolved Cd, Zn, Cu and Cr as a function of salinity in the Scheldt estuary in March 1993 (after Paucot and Wollast 1997)

tially deposited in the area of the turbidity maximum where they are often present as sulfides (Fig. 10). Organic contaminants are also associated preferentially to the fine fraction of the sediments (<20μm) but there is a considerable lack of data for these chemicals. The handling and disposal of contaminated muds dredged in estuaries is often a critical management problem. There is only a very few data on the speciation of the trace metals in these sediments and it is thus difficults to predict their behaviour during dredging and after disposal. The behaviour of trace elements and organic contaminants during estuarine mixing needs still large research efforts before an input-output budget can be established for most of the estuaries. Their transfer to the coastal zone remains often a critical, unknown flux.

Biogases

The production and exchange with the atmosphere of several biogases were investigated in various estuaries during the European project BIOGEST. Because of the respiration of detrital organic carbon in the upper stretch of the hydrographic basins, fresh waters entering the estuarine zone are often spectacularly oversaturated in CO_2 with respect to the atmosphere (Frankignoulle et al. 1998). In heavily polluted estuaries such as the Scheldt, the oversaturation exceeds by one to two orders of magnitude the equilibrium concentration, inducing intensive fluxes of CO_2 to the atmosphere (see Fig. 11). The exchange coefficients of the gases with the atmosphere are particularly high in estuaries due to the strong tidal currents. According to

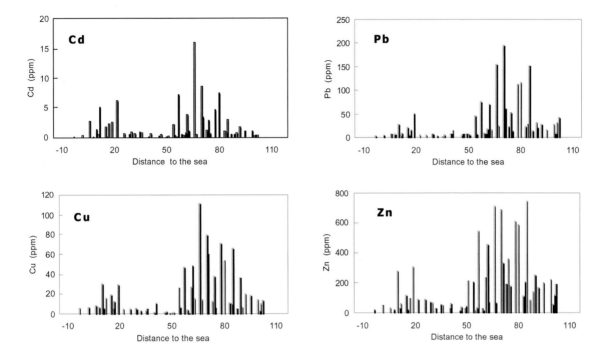

Fig. 10. Longitudinal distribution of trace elements in the sediments of the Scheldt showing their preferential accumulation in the area of the turbidity maximum (km 50-100) (Bouezmarni pers. com.).

Frankignoulle et al. (1998), the atmospheric flux of CO_2 generated by estuaries may represent a significant contribution to the global carbon cycle.

Nitrous oxide is also strongly oversaturated in most estuaries due to the nitrification and to a smaller extent to denitrification, in all the European estuaries investigated in the BIOGEST project (Fig. 11). This gas is not only known for its greenhouse effect but also for its involvement in photochemical reactions in the stratosphere where it destroys ozone. Here again the estuaries may be considered as a significant source of N_2O for the atmosphere and the fluxes are a hundred time those observed in oceanic systems.

In the upper estuary, the concentration of sulfate in fresh water is low compared to that in sea water and SO_4 does not contribute significantly to the respiration of organic carbon in the sediments. A large fraction of reactive organic matter is therefore available for methanogenesis. The methane produced in the sediments is only partially reoxidized in the overlying water column and most

of it is transferred to the atmosphere. High methane concentrations correspond usually to the occurrence of organic rich sediments, mainly in the fresh water part of the estuary Salt marshes associated with estuaries are also well known sources of methane for the atmosphere.

The biogas fluxes estimated for the European estuaries during the BIOGEST project are based on a limited amount of data which should be enlarged in the future. There are also uncertainties concerning the exchange coefficient with the atmosphere either measured with a Lagrangian bell jar or calculated from biogeochemical models (Vanderborght et al. 2002).

Modelling the estuarine processes

Many biogeochemical processes occuring in estuaries have been identified and often studied in detail. It remains nevertheless very difficult to obtain estimates of the overall effect of these processes on the fluxes of components through the mixing

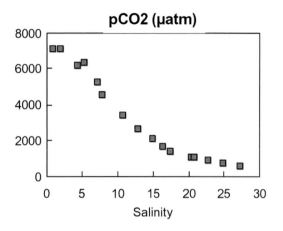

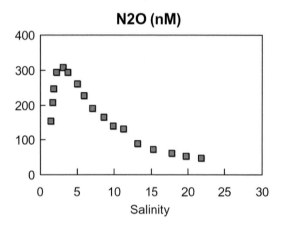

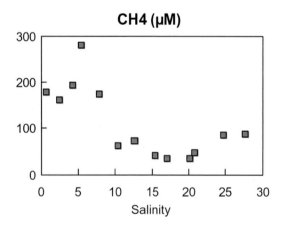

Fig. 11. Evolution of pCO$_2$ (Frankignoulle et al. 1998), of the concentration of N$_2$O (de Wilde and de Bie 2000), and of CH$_4$ (Iversen, BIOGEST database: web site: http://www.ulg.ac.be/oceanbio/biogest/biogest.htm) as a function of salinity in the Scheldt estuary.

zone and on their input to the coastal zone. In the best cases, long time series of measurements of the concentration of dissolved or particulate species in the river water allows to evaluate the fresh water input into the estuarine zone, if the river discharge is known. In most case, however, the frequency of sampling is too low with respect to the fluctuations of the concentration of a large variety of components. There is thus a need for long term, high frequency measurements of the composition of fresh water entering the estuaries.

As demonstrated here above, there are in addition a number of biogeochemical processes which may affect significantly the fluxes of riverborne material in the estuarine zone. The understanding and quantification of these processes is especially complicated in estuaries because of the changing physical, chemical and biological properties along the mixing zone of fresh water with salt water. The hydrodynamics is driven by external forces such as fresh water discharge and tidal amplitude at the mouth of the estuary. The balance among external forcing determines the water residence time, which in turn influences the degree of chemical or biological transformation of terrigenous material. The problem is further complicated by the fact that both river discharge and tidal amplitude are strongly variable at various time scales. They are however usually short with respect to the residence time of fresh water in the estuary and the system is therefore far from reaching steady state conditions. Unfortunately, many attempts to evaluate the transfer of components from the estuary to the coastal zone are based on the hypothesis of stationarity of the estuarine system.

It is the case for the widely used "apparent zero end-member" (AZE) method, (GESAMP 1987), where the concentrations observed in the high salinity range (> 30‰) are linearly extrapolated to zero salinity. The apparent zero end-member concentration is assumed to represent the concentration that would have been observed upstream if only conservative mixing was occurring. This approach assumes also that the distribution of any conservative dissolved element is completely determined by the dilution curve of river water with seawater. Time dependent transport models, which include tidal and river discharge fluctuations, indicate that the residual water flow to the coastal zone differs markedly from the fresh water discharge (Regnier et al. 1998) and that non linear property-salinity plots can be obtained for a conservative element under time varying conditions (tide, river discharge, fresh water composition). The results of the transport model indicate that the fluxes estimated by AZE method are meaningless.

Another method which has also been frequently applied is to use a simple one-dimensional model where it is assumed that the estuarine system is at quasi steady state. It is then easy to compute the longitudinal diffusion coefficient from the distribution of salinity, which is considered to be conservative, and the instantaneous river discharge. Here again the time dependant model shows that this approach introduces large errors especially for estuaries with long residence times, where steady-state conditions are almost never attained. Thus, long term contaminant flux estimates need to take into account the fluctuations of tidal amplitude, river discharge composition of river water.

A one-dimensional full transient model provides usually a satisfactory description of the residual estuarine circulation (Regnier et al. 1998). Implementation of the various physico-chemical and biological transformations allows to resolve the complex, non-linear behaviour of estuarine systems and evaluate much more carefully the fluxes to the adjacent coastal zone (Regnier et al. 1997; Vanderborght et al. 2002). The use of such models becomes an obligated path both to help in the quantification of biological processes occurring in the estuaries and for the evaluation of fluxes to the marine system.

Acknowledgments

This work is partly based on the results of the BIOGEST project supported by the European Union Environment and Climate program (contract nr ENV4-CT96-0213) and the Belgian State - Prime Minister's Services - Science Policy Office in the framework of the Marine Science Programme (Contract n°MS/11/060). I thank J.P. Vanderborght and one reviewer W. Balzer for their critical and very usefull comments. I also thank M. Loijens for editorial handling.

References

Abril G, Etcheber H, Le Hir P, Bassoulet P, Boutier P, Frankignoulle M (1999) Oxic/anoxic oscillations and organic carbon mineralization in an estuarine maximum turbidity zone (The Gironde, France). Limnol Oceanogr 44:1304-1315

Abril G, Riou S, Etcheber H, Frankignoulle M, De Wit R, Middelburg JJ (2000) Transient, tidal time scale nitrogen transformations in an estuarine turbidity maximum fluid mud system (The Gironde, France). Est Coast Shelf Sci 50:703-715

Berner R (1982) Burial of organic carbon and pyrite sulfur in the modern ocean and its geochemical and environmental significance. Amer J Science 282:451-47

Billen G, Lancelot C, Meybeck M (1991) N, P, and Si Retention along the Aquatic Continuum from land to ocean. In: Mantoura RFC, Martin J-M and Wollast R (eds) Ocean Margin Processes in Global Change Dahlem Workshop Report. J Wiley & sons, Chichester pp 19-44

Brenon I, Le Hir P (1999) Modelling the Turbidity Maximum in the Seine Estuary (France): Identification of Formation Processes. Est Coast Shelf Sci 49:525-544

Burchard H, Baumert H (1998) The formation of estuarine turbidity maxima due to density effects in the salt wedge. A hydrodynamic process study. J Phys Oceanogr 28 2:309-321

De Wilde HPJ, de Bie MJM (2000) Nitrous oxide in the Schelde esturay: Production by nitrification and emission to the atmosphere. Mar Chem 69:203-216

Elbaz-Poulichet F, Martin J-M, Huang WW, Zhu JX (1987) Dissolved Cd behaviour in some selected French and Chinese estuaries; consequences on Cd supply to the ocean. Mar Chem 22:125-136

Elbaz-Poulichet F, Garnier JM, Dao Ming Guan, Martin JM, Thomas A (1996) The conservative behaviour of trace matals (Cd, Cu, Ni and Pb) and As in the surface plume of stratified estuaries: Example of the Rhône river (France). Est Coast Shelf Sci 42:289-310

Feng H, Cochran JK and Hirschberg DJ (1999) 234[Th] and [7]Be as tracer for the sources of particles to the turbidity maximum of the Hudson river estuary. Est Coast Shelf Sci 49 5:629-645

Frankignoulle M, Wollast R, Bourge I (1996) Atmospheric CO_2 Fluxes in Highly disturbed estuary. Limnol Oceanogr 41 2:365-369

Frankignoulle M, Abril G, Borges A, Bourge I, Canon C, Delille B, Libert E and Théate JM (1998) Carbon dioxide emission from European estuaries. Science 282: 434-436

Gattuso J-P, Frankignoulle M, Wollast R (1998) Carbon and carbonate metabolism in coastal aquatic ecosystems. Ann Rev Ecol Syst 29:405-434

GESAMP (1987) Land/Sea boundary flux of contaminats: Contribution from river. Rep Stud GESAMP 32

Guan WB, Wolanski E, Dong L X (1998) Cohesive Sediment Transport in the Jiaojiang River Estuary, China. Est Coast Shelf Sci 46, 6:861-871

Heip C, Goosen N, Herman P, Kromkamp J, Middelburg J, Soetaert S (1995) Production and consumption of biological particles in temperate tidal estuaries. In: Ansell AD, Gibson RN and Barnes M (eds) Oceanography and Marine Biology: An Annual Review 33:1-149 UCL Press

Herzl V (2000) Biogeochemical behaviour of iron and manganese in the Scheldt estuary (Belgium) Thèse de doctorat, Université Libre de Bruxelles, 217 p

Humborg C, Conley DJ, Rahm L, Wulff F, Cociasu A, Ittekkot V (2000) Royal Colloquium - Silicon Retention in River Basins: Far-reaching Effects on Biogeochemistry and Aquatic Food Webs in Coastal Marine Environments. Ambio 29:45-50

Jouanneau JM, Latouche C (1981)The Gironde estuary. In: Fürchtbauer H , Lisitzyn AP, Millerman JD, Seibold E (eds) Contribution to Sedimentology. Stuttgart pp 10

Laane RW (1980) Conservative behaviour of dissolved organic in the EMS-Dollart estuary and the western Wadden Sea. Neth J Sea Res 14:192-199

Ludwig W, Probst JL, Kempe S (1996) Predicting the oceanic input of organic carbon by continental erosion. Glob Biogeochem Cycl 10:23-41

Martin J-M, Wollast R, Loijens M, Thomas A, Mouchel J-M, Nieuwenhuize J (1994) Origin and Fate of artificial Radionuclides in the Scheldt estuary. Mar Chem 46:189-202

Martin J-M, Windom HL (1991) Present and future roles of ocean margins in regulating marine biogeochemi-cal cycles of trace elements In: Mantoura RFC, Martin J-M, Wollast R (eds) Ocean Margin Processes in Global Change. Dahlem Workshop Report. J Wiley & Sons, Chichester pp 45-67

Mantoura RFC, Woodward EMS (1983) Conservative behaviour of riverine dissolved organic carbon in the Severn estuary: Chemical and geochemical implications. Geochim Cosmochim Acta 47:1293-1309

Meybeck M (1993) C, N, P and S in rivers: From sources to global inputs. In: Wollast R, Mackenzie FT, Chou L (eds) Interactions of C, N, P and S Biogeochemical Cycles and Global Change. NATO ASI Series, Vol. I4. Springer, Berlin pp 163-193

Meybeck M, Cauwet G, Dessery S, Somville M, Gouleau D, Billen G (1988) Nutrients (organic C, P, N, Si) in the eutrophic river Loire (France) and its estuary. Est Coast Shelf Sci 27:595-624

Middelburg J, Klaver G, Nieuwenhuize J, Markusse RM, Vlug T (1995) Nitrous oxide emissions from estuarine intertidal sediments. Hydrobiol 311:45-55

Nixon SW, Ammerran JW, Atkinson L.P, Berounsky VM, Billen G, Boicourt WC, Boynton WR, Church TM, DiToro DM, Elmgren R, Garber JH, Giblin AE, Jahnke RA, Owens NJP, Pilson MEQ, Seitzinger SP (1996) The fate of nitrogen and phosphorus at the land-sea margin of the North Atlantic Ocean. Biogeochem 35:141-180

Paucot H, Wollast R (1997) Transport and Transformation of Trace Metals in the Scheldt Estuary. Mar Chem Vol 58 1-2:229-244

Pritchard DW (1967) What is an estuary? Physical point of view. In: Lauff GH (ed) Estuaries. AAAS, 83, Washington DC, 158-179

Regnier P, Wollast R, Steefel CI (1997) Long-term fluxes of reactive species in macrotidal estuaries: Estimates from a fully transient, multi component reaction-transport model. Mar Chem 58 1-2:127-145

Regnier P, Mouchet A, Ronday F, Wollast R (1998) A discussion of methods for estimating residual fluxes in strong tidal estuaries. Cont Shelf Res 18:1543-1571

Rendel AR, Horrobin TM, Jickells TD, Edmunds HM, Brown J, Malcolm SJ (1997) Nutrient cycling in the Great Ouse estuary and its impact on nutrient fluxes in the Wash, England. Est Coast Shelf Sci 45, 5:653-668

Thornton SF, McManus J (1994) Application of organic carbon and nitrogen stable isotopes and C/N ratio as source indicators of organic matter provenance in estuarine systems. Evidence from the Tay estuary, Scotland. Est Coast Shelf Sci 38:219-231

Thouvenin B, Le Hir P, Romaña LA (1994) Dissolved oxygen model in the Loire estuary. In: Dyer KR and

Orth RJ (eds) Changes in Fluxes in Estuaries : Implications from Sciences to Management. Academic 169-178

Sholkovitz ER (1978) The flocculation of dissolved Fe, Mn, Al, Cu, Ni, Co and Cd during estuarine mixing. Earth Planet Sci Lett 41:77-86

Spitzy A, Ittekot V (1991) Dissolved and particulate organic matter in rivers. In: Mantoura RFC, Martin J-M, Wollast R (eds) Ocean Margin Processes in Global Change Dahlem Workshop Report. J Wiley and Sons, Chichester pp 5-17

Uncles R, Joint I, Stephens JA (1998) Transport and retention of suspended particulate matter and bacteria in the Himber-Ouse Estuary, UK, and their relationship with hypoxia and anoxia. Est 21:597-612

Vanderborght J-P, Wollast R, Loijens M, Regnier P (2002) Application of a transport-reaction model to the estimation of biogas fluxes in the Scheldt estuary. Biogeochem 59:115-145

Windom HL, Niencheski LF and Smith Jr. RG (1999) Biogeochemistry of nutrients and trace metals in the estuarine region of the Patos lagoon (Brazil). Est Coast Shelf Sci 48, 1:113-123

Trace Metals in the Oceans:
Evolution, Biology and Global Change

H.J.W. de Baar [1,2]* and J. La Roche [3]

[1] *Royal Netherlands Institute for Sea Research (NIOZ), PO Box 59, 1790 AB Den Burg, NL*
[2] *University of Groningen, PO Box 14, 9750 AA Haren, The Netherlands*
[3] *Institut für Meereskunde, Düsternbrooker Weg 20, 24105 Kiel, Germany*
* corresponding author (email): debaar@nioz.nl

Abstract: All living organisms require several essential trace metal elements. During biological evolution of prokaryotes and later on also eukaryotes several metals became incorporated as essential factors in many biochemical functions more or less in accordance with the abundance of these metals on the planet. As a result the biological importance of first row transition metals can be ranked roughly in the order Fe, Zn, Cu, Mn, Co, Ni. The second row metals Ag and Cd or third row metals like Hg and Pb appear to have no biological function, except possibly for Cd. Iron (Fe) being the fourth most abundant element in the crust of the planet has also played a role to temper the evolution of biogenic oxygen (O_2) in the atmosphere and oceans. Yet eventually O_2 has taken over the biosphere where now both atmosphere and ocean are strongly oxidizing. Inside every living cell the primordial reducing conditions have remained however. Therefore enzyme systems based on metal couples Fe-Mn and Cu-Zn are required to protect the cell interior from damage by reactive oxygen species. The key role of metals in these and many other enzymes as well as in protein folding is one of the major vectors in biological diversity at both the molecular and the species level. Abundance and biological role of metals in the oceans are being discovered only since 1976 and many questions remain. Until now many metals appear to be tightly linked with the large scale biological cycling. Recently the significance of very fine colloids as well as dissolved organic metal-complexes has been shown. Plankton ecosystems now appear to be governed by co-limitation of several essential metals, where the biological fractionation of their stable isotopes is expected to give rise to significant shifts in isotopic ratio for any given metal element. Co-limitation of plankton growth is consistent also with observed interactions between metals. Firstly substitution of for example Co or Cd for Zn is known for some but not all phytoplankton taxa. Secondly the cellular uptake of one metal, e.g. Cd, may respond to a complex matrix of other metals like Zn, Mn and possibly Fe. By mining and industrial use of metals, land use change and irrigation, mankind has greatly modified the abundances and mutual ratio's of metals in the biosphere. Biota can to some extent resist the ensuing external stress through cellular homeostasis. However at highly elevated levels or excessive metal ratio's several biological species cannot longer exist and major shifts of ecosystems and their diversity do occur. Likely such changes have already taken place for centuries and as such have largely gone unnoticed. The past decade was the onset of the iron age in oceanography. Nowadays Fe is known to be a severely limiting element in some 40 % of the world oceans. This limitation is sometimes relieved by aeolian supply of continental dust, but most Fe supply is actually from below emanating of reducing sediments. With adequate Fe there is a systematic response by the class of very large oceanic diatoms, their massive blooms then giving rise to uptake of CO_2 and export of both opal (SiO_2) and organic matter into the deep-sea. Hence the supply of Fe to ocean waters is one of the major controls of plankton ecosystems and ocean element cycling (C, Si, N, P) over time scales ranging from weeks to the 100,000 year periodicity of glaciations. Understanding the cycling and biological function of metals in the oceans is a prerequisite for understanding the role of the oceans in global change of past, present and future.

From WEFER G, LAMY F, MANTOURA F (eds), 2003, *Marine Science Frontiers for Europe.* Springer-Verlag Berlin Heidelberg New York Tokyo, pp 79-105

Introduction

At the onset of the twenty-first century it is commonly realized that the oceans, atmosphere and land are interacting within one integral biosphere. Moreover this biosphere consists of myriad interactions between physical, chemical, biological and geological processes. Integration among latter natural sciences is crucial for understanding the biosphere.

Variations of the global biosphere are taking place over short intervals of days (e.g. weather) and seasons, but also over longer climatic timescales from interannual to decadal and ultimately the 20,000-100,000 years periodicities of the glacial-interglacial variations. All these timescales of global change are receiving much attention as they are the background upon which the now confirmed global warming of the planet is superimposed (Mann et al. 1999; Houghton et al. 2001). The about 31 % increase of greenhouse gas CO_2 due to human activity is well known, and various other greenhouse gases are also changing.

This article is focusing on the less well known but crucial role of trace metals in the global biosphere and greenhouse, where changes in the abundance of several of these metals are also due to human activity. Major changes of metals are taking place on land and in the atmosphere. However below article is limited to the oceans. Land and atmosphere are mentioned only when directly relevant for the oceans.

Below we will find that the more abundant first row transition metals Mn, Fe, (Co), (Ni), Cu, Zn have played a key role in biological evolution. Nowadays these metals have functions in every cell of every living organism. Changes in the abundances and cycling of these metals in the oceans by human activity cause changes in the abundance and biodioversity of marine biota. This in turn is causing the re-distribution of major chemical elements like C, N and S between ocean and atmosphere. Therefore the air/sea exchange of greenhouse gases CO_2, CH_4, N_2O, DMS and a suite of biogenic organohalogen gases (e.g. CH_3Cl, CH_3Br, CH_2Br_2, CH_3I) depends also on the role of trace metals in marine ecosystems.

Origins

Origin of the transition metals

Following the Big Bang ~15 Gy (15 x 10⁹ years) ago, only the light elements H and He, together with small amounts of Li, were initially formed (Brownlee 1992). Yet during the ongoing formation and demise of stars, the nuclear fusion reactions within them gave rise to a sequence of heavier elements, in which certain masses which are multiples of ⁴He are favoured: these are ¹²C, ¹⁶O, ²⁴Mg, ³²S, ²⁸Si and ⁵⁶Fe. Some 4.6 Gy ago another star was born: the sun, and soon afterwards the planets evolved. The abundance of the chemical elements in our solar system shows a strong decrease with increasing atomic number, but ⁴He multiples such as Fe, C, O, Mg and Si stand out as distinct maxima in the curve (Fig. 1). Similar curves apply both for the bulk earth including its iron core, and for the outer crust which contains on average about 3.5% iron.

One notices the anomalous high abundance of the metal element Fe, about 1000-fold higher than a smooth curve would suggest (Turner et al. 2001). As a result Fe is the fourth most abundant element in the earth's crust. On the other hand the abundances of the heavier metal elements decrease strongly with increasing atomic number. Hence the first row transition metals are much more abundant than those of the second row, the latter more abundant again than the third row metals like Hg (Table 1).

Transition metals and the origin of life

Sometime after the formation of sun and planets, both the ocean and the atmosphere evolved on Earth. The primordial atmosphere most likely consisted mostly of CO_2 (~ 98%), with smaller amounts of N_2, CO and H_2O, and possibly reduced compounds like NH_3 and H_2; certainly there was no free O_2 present (Holland 1984; Falkowski and Raven 1997). As the first life developed, the chemical elements acquired distinct biochemical functions according to their chemistry and availability. The storage, transport and transfer of energy, often through

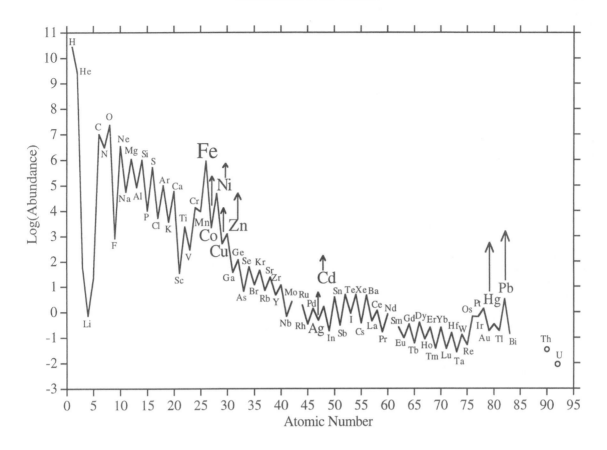

Fig. 1. Cosmic (solar) abundances of the chemical elements, also more or less representative for abundances in the biosphere of planet Earth. Vertical axis is relative to Si, which is given the arbitrary value of 10^6 (one million). Notice logarithmic scale such that each unit represents one order of magnitude. Most relevant here are the first row transition metals Mn, Fe, Ni, Co, Zn, the second row transition metals Ag and Cd, and third row metals Hg and Pb. Arrows indicate rising amounts in the biosphere of metals Co, Ni, Cu, Zn, Ag, Cd, Hg and Pb due to human activity. As a result toxic effects take place in ecosystems as well as human health. Figure drafted after data compilation of Anders and Grevesse (1989). Notice predominance of ^{4}He multiples C, O, Mg, S and Fe as well as even atomic number to odd atomic number predominance. Latter trend very obvious among lanthanide series La through Lu (de Baar et al. 1985) where odd-numbered element 61 (Pm not shown) is extreme case of unstable element not any longer existing on earth. Similarly element 43 (Tc not shown) is missing. Unstable daughters in radioactive decay series of Th and U do exist but in trivial atomic abundances and therefore not shown either (elements $_{87}$Fr, $_{88}$Ra, $_{89}$Ac and $_{91}$Pa).

redox reactions, are key functions for all living organisms. Several metals of the first transition series (Mn, Fe, Co, Cu) are particularly well suited to many of these functions, having pairs of oxidation states differing by only one electron and relatively close in terms of free energy (Frausto da Silva and Williams 1994). In the suboxic or mildly re-

ducing Archean ocean (3.8 – 2.5 Gy before present) these metals would have been readily soluble as Mn^{2+}, Fe^{2+}, Co^{2+} and $CuCl_2^-$ respectively (Saager et al. 1989; Lewis and Landing 1991) and would no doubt have been present at relatively high concentrations in an ocean where reduced sulphide is also present (Turner et al. 2001).

Mn	Fe	Co	Ni	Cu	Zn	
9550	900000	2250	49300	522	1260	
				Ag	Cd	
				0.49	1.61	
				Hg	Pb	
				0.34	3.15	

Table 1. In bold print the first row transition metals manganese, iron, cobalt, nickel, copper and zinc which are all involved in biological evolution and biochemical processes in living cells. Their biological importance can be ranked roughly in the order Fe, Zn, Cu, Mn, Co, Ni. (Frausto da Silva and Williams 1994). Iron is abundant on land but limiting for life in 40 % or more of the oceans. Numbers are the abundance at our planet relative to one million Si atoms. The second row metals silver and cadmium are much less abundant at our planet and not (Ag) or possibly (Cd) involved in biological evolution. Third row metals mercury and lead are also not abundant at our planet (with the somewhat higher Pb abundance due to ingrowth of several stable Pb isotopes as end-products of the slow U-Th radioactive decay chains). Mankind has increased the availability of all these metals in the biosphere, with ensuing ecological interferences among these metals causing re-organization of ecosystems with concomitant shifts in the global cycling of major elements such as carbon and nitrogen.

This allowed the key position of metals in many cellular functions, for example the ubiquity of the Fe-S rich ferredoxin (Wächtershäuser 1992; Russell et al. 1993) in all biological systems. Another key function is the Fe-Mn superoxide dismutase characteristic of prokaryotes. On the other hand Cu and Zn would be mostly bound as insoluble sulphides and not readily available for biological functions. This is even more true for the heavier second row metals Ag and Cd due to their much lower initial abundance in the solar system (Table 1)

The distribution of the three elements C, O and Fe underwent a slow but far-reaching change following the advent of photosynthesis. As more and more free O_2 was generated by photosynthesis, which also consumed atmospheric CO_2, some of the resulting organic carbon became fossilised and was stored in soils and sediments. Initially, for a period of more than 1 Gy, this oxygen from photosynthesis then was reduced by the available Fe(II), leading to the vast formations of iron oxide deposits of the Banded Iron Formations (BIF). However, eventually there was no Fe(II) remaining to cap-ture the oxygen and an oxygenated atmosphere and ocean evolved. From the time record of the BIF formation period it has been suggested that the first free O_2 in the atmosphere may have evolved about 2 Gy ago (Broecker 1985). Since then the oxygen in the atmosphere has increased and then stabilised at the current level of about 21 percent. The end result is a profound redistribution of the elements C, O and Fe.

The photosynthetic production of free oxygen (O_2) and somewhat concomitant evolution of eukaryotes (Margulis 1981) led to an oxygen-rich atmosphere and ocean where the metals Cu and Zn became more available, while Fe and Mn decreased by formation of insoluble oxides. As increasing biological evolution continued, Cu and Zn became increasingly important as co-factors in proteins as for example in cytoplasmatic superoxide dismutase of eukaryotes, while the mitochondria retained the ancestral Fe-Mn superoxide dismu-tases common in the prokaryotes. Also the many other biochemical functions of Fe and Mn already acquired at early stages remained. Moreover Cu in fact had a somewhat intermediate position as it was already a bit more available as $CuCl_2^-$ during primordial anoxia than Zn, Ag or Cd was at that time.

From our anthropocentric viewpoint we see the advent of photosynthesis some 2 Gy ago as a highly positive development which ultimately led to the development of our own species. However, the resulting redistribution of O and Fe cannot be interpreted as unreservedly positive. Oxygen is a toxic, reactive gas, and the intermediate products formed during its biological reduction to water (e.g. O_2^-, H_2O_2) are more reactive and dangerous still. Thus oxygen-tolerant organisms have developed elaborate defence mechanisms against these Reactive Oxygen Species (ROS), notably the super-oxide dismutases with metal atoms Cu-Zn or Fe-Mn as key components. Latter metals being crucial as anti-oxidants, the vitality and life expectancy of all organisms including man depends on supply of these metals in suitable ratio's. The second effect, whose consequences are central to past climate change, is the transfer of Fe from readily-soluble Fe(II) to sparingly-soluble Fe(III), which reduced the concentration of available iron by many orders of magnitude (de Baar and de Jong 2001). Similar changes affected several other redox-active and biologically-essential first row transition metals, but it is only in the case of iron that evidence has now been collected that the availability of this metal to marine organisms has become sub-optimal if not outright limiting (de Baar and Boyd 2000).

The complement of the modern O_2–rich atmosphere, combined with the oxygen fixed within the BIF's, is the fossil organic carbon accumulated within sedimentary deposits, being estimated at 12 million gigatonnes (12×10^{21} grams C). Thus very little CO_2 remains in the modern atmosphere, ranging from ~ 0.019% to ~ 0.027% at glacial and inter-glacial periods respectively (Raynaud et al. 1993; Petit et al. 1999). Due to modern burning of fossil carbon fuels, the atmospheric CO_2 concentration is now increasing again, reaching more than 0.036% in 1997 (Houghton et al. 1996), while a small but significant decrease of atmospheric O_2 is also observed (Keeling and Shertz 1992).

Metals in the biosphere

Surely the evolution of metal functions in all living organisms is much more complex than above sketched and only partly understood. However each time when a living cell had to perform a difficult piece of chemistry, it resorted to a transition metal as the natural diplomat intermediate (Hagen 2001) between hard and soft atoms (Pearson 1973) at either side of the periodic table (Stumm and Morgan 1981; Frausto da Silva and Williams 1994). Nowadays every living cell has Fe and Zn as the two most abundant and important metals in a large variety of crucial functions. Next Cu is almost as diverse in its biological roles as Zn and surely also quite abundant within biota. On the other hand the second row transition metals Ag and Cd, with analogous chemistries but much lower crustal abundances, were always deemed to have no biological functions at all.

Chemical speciation and size fractionation

Biological production has also led to high amounts of dissolved organic substances in rivers, estuaries and oceans, with extra organic input by mankind into many rivers and coastal waters worldwide. Metal-organic complexation in natural waters was long suspected but highly controversial due to both metal contamination and debate about applied electrochemical methods. Only in the 1980's the first generally accepted convincing evidence for organic complexation of a metal at realistic natural concentration in seawater was becoming available. Among the metals Cu was well known to have strong affinity for organic moieties (Mantoura et al. 1978) and first studies focusing on Cu-organics evolved from separation approaches (Mills et al. 1982; Moffett et al. 1990) to electrochemistry (van den Berg 1982, 1984; Buckley and Van den Berg 1986; Kramer 1986). This was followed by organic complexation of Cu and Zn in the Scheldt (van den Berg 1987). Since then we learned that Fe, Cu, Zn and Cd in oceanic waters are also strongly bound by dissolved organics (Zn, Bruland 1990; Cu, Coale and Bruland 1990; Cd, Bruland 1992; Fe, Gledhill and van den Berg 1994; van den Berg 1995; Nolting et al. 1998; Boyé et al. 1999, 2001). On the other hand silver

(Ag) appears to have only modest organic complexation in seawater (Miller and Bruland 1995). Initial suggestions of organic complexation of Co (Donat and Bruland 1988; Vega and Van den Berg 1997) were recently confirmed in oceanic waters (Ellwood and Van den Berg 2001; Saito and Moffett 2001). Meanwhile the organic complexation of Cu in the Scheldt river and estuary has been further unraveled also with respect to diatom growth (Gerringa et al. 1995, 1996, 1998). Within anoxic as well as oxic waters there is competition between the organic metal-complexes and the suite of proven or proposed metal-sulphide species in seawater (Lewis and Landing 1991; Rozan et al. 2000).

Matters are further complicated by the size spectrum of particles in rivers, estuaries and oceans. With classical filtration at about 0.2 or 0.4 micron a distinction used to be made between dissolved and particulate fractions. The river water upon mixing with seawater would experience massive flocculation with removal of dissolved Fe and other metals into particulate form. Yet recently by ultrafiltration at size cutoff of some 200 KDalton (relative molar mass) it was shown that most of the so-called dissolved pool consists of colloids, and it is these colloids which are removed by estuarine mixing (Dai and Martin 1995; Wen et al. 1999; Wells et al. 2000). Hence the truly dissolved fraction (<200 kD) is not removed at all and conservative throughout a regular oxic estuary. Obviously seasonal local anoxia would complicate this size fractionation. Moreover the organic complexation of this truly dissolved fraction still awaits assessment in rivers, after our first ever such combined ultrafiltration / organic complexes study (Fall 2000) in the Southern Ocean for ultrafiltered dissolved iron (Boyé et al. submitted a, submitted b). Latter study has shown that a significant portion (~20 %) of what hitherto was called dissolved organic Fe is in fact in the colloidal phase. This further extends from the observations of Nishioka et al. (2001) in the North Pacific Ocean where 13 to 50 % of the so-called dissolved Fe is actually in colloidal form.

Oceanography

Reliable oceanic distributions of metals have become available only since 1976. Remarkably both Cu, and Zn, as well as Ag and Cd are very closely correlated (Fig. 2) with the major nutrients silicate, phosphate and nitrate (Ni, Cu, Zn, Cd, Bruland 1980; Nolting et al. 1991, 2000; Nolting and de Baar 1994; Saager et al. 1992, 1997; Ag, Murozomi 1981; Martin et al. 1983; Zhang et al. 2001). This indicates strong involvement of these metals in the biological cycle, hence also biological functions, not only for Cu and Zn, but conceivably also for Cd. Initially this led to suggestion of cells substituting Zn by Cd when Zn levels were low (Price and Morel 1990). More amazingly the element ratio Zn/Cd was recently noticed (Nolting et al. 1999) to be only about 5-10 in the oceans, as compared to 600-800 on land or in deep-sea sediments.

Firstly the only 5-10 fold lower abundance of Cd in the oceans would after all make a true evolutionary usage more likely, rather than just a substitution when Zn is low. After all the many biochemical functions of Zn have been resolved on basis of studying only a few mostly land biota, and different biochemistry based on Cd for some of the marine biota is conceivable (Lee et al. 1995; Lane and Morel 2000). For example the tight oceanic relation of Zn with silicate could hint at a key role of Zn in one group of the eukaryotic diatoms, while the tight link of Cd with phosphate might then be ascribed to Cd biochemistry in other classes of bloom-forming phytoplankton.

Secondly the factor ~100 fractionation during geochemical cycling between land, seawater and eventually deep-sea sediments might well be taking place in estuaries, where known different solubilities of sulphides ZnS and CdS might play a role. Inorganic thermodynamic calculations did show a dramatic effect indeed as function of salinity and O_2/H_2S levels, and this is surprisingly consistent with a peak (Fig. 3) in the Zn/Cd ratio in the mid estuary of the Scheldt (Gerringa et al. 2001). However at the high salinity end member the ratio would again mimic that of the river end member, thus dramatic effects within the estuary but no apparent impact on global cycles or fractionations.

By analogy the Cu/Ag ratio in open oceanic waters is about 91 versus about 1060 in soils and rocks, and the apparent fractionation with factor of about 12 is also intriguing (Martin et al. 1983;

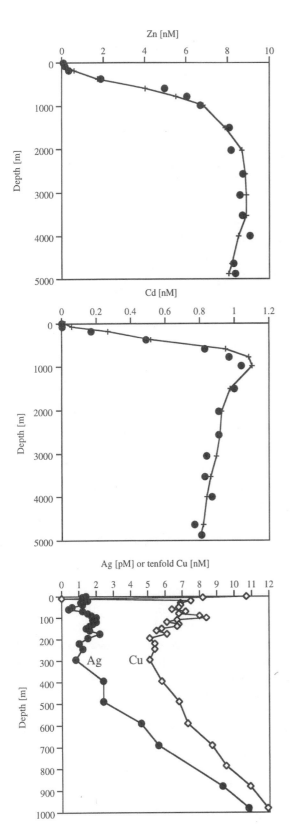

Fig. 2. Metal distributions in the North Pacific Ocean. Upper graph shows dissolved Zn (filled dots) versus depth (metres) with the similar dissolved silicate curve at adjusted scale. Covariance is $Zn[nM] = 0.053Si[\mu M] + 0.105$ ($r^2=0.988$, n=15). Middle graph shows dissolved Cd (filled dots) versus depth (metres) with the similar dissolved phosphate curve at adjusted scale. Covariance is $Cd[nM] = 0.331PO4[\mu M] + 0.041$ ($r^2=0.983$, n=15). Lower graph shows dissolved Ag (pM, filled dots) and similar tenfold dissolved Cu (nM, open diamonds) versus depth (metres). True Cu values in the 0.5 to 1.2 nM range are about 91-fold the dissolved Ag values with covariance $Ag[pM]= 9.52Cu[nM] - 4.46$ ($r^2=0.86$, n=30). Data of zinc, silicate, cadmium and phosphate of H-77 station 17 (33°N, 145°W) of Bruland (1980). Silver and copper data at 18°N, 108°W of Martin et al. (1983).

Zhang et al. 2001). Otherwise Ag does not have any known biological function due to its very low abundance at the planet. Nevertheless the oceanic distributions of Ag are much more closely following silicate (Zhang et al. 2001) than the distributions of Cu which does have known biological functions. Yet within polluted rivers and coastal waters anomalously high environmental ratio's of Ag/Cu or Ag/Zn may well give rise to Ag toxicity for biota (Sanudo-Wilhelmy and Flegal 1992). Similarly neither Hg nor Pb has a biological function due to their extremely low natural abundances, but nowadays are quite common in all waters of the Northern hemisphere and next also observed in the Southern hemisphere (Mason et al. 2001; Alleman et al. 2001a). Notably these metals are introduced from anthropogenic sources into the atmosphere and therefore distribute all over the mostly Northern hemisphere, as opposed to the other pollutant metals, which are largely introduced via streams and rivers into the coastal seas. The ensuing toxicity effects of Hg and Pb are notorious, where blocking of biochemical sites of biological metals (Cu, Zn) in for example cellular enzymes is often the toxic effect.

Plankton ecology

For more than one century marine ecologists have extensively studied the cellular requirements of

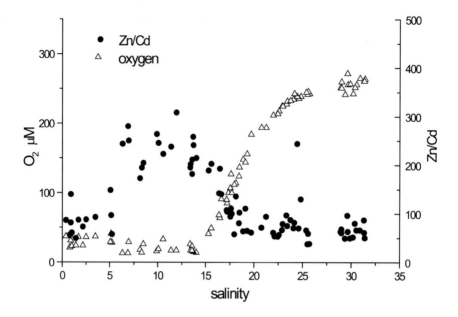

Fig. 3. The observed metal ratio Zn/Cd in the Scheldt estuary shows a distinct maximum at medium salinity ranges consistent with the theoretical calculation of ratio of solubility of each metal versus solid oxide and sulphide phases. Kindly drafted by Loes Gerringa as a summary of findings of Gerringa et al. (2001) on basis of previously reported field data (Nolting et al. 1999).

nitrogen (N), phosphorus (P), as well as silicon (Si) in the case of diatoms. These major nutrients exist in just one chemical state, such as phosphate or silicate, or at most some 2-4 chemical states (nitrate, nitrite, ammonia, urea) in seawater. Chemical conversions from one state to the other in seawater are not relevant for phosphate and silicate, and ignored for the various N-compounds. Limitation of phytoplankton growth has been studied for such single factor, for example either phosphate, or ammonia, or nitrate. Moreover this approach appeared consistent with the prevailing paradigm or supposed Liebig's Law of a single limiting factor, that is the suggestion that the growth of a phytoplankton cell is at one time always limited by only one nutrient, apart from light.

Only quite recently have inorganic CO_2 (Riebesell et al. 1993), and the trace nutrients iron (Fe) and zinc (Zn) been recognized (Martin and Fitzwater 1988; de Baar et al. 1990; Buma et al. 1991; Sunda and Huntsman 1992) as conceivably limiting nutrients. These 'new' nutrients have in common that they exist in a variety of chemical states in seawater, and strictly chemical interconversions have to be taken into account. It is this chemical speciation in multiple chemical forms in seawater which, both for various essential metals as well as CO_2, requires a new approach to unravel and understand their limiting role in marine ecosystems. Moreover it is realized that in dynamically changing plankton ecosystems, several chemical elements as a rule would simultaneously be limiting, or at least sub-optimal, for phytoplank-ton growth. In retrospect one century of adherence to the popular minimum rule in plankton ecology was found to result from incorrect extrapolations of Liebig's own writings (de Baar 1994). Slowly but surely the more realistic paradigm of simultaneous nutrient limitations becomes accepted (Kooyman 2000). After all within the cell the various essential elements are functioning in concert in the various biochemical pathways.

For example Zn is pivotal in carbonic anhydrase for utilization of bicarbonate which is by far the most ubiquitous chemical species of CO_2 in the ambient seawater (Morel et al. 1994; Buitenhuis et al. 1999). Another example is Fe which is not only crucial in the electron transport of the photosynthetic apparatus hence in supply of energy to the cell (Van Leeuwe et al. 1998; Van Leeuwe and de Baar 2000; Timmermans et al. 2001b), but also required in the enzymes nitrate reductase and nitrite reductase for utilization of major nutrient nitrate (Timmermans et al. 1994). Conversely major nutrient elements can also affect metabolism of metals, for example recently it was found that the CO_2 in seawater can modulate Cd uptake (Cullen et al. 1999). Next to these direct co-limitations of metals with major nutrients and light energy, the metals have numerous other functions in the cell, for example in cytochromes (Fe), DNA-replication (Zn-fingers), N_2-fixation (Fe and Mo), cytochrome oxidase (Cu) and superoxide dismutase (Cu), urease (Ni), and in various enzymes based on vitamin B-12 with Co as co-factor (Frausto da Silva and Williams 1994).

The eutrophication of coastal waters is commonly linked with increasing levels of previously limiting nutrients N and P due to Von Liebig's fertilizer enterprise in agriculture. Here we now realize this would not have been adequate without the simultaneous increase in coastal waters of limiting trace nutrient Zn as well, due to metal pollution.

Plankton organisms not only need metals for their growth but also have an important feedback on the distributions of metals in the sea. Above this was already noticed for the metals Cu, Zn, Ag and Cd which have an oceanic distribution similar to that of the major nutrients. The redox-metals Fe and Mn have a more complicated oceanic distribution. Nevertheless for Fe sometimes a covariance with nutrients is observed (e.g. Martin et al. 1989) while in coastal waters of the North Sea and an Arctic fjord the spring plankton bloom has dramatic impact on the ambient concentrations of both Fe and Mn (Schoemann et al. 1998, submitted).

Metal-metal interactions

Somewhat related to the above concept of co-limitations are the various concepts of metal-metal interactions in plankton growth. These can be both synergistic as antagonistic as well as by substitution. At low available ambient Zn it was found that both Cd and Co can substitute for growth of a marine diatom (Price and Morel 1990). Replace-

ment of Zn by Cd appeared to be a common option for phytoplankton species (Lee and Morel 1995). Moreover Sunda and Huntsman (1995) demonstrated that Co can interreplace Zn in the calcifying algae *Emiliania huxleyi* as well as in two diatom species. Timmermans et al. (2001c) confirmed these observations for *E. huxleyi* but on the other hand have shown that yet another diatom *Chaetoceros calcitrans* is not able to substitute Co for Zn. Various antagonistic and other interactions have been found between various combinations of the metals Mn, Fe, Cu, Zn, Cd and Pb (Bruland et al. 1991; Sunda and Huntsman 1983, 1996, 1998, 2000; Vasconcelos and Leal 2001).

Metals and the last glacial maximum

These metal-metal interactions will also affect oceanic cycling of any given metal. For example the oceanic cycling of Cd is affected by other metals (Sunda and Huntsman 2000). This further complicates the debate about whether or not the modern relationship between Cd and phosphate in seawater (Fig. 2) is applicable for the reconstruction of past phosphate distributions in the Last Glacial Maximum from the historic record of the Cd/Ca element ratio in fossil foraminifera (Saager et al. 1993; De Baar et al. 1994; Löscher et al. 1998a,b; Rickaby and Elderfield 1999; Elderfield and Rickaby 2000). Briefly Saager et al. (1993) demonstrated that the modern $Cd-PO_4$ relationship undoubtedly would shift to another relationship in the LGM ocean, such that one singular relationship does not exist throughout times. The underlying mechanism would be either preferential uptake of Cd by biota or fractionated deep water mineralization. Next Löscher et al. (1998b) were able to demonstrate preferential biological uptake in diatom blooms of the Polar Front. Recently such preferential uptake of Cd versus PO_4 was also observed during the SOIREE *in situ* iron enrichment experiment (Frew et al. 2001). This demonstrated preferential uptake of Cd can be parameterized, for example in the recent simple model for Cd/PO_4 fractionation by Elderfield and Rickaby (2000). This may or may not eventually help to settle the current debate (Sigman and Boyle 2000) about the nutrient status of the Southern Ocean during the

LGM. Here some evidence in support of more efficient nutrient utilization (Francois et al. 1997) hence more uptake of CO_2 from the atmosphere (Moore et al. 2000), tends to contradict other evidence (Rosenthal et al. 2000; De la Rocha et al. 1998) in favor of less nutrient utilization and a relative loss of CO_2 to the atmosphere. Thus various metals may play a role either as a major trigger (Fe; Martin 1990; Lancelot et al. 2000; Hannon et al. 2001; Watson et al. 2000) or as a tracer (Cd) of carbon dioxide uptake by the Southern Ocean (CARUSO) in the LGM versus the modern era.

Homeostatis and pollutant metals

Living organisms are capable of maintaining constant metal concentrations and ratios (e.g. Zn/Cu) inside the cell, despite large variations of metal concentration and metal speciation (biological availability) in the outside environment or food supply. This is achieved by metal-selective uptake sites at the cell wall, but also by neutralization of excess metal into metallothionein bonds (Frausto da Silva and Williams 1994). Nevertheless at extreme concentrations or element ratios the cell cannot cope anymore. For example at extremely high or low external Cu/Zn ratio it becomes impossible to maintain the required ratio Cu/Zn=1 in superoxide dis-mutase.

Due to mining and processing of metals in various industrial applications the emissions of many metals into the biosphere have increased. Hence the levels of Cu, Zn, Ag, Cd, Hg as well as Pb in the environment have not only increased, but also the element ratio's have changed. The organic loading in rivers and coastal waters has moreover given rise to both organic complexation and shifts in the O_2/H_2S conditions, both affecting the chemical speciation hence biological availability of metals. Many organisms are no longer capable of maintaining homeostatis. Therefore the composition and biodiversity of biological species in riverine and coastal ecosystem likely has shifted over the past centuries.

Land use change in past decades and the future is also affecting metal abundances in the biosphere. Desertification in for example the Sahel region has given rise to enhancement of dust storms over the ocean, where increased Fe input into the central Atlantic likely has stimulated plankton productivity (Jickells and Spokes 2001). For the global ocean it has been estimated that about half of the Fe input with aeolian dust is in fact due to land use change by mankind. Changes in water management for irrigation, drinking water supply or waste water disposal do affect groundwater levels and rivers and lakes. One obvious results is a shift in the transition between oxic and deeper anoxic groundwaters, with an accompanying shift in dissolution of some metals and precipitation of others, notably as oxides or sulphides. For example when pumping up deep anoxic ground waters from aquifers the reduced ZnS and CdS do become oxidized into more soluble state, leading to toxic levels in drinking and irrigation waters. This is akin to the well known arsenic (As) poisoning in water supplies of Bangla Desh. Finally increased agriculture has led to eutrophication of coastal waters by washout of fertilizer nutrients N and P. This is also causing (seasonal) oxygen depletion in subsurface waters, with concomitant changes in availability of trace metals.

Light spectrum

Incoming sunlight is beneficial for photosynthesis. However recent increases of UV irradiation in Southern but also Northern hemisphere due to a thinner ozone layer have several effects. Inside the cell there occurs DNA damage. For repair of damage Zn plays a role in zinc-fingers crucial for DNA replication. Moreover reactive oxygen species (ROS, notably hydrogen peroxide and superoxide radical) become more abundant causing further damage inside the cell. Here the counteracting of such very damaging ROS by the Fe-Mn and Cu-Zn superoxide dismutases comes into play. Moreover the visible and UV light also reacts with chromophores in the dissolved organic matter in seawater, this leading to shifts in the metal-organic complexation. For some metals like Cu and Fe this is further accompanied by their photoreduction from Cu(II) to Cu(I) and Fe(III) to Fe(II) respectively. Interactions of light with metal-organic complexes and biological availability of metals is quite a new focus of attention.

Biological availability in EDTA-free cultures

Almost all investigations of metal requirement or metal toxicity of phytoplankton have been done in seawater medium where large amounts of EDTA was added. The EDTA had been introduced as a buffer for controling metal concentrations. However it modifies the composition of seawater considerably. Therefore it masks the biological availability of different natural chemical forms of a metal in seawater.

Recently Gerringa et al (2000) assessed the disturbance of natural chemistry in the case of Fe. The limiting role of Fe for algal growth has until recently almost always been studied in seawater medium controlled by large amounts of EDTA chelator. These studies led to the paradigm that only the dissolved inorganic Fe(III) chemical forms or species, together known as Fe', act as the rate-limiting factor for growth, i.e represent 'biological availability'. For each growth experiment the Fe' could be exactly calculated by known stability constants of the Fe(III)-EDTA, and the stability constants of assumed existing inorganic species, mostly the hydrolyzed complexes (Millero 1998). However, some 99% or more of the dissolved Fe in oceanic waters exists in some organic form at the thermodynamic equilibrium, the Fe(III) being very strongly bound by at least one organic ligand of unknown molecular structure (Van den Berg 1995). Experimental study has shown that certain algae such as *Emiliania huxleyi* are capable of producing organic ligand of Fe(III), providing long term availability of iron directly or indirectly (Boyé and van den Berg 2000). Moreover the reduced inorganic Fe(II) oxidation state in surface waters can sometimes account for half or more of the overall dissolved Fe pool (Waite et al. 1995; Boyé et al. submitted c), where photoreduction apparently plays a major role. Moreover cycling of Fe through the microbial loop will also change its chemical form hence its availability for plankton assimilation (Barbeau et al. 1996). We now realize there in fact exist three pools: Fe'(III), Fe(III)-organic and Fe(II) in seawater (Fig. 4). Among these the steady-state Fe' is extremely minor, and at an overall dissolved Fe in oceanic surface waters on the order of 0.1 to 1 nM, the calculated Fe' is less than 1 pM and would

be far too low to allow any phytoplankton growth at all. Hence either any Fe' taken up by the cell is very rapidly replenished by fast dissociation of the organic Fe(III)-ligand complex, or the complete Fe(III)-ligand serves as an organic substrate, or the reduced Fe(II) state is the preferred form for uptake. Either way, the common paradigm of Fe' as the growth-rate controling variable is no longer valid.

The strong organic complexation of Zn in natural surface seawater was demonstrated in the late 1980's (Van den Berg et al. 1987; Bruland 1989). Otherwise Zn is a relatively simple metal nutrient as it exists in only one oxidation state Zn(II) in seawater. Nevertheless the prevailing paradigm of the inorganic Zn' control of plankton growth has to be abandoned. Similarly for other nutrient metals, like Cu, strong organic complexation has been found (Cu, Coale and Bruland 1988) and the paradigm of free Cu (calculated from EDTA cultures) being the control of biological availability is also no longer valid.

More recently our group has pioneered culturing in strictly natural seawater (Timmermans et al. 2001a). Here it is furthermore possible to assess biological availability not by chemical approaches, but simply by the phytoplankton organisms themselves. Let the biota themselves define what is bioavailable for them. By growing different, large and small, phytoplankton species, in the same natural seawater, one obtains data on what ambient dissolved concentration is required for half-optimal (Michaelis Menten) and optimal growth (Fig. 5). Here it was found the very smallest size class of diatoms could never be grown into Fe limitation, unless by addition of a minute amount of natural siderophore, an organic molecule with exceedingly high affinity for Fe. Thus it was concluded that the ambient Fe concentration necessary for a small Antarctic diatom is about four orders of magnitude less than that of a large diatom.

Molecular biology and functional diversity

Molecular biological techniques have revolutionized all fields of biology, including biological oceanography. However, most of the marine research carried out with molecular techniques has been so

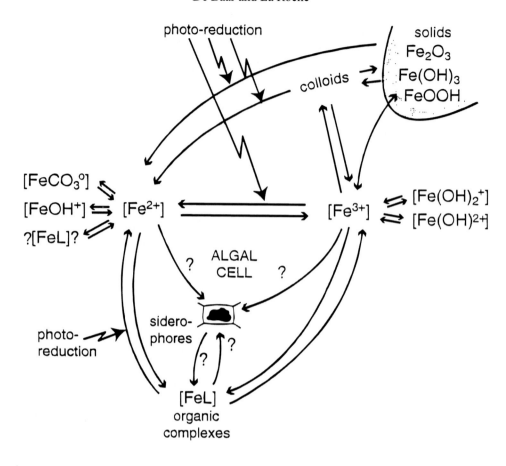

Fig. 4. The different chemical forms of Fe in seawater, interacting with the phytoplankton in various ways. Simplified scheme after Gerringa et al. (2000) with permission of Elsevier Scientific Publishers.

far in the area of molecular taxonomy and phylogeny. Alignment and comparison of DNA sequences by a variety of techniques (e.g. DNA sequencing, finger printing, denaturing gradient gel electrophoresis or DGGE) have been used to infer the taxonomic and phylogenetic affiliation of marine organisms ranging from bacteria, archea, phytoplankton, macrophytes and fish. Although the study of diversity has some intrinsic merit, it is time to put taxonomic diversity into a more meaningful context of functional diversity. The underlying question, at least for marine biogeochemistry, is how taxonomic diversity relates to functional redundancy (i.e. how many different species are capable of performing a function) of metabolic transformations within an environment.

Taxonomic and phylogenetic studies using rRNA genes as target alone will tell us very little about the important role that metals have played in the evolution of protein structure and function. The importance of metals as cofactor is ubiquitous among all classes of enzymes and proteins. Metals can thus play roles in synthetic pathways, gene regulation and cellular signaling. Active sites of proteins, which often contain metals as co-factors, are generally highly conserved and the overall diversity of homologous metallo-proteins sharing a similar function, will be to some extent subject to constraints imposed by the need to preserve the metal binding sites and other structure that confer activity.

Within a broader biological context, the advent of genomics and high throughput DNA sequencing techniques has greatly accelerated the pace at which we can now obtain DNA sequence data. The genomes of three important open-ocean photo-

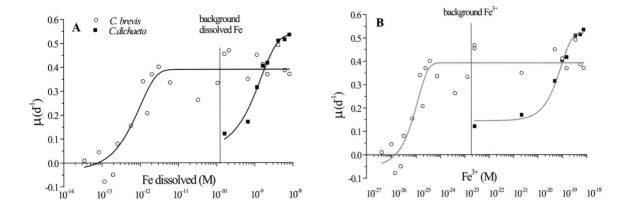

Fig. 5. Growth rates of *Chaetoceros dichaeta* and *Chaeotoceros brevis*. versus **a)** disssolved Fe and **b)** free Fe^{3+}-ion. Experiments of single strains of diatoms incubated in natural Antarctic seawater (Fe = 0.16 nM) without addition of EDTA. The large diatom *C. dichaeta* is Fe-limited in the natural seawater but its growth rate increased upon addition of extra dissolved Fe. The small diatom *C. brevis* showed optimal growth both at the natural Fe concentration as well as after Fe additions. Only by addition of a strong Fe-binding siderophore ligand desferrioxamine-B was it possible to achieve Fe limitation of *C. brevis*. From these cultures with increasing amounts of siderophore the free Fe^{3+} ion was calculated, and next calculated what the total dissolved Fe concentration should be in natural seawater to achieve limitation of *C. brevis*. These concentrations in the 10^{-12} M range are orders of magnitude below the natural background value (0.16×10^{-9} M), i.e. in natural waters *C. brevis* is never Fe-limited. Notice the about four orders difference with *C. dichaeta* which as a rule will be Fe-limited in natural waters. After Timmermans et al. (2001a) with permission of the American Society of Limnology and Oceanography, Inc.

synthetic prokaryotes (*Synechococcus* WH8102, 2 strains of *Prochlorococcus*) have been sequen-ced and we can expect genome sequences from more key organisms in the next few years. A quick pe-rusal of the genome databank for these three ma-rine organisms (http://spider.jgi-psf.org/JGI_mi-crobial/html/) already tells us something about presence and absence of genes that are iron-regu-lated. All have genes coding for two different, probably high and low affinity, Fe uptake systems, along with the ferric uptake regulatory protein (Fur) (Hantke 2001). Conversely, the gene coding for the chlorophyll-binding iron-stress-induced protein (IsiA) which works in energy dissipation and thereby protects the photosystem II during iron-stress (Park et al. 1999) is present in *Synechococcus* but has diverged to become the major light harvesting chlorophyll binding protein in the Prochlorophyte (La Roche et al. 1996b). Genes for flavodoxin (La Roche et al. 1996a) and ferritin (an iron storage protein) while present in *Prochlorococcus* have not yet been identified in the

genome of *Synecho-coccus* WH8102. However, we know nothing yet about the identity of the genes encoding for iron uptake and storage in eukaryotic phytoplankton species. Whole genome sequencing of marine microorganisms offers the possibility to look at the global gene expression within an organism using DNA micro-array technology, and one can envision working both with field samples or controlled laboratory experiments targeted at understanding, for example, the coordinated expression of iron-regulated genes.

Although DNA sequences and whole genome information will be invaluable for our understanding of gene expression in marine organisms, sequence data alone cannot explain the complexity and functioning of an organism. It is now apparent that number of genes in an organism does not increase proportionally to genome size. For example, although the human genome (3,400 Mb) is 3 orders of magnitude larger then the average bacterial genome (3,5 Mb), the estimated number of genes in humans (40,000) is only 10 times higher then that of bacteria. Even more puzzling, is the realization that many plants and protozoans have much larger genomes then humans, with some species of *Amoeba* reaching 670,000 Mb (http://www.cbs.dtu.dk/databases/DOGS/abbr_table.bysize.txt). It is also apparent that the increased complexity of eukaryotic genomes is expressed not only by larger genome size but also by complex patterns of gene expression, regulatory sequences and RNA processing.

In fact, the plethora of new molecular techniques reaches far beyond molecular taxonomy and phylogeny. In the post-genomic era, this is clearly demonstrated by the development of new fields of research in the area of proteomics, related to the analysis of protein structure, function and regulation. The amino acid sequence of a protein or its 'primary structure' contains most of the necessary information required for the manifestation of the secondary and tertiary structures. Each amino acid has a side chain conferring specific chemical properties that dictate the type of chemistry it can be involved in. For example, about a third of the amino acids can bind metals, cysteine (Zn, Cu, Fe, Ni, Mo), histidine (Zn, Cu, Mn, Fe, Ni) being the most versatile. The initial folding of a polypeptide chain or its secondary structure, results in only a handful of structures namely alpha helices, beta sheets, and turns giving it a three-dimensional shape, the tertiary structure. However, a functional enzyme is often made up of several subunits, and the assembly of identical or different protein subunits into quarternary structure adds a level of complexity that cannot be resolved simply by analysing DNA sequence or even genome information. Metal associated to polypeptides may also play a role in the three-dimensional shape of an enzyme. For example in the above mentioned Cu-Zn superoxide dismutase the Zn atom is functioning as to allow optimal folding, while only the Cu atom is the co-factor in the enzymatic function (Frausto da Silva and Williams 1994). Obviously then the external abundance and biological availability of metals will affect the proper assembly of a protein into their active form. For example in a marine daitom the oxidative stress is interrelated with availabilities of both Cu and Zn (Rijstenbil et al. 1994, 2002).

It is clear that the rapidly developing field of proteomics has also a lot of potential in the study of metal binding proteins, and that the rapidly growing data bases on protein signature sequences and three-dimensional structure will be invaluable for the study of metalloproteins. However, it is also clear that marine scientists have to also take initiative to promote the study of marine organisms such as unicellular algae with new high throughput molecular methods for the study of DNA, protein and regulation at molecular level.

Isotope biology of metals

The first row transition metals being involved in biological functions and cycling, mass fractionations akin to major biological elements $^{13}C/^{12}C$ and $^{15}N/^{14}N$ are to be expected. Conversely observation of such fractionations either in the field or in controlled laboratory experiments would be useful for unraveling biochemical mechanisms of metals.

Upon many fruitless attempts by Thermal Ionization Mass Spectrometry for among others Cd (own unpublished work), and recent TIMS findings for Fe which are still subject of debate (Beard et al. 1999; Anbar et al. 2000), the development of High Resolution ICP-MS (Wells and Bruland 1998; Field et al. 1999; Halliday et al. 2000) now would allow accurate assessment of isotopic ratio's of various essential metals such as Fe (Zhu et al. 2000a), Cu and Zn (Zhu et al. 2000b; Maréchal et al. 1999). In our laboratory methods are now being set up for the HR ICP-MS in order to combine its powerful analytical/isotopic capabilities with our expertise in ultraclean sampling and biological experiments for trace metals.

The iron age in oceanography

The first suggestion of Fe limitation of oceanic plankton by Gran (1931) was demonstrated convincingly only in August 1987 (Martin and Fitzwater 1988). The scientific history of limitation by Fe in context of plankton ecology has been synthesized elsewhere (de Baar 1994). Throughout the 1990's the subject of Fe limitation has grown in attention into what now may be called an iron age in oceanography (Coale et al. 1999).

Iron limitation

The role of iron in plankton ecology and carbon dioxide transfer of the oceans has recently been reviewed by de Baar and Boyd (2000). Briefly in about 40 % of the world oceans there is adequate supply of major nutrients N, P and Si. In these High Nutrient Low Chlorophyll (HNLC) waters the development of phytoplankton blooms is limited by trace element Fe, next to light limitation and top-down control by grazing pressure. The three major HNLC regions are the Antarctic Ocean, the Equatorial Pacific Ocean and the Subarctic North Pacific Ocean.

In the central gyres of the oceans, constituting the remaining about 60 % of surface waters, the major nutrients N, P and Si are limiting either on their own or in concert. However anytime this limitation is alleviated the plankton ecosystem was found to experience Fe limitation as well. The overall findings of such experiments in temperate, low-nutrient, waters are consistent with various degrees of co-limitation by several nutrients including Fe. An adequate supply of all nutrients, N, P, Si and Fe, yields diatom blooms as also found in HNLC waters; obvious exception to the rule occurs where Si supply is inadequate.

The regular supply of iron and the microbial foodweb

The sources, sinks and distributions of dissolved Fe in oceans, coastal seas and restricted basins have recently been reviewed in an extensive synthesis comprising all available data of the literature (de Baar and de Jong 2001).

In the three major HNLC regions the supply of dissolved iron from below (e.g. Antarctic Circumpolar Current) is adequate for sustaining the recycling community of mostly small sized algae being efficiently grazed upon by microzooplankton. In general, the small algae appear to be growing well, i.e. they appear not iron-limited per se and grazing largely controls their biomass. Nevertheless in all three HNLC regions, there are indications from various physiological parameters that even these small algae experience at least some Fe stress.

The earlier suggestion of dominant aeolian Fe input (Martin 1990, 1991) into both equatorial Pacific and Southern Ocean, has now given way to the coastal margin sediments and upwelling, or the combination thereof, being the major sources of Fe supply (de Baar et al. 1995; Coale et al. 1996a; de Baar and De Jong 2001). Thus only for the subarctic Pacific Ocean, aeolian input still remains to be considered the major local source term (de Baar et al. 1995).

For example for the Atlantic-Antarctic Ocean south of the Polar Front the new production sustained by Fe upwelling was found to be 14 mmole C m^{-2} d^{-1} (for Fe:C = 1:100,000), about half of the potential new production as assessed on basis of simultaneous upwelling of nitrate (de Baar et al. 1995). By analogy in the equatorial Pacific the upwelled Fe can support new production of about 20 mmole C m^{-2} d^{-1} (for Fe:C = 1:167,000), which would only be about 20% of the potential of simultaneous nitrate supply from below (Coale et al. 1996a).

The upwelling supply of Fe in the northeast Pacific Ocean is relatively small compared to the two other regions. Here the episodic Fe input from aerosols might lead to significant blooms with uptake of CO_2 and major nutrients, but this has never been directly observed in the region. Such aeolian Fe input is conceivable though, judging from direct observations and subsequent transient stimulation of growth in the central North Pacific Ocean (Young et al. 1991; DiTullio and Laws 1991). Recently a major wet deposition event was also reported for the equatorial Pacific (Hanson et al. 2001). Moreover the downward flux of frustules of large diatom species, as collected in sediment traps, would suggest that diatom blooms are regular events in the subarctic Pacific region (Takahashi et al. 1990).

The enhanced supply of iron and diatom blooms

Upon addition of iron, either naturally in the Polar Front (de Baar et al. 1995; Löscher et al. 1997) and neritic nearshore Antarctic waters (de Baar et al.

1990; Nolting et al. 1991), or in shipboard incubations (Buma et al. 1991; Van Leeuwe et al. 1997; Scharek et al. 1997; Timmermans et al. 1998), or by *in situ* perturbations (Coale et al. 1996a; Boyd et al. 2000; Smetacek et al. 2001), the larger size class of mostly diatoms is able to outgrow the background population of small algae. Towards the Antarctic continent the increasing dissolved Fe levels will support more intense diatom blooms (de Baar et al. 1999). Apparently the larger mesozooplankton are not capable of keeping the larger diatoms fully in check, this in keeping with the notion of longer generation times of mesozooplankton as compared to microzooplankton. Obviously diatoms also need silicate for their frustules, and in low silicate waters of the subantarctic the diatoms are absent (de Baar et al. 1999).

The almost selective stimulation of larger diatoms is a strikingly consistent trend in all Fe studies done thus far. It appears in keeping with the concept of the lower surface/volume ratio of the large diatom cells (Timmermans et al. 2001a). Here Muggli et al. (1996) have shown that the morphology of the opaline skeleton largely prevents a large oceanic diatom from decreasing its cell size in response to Fe stress. In contrast, other algae are strongly responding to Fe stress by reducing their size (Muggli and Harrison 1996; van Leeuwe and Stefels 1998). Apart from cell size, it is also important to realise that coastal *neritic* diatoms likely have higher cellular Fe requirements per unit biomass than some of their *oceanic* counterparts (Sunda et al. 1991). On the other hand, the largest *oceanic* diatoms, whether pennates in the equatorial Pacific (Zettler et al. 1996), centric *Actinocyclus* sp. in the subarctic Pacific (Muggli et al. 1996) or chain-forming *Fragilariopsis kerguelensis* in the Polar Front, also appear to have a high Fe requirement for optimal growth, this corresponding to ambient levels of dissolved Fe apparently equal to or exceeding the 0.2- 2.0 nM range. Recently the half-saturation constants were determined in natural seawater medium (without EDTA additions) for Antarctic large diatoms *Chaetoceros dichaeta* and and *Fragilariopsis kerguelensis* found to be 1.2 nM (Fig. 5) and 0.4 nM respectively (NIOZ, unpublished results). Thus the ruling paradigm of the 1990's of high and low Fe require-

ments in coastal versus oceanic phytoplankton, is now replaced by a concept where a third major class of large bloom-forming *oceanic* diatoms with high Fe requirement exists as well. It is latter class of phytoplankton which likely is crucial for the big blooms and concomitant export of organic matter in open oceans. Thus for global cycling of major biological elements (C, N, P, Si) including uptake of CO_2 from the atmosphere, this third class of large oceanic diatoms is crucial.

In studying these large diatoms it was furthermore found that large deviations of the classical Redfield ratio (C:N:P) can occur in the field (de Baar et al. 1997). From water column data the uptake ratio of nitrate to phosphate was found to be as low as 4, as compared to a general Antarctic ratio of about 14 and an average global ratio of about 15. The suggestion that suboptimal Fe supply is affecting this via the suppressed activity of Fe-containing enzymes nitrate reductase and nitrite reductase was recently confirmed for the chain forming *Fragilariopsis kerguelensis* in the laboratory (de Baar et al. 2000). At low Fe supply its N/P uptake ratio was about 4-6 as compared to about 12-14 at adequate Fe supply.

Iron and the geochemical budgets of carbon dioxide and nutrients

In the Antarctic Polar Front, the natural effect of Fe on the biological uptake of CO_2 and nutrients in major diatom blooms has been observed (de Baar et al. 1995). The corresponding CO_2 budget is consistent with the other JGOFS observations at this front (Bakker et al. 1997). Moreover the Polar Front is a site where formation of deep water (AAIW) is taking place such that a combined biological / physical pump for drawdown of CO_2 may be at work. These findings are consistent with plankton ecosystem modeling (Lancelot et al. 2000). Once having validated this model it was also possible to demonstrate that higher Fe supply as during the Last Glacial Maximum, would indeed stimulate diatom blooms and oceanic CO_2 uptake (Hannon et al. submitted).

The Southern Ocean Iron Enrichment Experiment (SOIREE) in February 1999 largely confirmed the stimulation by Fe addition of the *in situ* phyto-plankton community (Boyd et al. 2000). At the final day 13 the abundance of large diatom *Fragi-lariopsis kerguelensis* had increased significantly (Gall et al. 2001). Moreover a significant decrease of CO_2 was observed inside the Fe enriched patch (Bakker et al. 2001).

In the equatorial Pacific, an enhanced phytoplankton standing stock downstream of the Galapagos Islands has been reported (Martin et al. 1994), but not accompanied by anomalies of CO_2 and major nutrients. However an intentional *in situ* perturbation experiment with repeated iron additions did yield significant anomalies of CO_2 and nutrients in the equatorial Pacific region (Coale et al. 1996b). Within this Fe-enriched patch of IronEx II, the fugacity of CO_2 in surface water was observed to decrease from a background value of about 510 µatm to approximately 420 µatm (Cooper et al. 1996). This evidence is in keeping with the original direct observations of a natural decrease from about 356 µatm to values as low as approximately 310 µatm exactly coinciding with the Fe-rich Polar Front blooms (de Baar et al. 1995; Bakker et al. 1997). An important difference is that the equatorial surface waters shifted from strongly oversaturated (relative to the atmosphere) to less oversaturated, whereas the Polar Front had shifted from approximate equilibrium to distinct undersatu-ration. Hence in the IronEx II patch the *in situ* perturbation led to a transient suppression of natural outgassing at the equator, whereas the natural *in situ* bloom at the Polar Front led to an influx of CO_2 into the Southern Ocean.

The evolution of diatom blooms appears to be important for understanding the export into the deep-sea. Apparently at the end of a bloom period, when Fe may have become depleted, the Antarctic diatom *Corethron criophilum* reaches a stage of sexual reproduction accompanied by shedding and settling of diatom frustules (Crawford 1995). This would strongly affect the Si budget. In the laboratory, Muggli et al. (1996) observed a fivefold increase of settling velocity of a subarctic diatom when grown under Fe-stress. In the *in situ* situation this would affect the budgets of not only Si, but also C, N and P.

Common trends and differences between HNLC regions and oligotrophic gyres

In general at the three major HNLC regions the same mechanisms are at play, only their prevalence varies. In HNLC waters the large oceanic diatoms increase upon Fe enrichment; blooms develop and suffer little from grazing due to the different generation times of diatoms (days) and mesozooplankton (months). In all three HNLC regions, the microzooplankton grazers (e.g. protozoa) have similar generation times as the algae and appear well capable of keeping the pico- and nanoplankton in check. In the Antarctic Ocean, the supply of iron from below is significant. This is true also in the equatorial Pacific Ocean. The upward supply is relatively small in the strongly stratified NE subarctic Pacific. In all three HNLC regions, this upward supply from below is deemed adequate for sustaining the efficiently recycling microbial foodweb. On the other hand, in the Antarctic Ocean the deep wind mixed layer and seasonal sea-ice cover make the Antarctic severely light-limited as compared to the two other regions. Additional iron supply in coastal waters and those frontal systems downstream from land sources, is supporting most Antarctic blooming, whereas intermittent aeolian supply would be mostly responsible for blooming in the open equatorial Pacific and Subarctic Pacific. For the Southern Ocean at large, an aeolian supply may also contribute to the Atlantic Subantarctic region and even further south into the Polar Frontal Zone, but this conceivable source term needs further quantification.

In the oligotrophic gyres of temperate oceans there appears to exist co-limitation of not only major nutrients but also Fe (DiTullio et al. 1991) and likely other metals as well such as notably Zn.

Iron ages

The supply of Fe to oceanic surface waters does vary over a suite of time scales. Over the longer time scale of the 420,000 years Vostok ice core record (Petit et al. 1999) there exist distinct maxima of Fe dust at each glacial maximum, coincident with minima in both atmospheric CO_2 and temperature. The elevated Fe dust input is confirmed for the Last Glacial Maximum (about 12,000 years BP) by high resolution studies of Antarctic snow and ice (Edwards et al. 1998; Edwards 2000). Latter author found on average about fifteenfold higher Fe input during the LGM than nowadays. The about 120 metres lower sea level during the LGM would be consistent with a much larger region of the extended Patagonia desert supplying more Fe dust to a more nearby deep ocean region (Iriondi 2000). These lines of evidence strongly suggest that enhanced Fe supply towards the Southern Ocean, via stimulation of diatom growth, is causing a decrease of atmospheric CO_2 (Moore et al. 2000; Watson et al. 2000). Indeed in plankton ecosystem modeling this was confirmed by running of various LGM scenario's with enhanced Fe supply and modified wind velocities (Lancelot et al. 2000; Hannon et al. submitted).

Decadal to interannual variations of desertification on land can either be natural (Southern Oscillation component of ENSO, Sahel drought ?), or by land use change (agriculture diverting river flows, e.g. Aral Lake) or by the now confirmed global warming (IPCC 2001). In the latter case an intensification of the Patagonia desert region is quite likely. Aeolian transport from South Africa stimulating productivity in the South Indian Ocean has also been suggested (Piketh et al. 2000). Any such changes will modify the dust input into the oceans, hence the controling role of Fe in plankton blooms and CO_2 drawdown.

Over very short time scales of typically one week satellite observations have shown that strong dust events, followed by wet deposition (i.e. visible clouds) often leads to an observed phytoplankton bloom. After one week the signal of either ocean color (i.e Chl a) or reflectance (i.e. coccoliths of an *E. huxleyi* bloom) can often be detected (Saydam and Polat 1999).

Mitigation of fossil fuel CO_2 ?

An intentional Fe fertilization scenario for removal of fossil fuel CO_2 from the atmosphere was first discussed in late 1989 (see review de Baar 1994). Soon it was rejected as a policy option in a con-

sensus statement unanimously agreed at a special ASLO symposium in 1991 on High Nutrient Low Chlorophyll regions (Chisholm and Morel 1991). Since then neither this scenario nor other schemes for enhancing the storage of CO_2 in the oceans (de Baar 1992) have received much attention in oceanography journals. However at the end of the twentieth century new commercial interest in large scale Fe fertilization once again led to major concern. Therefore Engelbrecht (2000, 2001) extensively reviewed the scientific and technical feasibility, environmental consequences, socio-economic implications including international relations and treaties, as well as ethics of the scheme. For each of these aspects the fertilization scheme appears to be very controversial.

Once again under aegis of ASLO a meeting was held (April 2001) mostly for sorting out these various issues at stake among a small group of experts in ocean science as well as technology and international law, and some proponents of the fertilization scheme. This time a consensus statement could not be realized. Nevertheless all about 40-50 participants gained considerably from each others expertise in open presentations and discussions. Upon further communications a summary statement was released by ASLO in August 2001, with consensus on at least one item, the desire to organize an open international conference on all aspects of large scale fertilization (Phinney 2001). This goal is now being pursued both by ASLO and SCOR with support of several scientists within the European Union, North America and elsewhere in the world. On the one hand, the issue does provide a focal point for bringing together ocean science, ocean technology and ocean policy aiming at good stewardship of the ocean common, a property of all mankind. Here an integrated research project, quite conceivable e.g. in European context, would appear most timely. On the other hand the controversial, if not fashionable (and sometimes hilarious) issue of oceanic iron fertilization is a detraction from the more valid and interesting scientific quest of understanding the natural role of Fe in the oceans as described elsewhere in this chapter. (Note added in proof: recently three individual participants of the April 2001 ASLO workshop reported a personal account by discrediting ocean fertilization; Chisholm et al. 2001).

Summary

Biological evolution hinges on availability of transition metals on the planet, where evolution of an oxygen-rich atmosphere plays a major role.

Within the oceans the vertical and interoceanic distributions of trace metals have strong co-variance with major nutrients, also for trace metals which do not have a known biochemical function.

Plankton ecosystems experience multiple nutrient limitations by both major (N, P, Si) and trace (Fe, Mn, Cu, Zn) elements.

Increasing UV light is modifying the organic complexation as well as redox state of metals in seawater, hence the biological availability for uptake by plankton.

Throughout the evolution of life, metals have been incorporated into protein structures as a function of their abundance and availability at the time a function evolved. Once a successful biochemical pathway has evolved, the metal has been retained throughout geological times despite changes in the abundance of these metals due to change in the redox state of the atmosphere and of the ocean. In the future, the evolution of new metalloprotein will continue to reflect the abundance and availability of metals at the time the new function arises.

Iron enrichment of ocean waters invariably stimulates the largest class of bloom-forming oceanic diatoms.

This class of large bloom-forming diatoms is a major driver of ocean uptake of CO_2 and export of organic carbon and opal into the deep-sea.

Working hypotheses

Biological availability of metals for plankton and bacteria can best be assessed in natural seawater without addition of EDTA or other metal buffers.

The close relation of Cd and phosphate in the oceans together with the low oceanic ratio Zn/Cd = 5-10 suggest that Cd is a true biological element and not just a substitute for Zn.

Eutrophication of coastal waters by fertilizer nutrients N and P could only have caused nuisance plankton blooms because the essential metal Zn had also increased by pollution.

Pollution input of metals has strongly modified the plankton ecosystem of coastal waters over preceding centuries.

Most supply of Fe into surface waters of the oceans is from below emanating of reducing marine sediments.

Aeolian Fe supply from land sources varies over all time scales and is an integral cause as well as result of global climate change.

Acknowledgements

The authors are grateful to the organizers profs. Mantoura and Wefer for the kind invitation to take part in the Hanse Symposium and contribute the advance position paper. This article does not contain original data. It is merely an effort to combine existing knowledge of separate disciplines and recent findings of various colleagues into an overall view of the role of metals in the oceans. For this synthesis we relied on the insights gained in recent European projects MERLIM (Marine Ecosystem Regulation: Trace Metal and Carbon Dioxide Limitation; 1996-1999; EU contract MAS3-CT95-0005) and CARUSO (Carbondioxide Uptake Southern Ocean, 1998-2001; EU contract ENV4-CT97-0472). Some paragraphs were based on recent review chapters published elsewhere with permission of their authors (de Baar and Boyd 2000; de Baar and de Jong 2001; Turner et al. 2001). Colleagues Marie Boyé, Anita Buma, Peter Croot, Loes Gerringa and Klaas Timmermans at NIOZ and University of Groningen kindly provided advice, comments and various figures of their recent articles. Both authors HDB and JLR are currently involved in the European program IRONAGES (Iron Resources and Oceanic Nutrients - Advancement of Global Environment Simulations; EU contract EVK2-CT1999-00031; see further at the website http://www.nioz.nl/ projects/ironages/index.htm). Here twelve institutes are collaborating for including four limiting nutrients (Fe, Si, N, P) and five major bloom-forming algae in ocean-atmosphere Biogeochemical Climate Models for predicting air/sea transfers of

CO_2 and DMS. Once completed these more realistic climate models would then be suitable for expansion with other trace metal nutrients, notably Cu and Zn, as well as the emissions of additional biogenic gases into the atmosphere. This article and its underlying research projects are contributing to the international Surface Ocean Lower Atmosphere Study (SOLAS of SCOR/IGBP). This is Royal NIOZ contribution number 3650.

References

Alleman LY, Church TM, Véron AJ, Kim G, Hamelin B and Flegal AR (2001) Isotopic evidence of contaminant lead in the South Atlantic troposphere and surface waters. Deep-Sea Res II 48:2811-2828

Anbar AD, Roe JE, Barling J, Nealson KH (2000) Non-biological fractionation of iron isotopes. Science 288:126-127

Anders E and Grevesse N (1989) Abundances of the elements: Meteorite and solar. Geochim Cosmochim Acta 53:197-214

Baar de HJW (1992) Options for enhancing the storage of carbon dioxide in the oceans: A review. Energy Convers Mgmt 33(508):635-642

Baar de HJW (1994) von Liebig's Law of the Minimum and Plankton Ecology (1899-1991). Progress in Oceanography 33:347-386

Baar de HJW, Boyd PM (2000) The Role of Iron in Plankton Ecology and Carbon Dioxide Transfer of the Global Oceans. Chapter 4 In: Hanson RB, Ducklow HW and Field JG (eds) The Dynamic Ocean Carbon Cycle: A Midterm Synthesis of the Joint Global Ocean Flux Study. International Geosphere Biosphere Programme Book Series, Vol. 5, Cambridge University Press pp 61-140

Baar de HJW, Jong de JTM (2001) Distributions, Sources and Sinks of Iron in Seawater. Review chapter 5 In: Turner D and Hunter KA (eds) Biogeochemistry of Iron in Seawater. IUPAC Book Series on Analytical and Physical Chemistry of Environmental Systems 7:123-253

Baar de HJW, Bacon MP, Brewer PG and Bruland KW (1985) Rare earth elements in the Atlantic and Pacific Oceans. Geochim Cosmochim Acta 49:1943-1959

Baar de HJW, Buma AGJ, Nolting RF, Cadée GC, Jacques G and Tréguer PJ (1990) On iron limitation of the Southern Ocean: Experimental observations in the Weddell and Scotia Seas. Mar Ecol Prog Ser 65:105-122

Baar de HJW, Saager PM, Nolting RF, Van der Meer J (1994) Cadmium versus Phosphate in the World Ocean. Mar Chem 46:261-281

Baar de HJW, de Jong JTM, Bakker DCE, Löscher BM, Veth C, Bathmann U and Smetacek V (1995) Importance of Iron for Phytoplankton Spring Blooms and CO2 Drawdown in the Southern Ocean. Nature 373: 412-415

Baar de HJW, van Leeuwe MA, Scharek RA, Goeyens L, Bakker K and Fritsche P (1997) Nutrient anomalies in *Fragilariopsis kerguelensis* blooms, iron deficiency and the nitrate/phosphate ratio (AC Redfield) of the Antarctic Ocean. Deep-Sea Res II 44(1/2):229-260

Baar de HJW, de Jong JTM, Nolting RF, Van Leeuwe MA, Timmermans KR, Bathmann U, Rutgers van der Loeff M, Sildam J (1999) Low dissolved Fe and the absence of diatom blooms in remote Pacific waters of the Southern Ocean. Mar Chem.66:1- 34

Baar de HJW, Croot PL, Stoll MHC, Kattner G, Pickmere S, Freyer U, Boyd P and Smetacek V (2000) Nutrient anomalies of *Fragilariopsis kerguelensis* blooms revisited. Abstract Southern Ocean JGOFS Symposium, Brest, July 2000

Bakker DCE, de Baar HJW and Bathmann U (1997) Changes of carbon dioxide in surface waters during the spring in the Southern Ocean. Deep-Sea Res II 44(1/2):91-127

Bakker DCE, Watson AJ and Law CS (2001) Southern Ocean iron enrichment promotes inorganic carbon drawdown. Deep-Sea Res II 48:2483-2508

Barbeau K, Moffett JW, Caron DA, Croot PL and Erdner DL (1996) Protozoan Grazing May Play a Role in Relieving Iron Limitation of Phytoplankton. Nature 380:61-64

Beard BL, Johnson CM, Cox L, Sun H, Nealson KH, Aguilar C (1999) Iron isotopes biosignatures. Science 285:1889-1892

Bowie AR, Maldonado MT, Frew RD, Croot PL, Law CS, Achterberg EP, Mantoura RFC, Worsfold PJ, Boyd PW(2001) The fate of added iron during a mesoscale fertilisation experiment in the Southern Ocean. Deep-Sea Res II 48:2703-2743

Boyd PW, Watson AJ, Law CS, Abraham E, Trull T, Murdoch R, Bakker DCE, Bowie AR, Buesseler K, Chang H, Charette M, Croot P, Downing K, Frew R, Gall M, Hadfield M, Hall J, Harvey M, Jameson G, La Roche J, Liddicoat M, Ling R, Maldonado M, McKay RM, Nodder S, Pickmere S, Pridmore R,

Rintoul S, Safi K, Sutton P, Strzepek R, Tanneberger K, Turner S, Waite A, Zeldis J (2000) Phytoplankton bloom upon mesoscale iron fertilisation of polar southern ocean waters. Nature 407:695-702

Boye M and van den Berg CMG (2000) Iron availability and the release of iron complexing ligands by *Emiliania huxleyi*. Mar Chem 70(4):277-287

Boyé M, Sarthou G and van den Berg CMG (1999) Organic speciation of iron and zinc in the upper water column of the north-eastern Atlantic ocean. Deep-Sea Res II, Accepted

Boyé M, van den Berg CMG, de Jong JTM, Leach H, Croot P and de Baar HJW (2001) Organic complexation of iron in the Southern Ocean. Deep-Sea Res 48:1477-1497

Boyé M, Nishioka J, Croot P, Laan P, Timmermans KR and de Baar HJW (submitted a) Colloidal Fe accounts for a significant part of the dissolved organic Fe-complexes in the Southern Ocean

Boyé M, Aldrich AP, van den Berg CMG, de Jong JTM, Timmermans KR and de Baar HJW (submitted b) Horizontal gradient of the organic complexation of iron and its redox-speciation in surface waters of the northeast Atlantic Ocean

Broecker WS (1985) How to Build a Habitable Planet. Eldigio Press, Palisades, New York 291 p

Brownlee DE (1992) The origin and early evolution of the Earth. In: Butcher SS, Charlson RJ, Orians GH, Wolfe GV (eds) Global Biogeochemical Cycles. Academic Press, London pp 9-20

Bruland KW (1980) Oceanographic distributions of cadmium, zinc, nickel and copper in the North Pacific. Earth Planet. Sci Lett 47:176-198

Bruland KW (1989) Oceanic zinc speciation: Complexation of zinc by natural organic ligands in the central North Pacific. Limnol Oceanogr 34:267-283

Bruland KW (1992) Complexation of cadmium by natural organic ligands in the central North Pacific. Limnol Oceanogr 37:1008-1017

Bruland KW, Donat JR and Hutchins DA (1991) Interactive influences of bioactive trace metals on biological production in oceanic waters. Limnol Oceanogr 36:1555-1577

Buckley PJM, van den Berg CMG (1986) Copper complexation profiles in the Atlantic Ocean. Mar Chem 19:281-296

Buitenhuis ET, Timmermans KR and de Baar HJW (1999) Zinc-bicarbonate co-limitation of *Emiliania huxleyi* (Prymnesiophyceae). Chapter 3 In: Ph.D. thesis E.T. Buitenhuis (ed) Interactions between *Emiliania huxleyi* and the dissolved inorganic carbon system. University of Groningen pp 42-49

Buma AGJ, de Baar HJW, Nolting RF and van Bennekom AJ (1991) Metal enrichment experiments in the Weddell-Scotia Seas: Effects of Fe and Mn on various plankton communities. Limnol Oceanogr 36(8):1865-1878

Chisholm SW and Morel FMM (1991) What controls phytoplankton production in nutrient-rich areas of the open sea ? Limnol Oceanogr 36(8):1507-1970 with preface by the editors

Chisholm SW, Falkowski PG and Cullen J (2001) Discrediting Ocean Fertilization. Science 294:309-310

Coale KH and Bruland KW (1988) Copper complexation in the northeast Pacific. Limnol Oceanogr 33:1084-1101

Coale KH, Fitzwater SE, Gordon RM, Johnson KS, Barber RT (1996a) Control of community growth and export production by upwelled iron in the equatorial Pacific Ocean. Nature 379:621-624

Coale KH et al. (1996b) A massive phytoplankton bloom induced by an ecosystem-scale iron fertilisation experiment in the equatorial Pacific Ocean. Nature 383:495-501

Coale K, de Baar HJW, Worsfold P (1999) The iron age of oceanography. EOS, Trans Amer Geophys Union 80:377-382

Cooper DJ, Watson AJ and Nightingale PD (1996) Large decrease in ocean-surface CO_2 fugacity in response to *in situ* iron fertilisation. Nature 383:511-13

Crawford RM (1995) The role of sex in the sedimentation of a marine diatom bloom. Limnol Oceanogr 40:200-04

Cullen JT, Lane TW, Morel FMM and Sherrell RM (1999) Modulation of cadmium uptake in phytoplankton by seawater CO_2 concentration. Nature 402:165-167

Dai M and Martin JM (1995) First data on trace metal level and behaviour in two major Arctic river-estuarine systems (Ob and Yenisey) and in the adjacent Kara Sea. Earth Plan Sci Lett 131:127-141

De la Rocha CL, Brzezinski MA, De Niro MJ and Shemesh A (1998) Silicon-isotope composition of diatoms as an indicator of past oceanic change. Nature 395:680-683

DiTullio GR and Laws EA (1991) Impact of an atmospheric-oceanic disturbance on phytoplankton community dynamics in the North Pacific central gyre. Deep-Sea Res 38:1305-29

Donat R and Bruland KW (1988) Direct determination of dissolved cobalt and nickel in seawater by differential pulse cathodic stripping voltammetry preceded by adsorption of their nioxime complexes.

Anal Chem 60:240-244

Edwards PR (2000) Iron in modern and ancient East Antarctic snow: implications for phytoplankton production in the Southern Ocean. PhD Thesis, University of Tasmania, Australia 177p

Edwards PR, Sedwick PN, Morgan VI, Boutron CF, Hong S (1998) Iron in ice cores from Law Dome, East Antarctica: Implications for past deposition of aerosol iron. Ann Glaciol 27:365-370

Elderfield H and Rickaby REM (2000) Oceanic Cd/P ratio and nutrient utilization in the glacial Southern Ocean. Nature 405:305-310

Ellwood MJ and van den Berg CMG (2001) Determination of organic complexation of cobalt in seawater by cathodic stripping voltammetry. Mar Chem 75:33-47

Engelbrecht D (2000) Is intentional ocean fertilization an allowable CO_2 mitigation strategy. Student Essay, Departments of Philosophy and Theoretial Biology, Vrije Universiteit, Amsterdam 73p

Engelbrecht D, Zonneveld C, de Baar H and Kooyman S (2001) Ocean fertilization combats global warming: ingenious ploy or fata morgana ? Abstract at IGBP Open Science Conference on Challenges of a Changing Earth, Amsterdam 10-13 July 2001

Falkowski PG, Raven JA (1997) Aquatic Photosynthesis. Blackwell Science Inc, Malden, MA 375 p

Field MP, Cullen JT and Sherrell RM (1999) Direct determination of 10 trace metals in 50 μL samples of coastal seawater using desolvating micronebulization sector field ICP-MS. J Anal Atomic Spectros 14:1425-1432

Francois R, Altabet MA, Yu EF, Sigman DM, Bacon MP, Frank M, Bohrmann G, Bareille G and Labeyrie LD(1997) Contribution of Southern Ocean surface-water stratification to low atmospheric CO_2 concentration during the last glacial period. Nature 289:929-935

Frausto da Silva JJR and Williams RJP (1994) The biological chemistry of the elements. Clarendon Press, Oxford 561p

Frew R, Bowie A, Croot P and Pickmere S (2001) Macronutrient and trace-metal geochemistry of an *in situ* iron-induced Southern Ocean bloom. Deep-Sea Res II 48:2467-2482

Gall MP, Boyd PW, Hall J, Safi KA and Chang H (2001) Phytoplankton processes. Part 1: Community structure during the Southern Ocean Iron RElease Experiment (SOIREE). Deep-Sea Res II 48:2551-2570

Gerringa LJA, Poortvliet TCW, Rijstenbil JW, van Drie J and Schot MC (1995) Speciation of copper and responses of the marine diatom *Ditylum brightwellii*

upon increasing copper concentrations. Aquatic Toxicol 31:77-90

Gerringa LJA, Poortvliet TCW and Hummel H (1996) Comparison of chemical speciation of Cu in the Oosterschelde and Westerschelde estuary, the Netherlands. Est Coast Shelf Sci 42:629-643

Gerringa LJA, Hummel TCW, Moerdijk-Poortvliet H (1998) Relations between free copper and salinity, DOC and POC in the Oosterschelde and Westerschelde, the Netherlands. J Sea Res 40/3-4:193-203

Gerringa LJA, de Baar HJW and Timmermans KR (2000) Iron limitation of phytoplankton growth in natural waters versus artificial laboratory media with EDTA. Mar Chem 68:335-346

Gerringa LJA, de Baar HJW, Nolting RF, Paucot H (2001) The influence of salinity on the solubility of sulphides of Zn and Cd in estuaries, with the Scheldt estuary as an example. J Sea Res 46:201-211

Gledhill M and van den Berg CMG (1994) Determination of complexation of iron (III) with natural organic ligands in seawater using cathodic stripping voltametry. Mar Chem 47:41-54

Gledhill M, van den Berg CMG, Nolting RF, Timmermans KR (1998) Variability in the speciation of iron in response to an algal bloom in the northern North Sea. Mar Chem 59: 283-300

Gran HH (1931) On the conditions for the production of plankton in the sea. Rapp Proc Verb Réun Cons Int Expl Mer 75:37-46

Hagen (2001) Chemisch Weekblad, 2: 8

Halliday AN, Christensen JN, Lee DC, Hall CM, Luo X and Rehkamper (2000) Multiple-Collector Inductively Coupled Plasma Mass Spectrometry. Chapter 8 In: Inorganic Mass Spectrometry, Fundamentals and Applications. Marcel Dekker Inc, New York pp 291-328

Hannon E, Boyd PW, Silvoso M and Lancelot C (2001) Modeling the bloom evolution and carbon flows during SOIREE: Implications for future *in situ* iron-enrichments in the Southern Ocean. Deep-Sea Res II 48:2745-2773

Hannon E, Stoll MHC, de Baar HJW, Veth C and Lancelot C (submitted) Control of CO_2 drawdown in the Southern Ocean by iron and wind, a modeling study. Global Biogeochem Cycl

Hanson AK, Tindale NW and Abdel-Moati MAR (2001) An Equatorial Pacific rain event: influence on the distribution of iron and hydrogen peroxide in surface waters. Mar Chem 75:69-88

Hantke K (2001) Iron and metal regulation in bacteria.Curr Opin Microbiol 172-7

Holland HD (1984) The Chemical Evolution of the Atmosphere and Oceans. Princeton University Press 582 p

Houghton JT, Meria Filho LG, Callender BA, Harris N, Kattenberg A, Maskell K (1996) Climate Change 1995: The Science of Climate Change. Contribution of Working Group One to the Second Assessment Report of the Intergovernmental Panel on Climate Change. Cambridge University Press, Cambridge 572 p

Houghton JT, Ding Y, Griggs DJ, Noguer M, van der Linden PJ and Xiaosu D (2001) Climate Change 2001; The Scientific Basis. Intergouvernmental Panel on Climate Change, Cambridge University Press 944p

Iriondi M (2000) Patagonian dust in Antarctica. Quaternary International pp 68-71, pp 83-86

IPPC Climate Change (2001) The Scienfitic Basis. Cambridge University Press 98 p

Jickells TD and Spokes LJ (2001) Atmospheric Iron Inputs to the Oceans, Chapter 4 In: Turner D and Hunter KA (eds) Biogeochemistry of Iron in Seawater. IUPAC Book Series on Analytical and Physical Chemistry of Environmental Systems, Volume 7 85-122

Keeling RF, Shertz SR (1992) Seasonal and interannual variations in atmospheric oxygen and implications for the global carbon cycle. Nature 358:723-727

Kooyman SALM (2000) Dynamic energy and mass budgets in biological systems. Cambridge University Press 424p

Kramer CJM (1986) Apparent copper complexation capacity and conditional stability constants in North Atlantic waters. Mar Chem 18:335-349

Lancelot C, Hannon E, Becquevort S, Veth C and de Baar HJW (2000) Modeling phytoplankton blooms and carbon export production in the Southern Ocean: Dominant controls by light and iron in the Atlantic sector in Austral spring 1992. Deep-Sea Res 47:1621-1662

Lane TW and Morel FMM (2000) A biological function for Cd in marine diatoms. Proc Nat Acad Sci 97:4627-4631

La Roche J, Boyd PW, McKay RML and Geider R J (1996a) Flavodoxin as an *in situ* marker for iron stress in phytoplankton. Nature 382:802-805

La Roche J, van der Staay GWM, Partensky F, Ducret A, Aebersold R, Li R, Golden SS, Hiller RG, Wrench PM, Larkum AWD, Green B (1996) Independent evolution of the prochlorophyte and the green plant chlorophyll a/b light-harvesting proteins. Proc Nat Acad Sci 93:15244-15248

Lee JG and Morel FMM (1995) Replacement of zinc by cadmium in marine phytoplankton. Mar Ecol Prog Ser 127:305-309

Lee JG, Roberts SB and Morel FMM (1995) Cadmium: A nutrient for the marine diatom *Thalassiosira weissflogii*. Limnol Oceanogr 40:1056-1063

Leeuwe van MA and Stefels J (1998) Effects of iron and light stress on the biochemical composition of Antarctic *Phaeocystis* sp. (Prymnesiophyceae). II. Pigment composition and fucoxanthin markers. J Phycology 34:469-503

Leeuwe van MA and de Baar HJW (2000) Photo acclimation by the Antarctic flagellate *Pyramimonas* sp. (Praesinophyceae) affected by iron limitation. Euro J Phycology 35:207-215

Leeuwe van MA, Scharek R, de Baar HJW, de Jong JTM and Goeyens L (1997) Iron enrichment experiments in the Southern Ocean: physiological responses of plankton communities. Deep-Sea Res II 44(1/2):189-207

Leeuwe van MA, Timmermans KR, Witte HJ, Kraay GW, de Baar HJW (1998) The influence of iron limitation on chromatic adaptation by natural phytoplankton populations in the Pacific region of the Southern Ocean. Mar Ecol Prog Ser 166:43-52

Lewis BL, Landing WM (1991) The biogeochemistry of manganese and iron in the Black Sea. Deep-Sea Res II 38:773-803

Löscher BM, de Jong JTM, de Baar HJW, Veth C and Dehairs F (1997) The distribution of Fe in the Antarctic Circumpolar Current. Deep-Sea Res II, 44(1/2):143-187

Löscher BM, de Jong JTM and de Baar HJW (1998a) The distribution and preferential biological uptake of cadmium at 6°W in the Southern Ocean. Mar Chem 62:259-286

Löscher BM, van der Meer J, de Baar HJW, Saager PM and de Jong JTM (1998b) The global Cd/phosphate relationship in deep ocean waters and the need for accuracy. Mar Chem 59:87-93

Mann ME, Beadley SR and Hughes MK (1999) Northern hemisphere temperatures during the past millenium: inferences, uncertainties and limitations. Geophys Res Letts 26:759-762

Mantoura RFC, Dickson A and Riley JP (1978) The complexation of metals with humic materials in natural waters. Est Coast Mar Sci 6:387-408

Maréchal CN, Telouk P and Albarede F (1999) Precise analysis of copper and zinc isotopic compositions by plasma-source mass spectrometry. Chem Geol 156:251-273

Margulis L (1981) Symbiosis in Cell Evolution. WH Freeman and Company, New York.

Martin JH (1990) Glacial - interglacial CO_2 change: The

iron hypothesis. Paleoceanogr 5:1-13

Martin JH (1991) Iron still comes from above. Nature 353:123

Martin JH, Fitzwater SE (1988) Iron deficiency limits phytoplankton growth in the northeast Pacific subarctic. Nature 331: 341-343

Martin JH, Knauer GA and Gordon RM (1983) Silver distributions and fluxes in north-east Pacific waters. Nature 305:306-309

Martin JH, Gordon RM, Fitzwater SE, Broenkow WW, (1989) VERTEX: Phytoplankton/iron studies in the Gulf of Alaska. Deep-Sea Res 36:649-680

Martin JH et al. (1994) Testing the iron hypothesis in ecosystems of the equatorial Pacific Ocean. Nature 371:123 - 129

Mason RP, Lawson NM and Sheu GR (2001) Mercury in the Atlantic Ocean: Factors controlling air-sea exchange of mercury and its distribution in the upper waters. Deep-Sea Res II 48:2829-2854

Miller LA and Bruland KW (1995) Organic speciation of silver in marine waters. Environ Sci Technol 29: 2616-2621

Millero FJ (1998) Solubility of Fe(III) in seawater. Earth Plan Sci Lett 154:323-330

Mills GL, Hanson Jr AK, Quinn JG, Lammela WR, Chasteen ND (1982) Chemical studies of copper-organic complexes isolated from estuarine waters using C_{18} reverse phase liquid chromatography. Mar Chem 11:355-377

Moffett JW, Zika RG and Brand LE (1990) Distribution and potential sources and sinks of copper chelators in the Sargasso Sea. Deep-Sea Res 37:27-36

Moore JK, Abbott MR, Richman JG and Nelson DM (2000) The Southern Ocean at the last glacial maximum: A strong sink for atmospheric carbondioxide. Glob Biogeochem Cycl 14:455-475

Morel FMM, Reinfelder JR, Roberts SB, Chamberlain CP, Lee JG and Yee D (1994) Zinc and carbon co-limitation of marine phytoplankton. Nature 369:740-742

Muggli DM and Harrison PJ (1996) Effects of nitrogen source on the physiology and metal nutrition of *Emiliania huxleyi* grown under different iron and light conditions. Mar Ecol Prog Ser 130:255-67

Muggli DM, LeCourt M and Harrison PJ (1996) The effects of iron and nitrogen source on the sinking rate, physiology and metal composition of an oceanic diatom from the subArctic Pacific. Mar Ecol Prog Ser 132:215-27

Murozomi M (1981) Isotope dilution surface ionization mass spectrometry of trace constituents in natural environments and in the Pacific. Jap Soc Analyt Chem 30:19-26

Nishioka J, Takeda S, Wong CS, Johnson WK (2001) Size-fractionated iron concentrations in the northeast Pacific Ocean: Distribution of soluble and small colloidal iron. Mar Chem 74:157-179

Nolting RF and de Baar HJW (1994) Behavior of nickel, copper, zinc and cadmium in the upper 300m of a transect in the Southern Ocean. Mar Chem 45:225-242

Nolting RF, de Baar HJW, van Bennekom AJ and Masson A (1991) Cadmium, Copper and Iron in the Scotia Sea, Weddell Sea and Weddell/Scotia Confluence (Antarctica). Mar Chem 35:219-243

Nolting RF, Gerringa LJA, Swagerman MJW, Timmermans KR, de Baar HJW (1998) Fe (III) speciation in the High Nutrient, Low Chlorophyll Pacific region of the Southern Ocean. Mar Chem 62: 335-352

Nolting RF, de Baar HJW, Timmermans KR, Bakker K (1999) Chemical fractionation of zinc versus cadmium among other metals nickel, copper and lead in the northern North Sea. Mar Chem 67:267-287

Nolting RF, Heijne M, Jong JTM de, Timmermans KR, Baar HJW de (2000) The determination and distribution of Zn in surface water samples collected in the northeast Atlantic Ocean. J Environ Monit 534 – 538

Nolting RF, Helder W, de Baar HJW, Gerringa LJA (1999) Contrasting behaviour of trace metals in the Scheldt estuary in 1978 compared to recent years. J Sea Res 42:275-290

Park Y-I, Sanström S, Gustafsson P, Öquist G (1999) Expression of the isiA gene is essential for the survival of the cyanobacterium *Synechococcus* sp. PCC 7942 by protecting photosystem II from excess light under iron limitation. Mol Microbiol 32(1):123-129

Pearson RD (1973) Hard and Soft Bases. Dowden, Hutchinson and Ross, Stroudsburg, Pennsylvania 480 p

Petit JR, Jouzel J, Raynaud D, Barkov NI, Barnola J-M, Basile I, Bender M, Chapellaz J, Davis M, Delaygue G, Delmotte M, Kotlyakov VM, Legrand M, Lipenkov VY, Lorius C, Pépin L, Ritz C, Saltzman E, Stievenard M (1999) Climate and atmospheric history of the past 420,000 years from the Vostok Ice Core, Antarctica. Nature 399:429-436

Phinney J (2001) The Scientific and Policy Uncertainties Surrounding the Use of Ocean Fertilization to Transfer Atmospheric Carbon Dioxide to the Oceans. A summary statement drafted by participants in a workshop sponsored by the: American Society of Limnology and Oceanography (ASLO) (www.aslo. org), April 25, 2001 Washington DC

Piketh SJ, Tyson PD and Steffen W (2000) Aeolian transport from southern Africa and iron fertilization of marine biota in the South Indian Ocean. South African J Sci 96:244-246

Price NM and Morel FMM (1990) Cadmium and cobalt substitution for zinc in a marine diatom. Nature 344:658-660.

Raynaud D, Jouzel J, Barnola JM, Chappellaz J, Delmas RJ, Lorius C (1993) The ice record of greenhouse gases. Science 259:926-934

Rickaby REM and Elderfield H (1999) Planktonic Cd/Ca: paleonutrients or paleotemperature. Paleoceanogr 14:293-303

Riebesell U, Wolf-Gladrow DA and Smetacek V (1993) Carbon dioxide limitation of marine phytoplankton growth rates. Nature 36:249-251

Rosenthal Y, Boyle EA and Labeyrie L (1997) Last glacial maximum paleochemistry and deep water circulation of the Southern Ocean: Evidence from foraminiferal cadmium. Paleoceanogr 12:787-796

Rijstenbil JW, Gerringa LJA (2002) Interactions of algal ligands, metal complexation and availability and cell responses of the marine planktonic diatom *Ditylum brightwellii*, throughout a gradual increase of copper to toxic levels. Aquat Toxicol 56:115-131

Rijstenbil JW, Derksen JWM, Gerringa LJA, Poortvliet TCW, Sandee A, van den Berg M, van Drie J, Wijnholds JA (1994) Oxidative stress induced by copper: Defence and damage in the marine planktonic diatom *Ditylum brightwellii* (Grunow) West, grown in continuous cultures with high and low zinc levels. Mar Biol 119:583-590

Rozan TF, Lassman ME, Ridge DP, Luther III GW (2000) Evidence for iron, copper and zinc complexation as multi nuclear sulphide clusters in oxic rivers. Nature 406:879–882

Russell MJ, Daniel RM, Hall AJ (1993) On the emergence of life via catalytic iron-sulphide membranes. Terra Nova 5:343-347

Saager PM and de Baar HJW (1993) Limitations to the quantitative application of Cd as a paleoceanographic tracer, based on results of a multi-box model (MENU) and statistical considerations. Glob Planet Change 8:69-92

Saager PM, de Baar HJW and Burkill PH (1989) Manganese and Iron in Indian Ocean Waters. Geochim Cosmochim Acta 53:2259-2267

Saager PM, de Baar HJW and Howland RJ (1992) Cd, Zn, Ni and Cu in the Indian Ocean. Deep-Sea Res 39(1):9-35

Saager PM, de Baar HJW, de Jong JTM, Nolting RF and Schijf J (1997) Hydrography and local sources of dissolved Mn, Ni, Cu and Cd in the northeast Atlantic Ocean. Mar Chem 57:195-216

Saito MA and Moffett JW (2001) Complexation of cobalt by natural organic ligands in the Sargasso Sea as determined by a new high-sensitivity electrochemical cobalt speciation method suitable for open ocean work. Mar Chem 75:49-68

Sanudo-Wilhelmy S, Flegal AR (1992) Anthropogenic silver in the Southern California Bight: A new tracer of sewage in coastal waters. Environ Sci Technol 26:2147-2151

Saydam AC and Polat I (1999) The Impact of Saharan Dust on the Occurence of Algae Blooms. Proceedings of EUROTRAC Symposium (eds Borrell PM and Borrell P), WIT Press, Southampton pp 656-663

Scharek R, van Leeuwe MA and de Baar HJW (1997) Responses of Southern Ocean phytoplankton to addition of trace metals. Deep-Sea Res II 44:209-227

Schoemann V, de Baar HJW, de Jong JTM and Lancelot C (1998) Effects of phytoplankton blooms on the cycling of manganese and iron in coastal waters. Limnol Oceanogr 43:1427-1441

Schoemann V, de Jong JTM, de Baar HJW, Lancelot C, Reigstad M and Wassmann P (submitted) The cycling of Mn and Fe during a *Phaeocystis* bloom in an Arctic fjord. Limnol Oceanogr

Sigman DM and Boyle EA (2000) Glacial/interglacial variations in atmospheric carbondioxide. Nature 407:859-869

Smetacek V, Watson AJ, de Baar HJW, Bathmann U, Strass V, Riebesell U (2001) Polarstern ANT 18/2 Cruise Report, Ber Polarforschung

Stumm W and Morgan JJ (1981) Aquatic Chemistry. J Wiley and Sons 780 p

Sunda WG and Huntsman SA (1983) Effect of competitive interactions between manganese and copper on cellular manganese and growth in estuarine and oceanic species of the diatom *Thalassiosira*. Limnol Oceanogr 28:924-934

Sunda WG and Huntsman SA (1992) Feedback interactions between zinc and phytoplankton in seawater. Limnol Oceanogr 37:25-40

Sunda WG and Huntsman SA (1995) Cobalt and zinc interreplacement in marine phytoplankton: Biological and geochemical implications. Limnol Oceanogr 40:1404-1417

Sunda WG and Huntsman SA (1996) Antagonisms between cadmium and zinc toxocity and manganese limitation in a coastal diatom. Limnol Oceanogr 41:373-387

Sunda WG and Huntsman SA (1998) Interactions among Cu^{2+}, Zn^{2+} and Mn^{2+} in controlling cellular

Mn, Zn, and growth rate in the coastal alga *Chlamydomonas*. Limnol Oceanogr 43:1055-1064

Sunda WG and Huntsman SA (2000) Effect of Zn, Mn and Fe on Cd accumulation in marine phytoplankton: Implications for oceanic Cd cycling. Limnol Oceanogr 45:1501-1516

Sunda WG, Swift D and Huntsman SA (1991) Iron growth requirements in oceanic and coastal phytoplankton. Nature 351:55-57

Takahashi K, Billings JD and Morgan JK (1990) Oceanic province: Assessment from the time-series diatom fluxes in the northeastern Pacific. Limnol Oceanogr 35:154-65

Timmermans KR, Stolte W and de Baar HJW (1994) Iron-mediated effects on nitrate reductase in marine phytoplankton. Mar Biol (121):389-396

Timmermans KR, van Leeuwe MA, de Jong JTM, McKay RML, Nolting RF, Witte HJ, van Ooyen J, Swagerman MJW, Kloosterhuis H and de Baar HJW (1998) Iron limitation in the Pacific region of the Southern Ocean: Evidence from enrichment bioassays. Mar Ecol Prog Ser 166:27-41

Timmermans KR, GerringaLJA, de Baar HJW, van der Wagt B, Veldhuis MJW, de Jong JTM, Croot PL and Boye M (2001a) Growth rates of large and small Southern Ocean diatoms in relation to availability of iron in natural seawater. Limnol Oceanogr 46(2):260–266

Timmermans KR, Davey MS, van der Wagt B, Snoek J, Geider RJ, Veldhuis MJW, Gerringa LJA, de Baar HJW (2001b) Co-limitation by iron and light of *Chaetoceros brevis*, *C. dichaeta* and *C. calcitrans* (bacillariophyceae). Mar Ecol Prog Ser 217:287-297

Timmermans KR, Snoek J, Gerringa LJA, Zondervan I and de Baar HJW (2001c) Not all eukaryotic algae can interreplace cobalt and zinc: *Chaetoceros calcitrans* (Bacillariophyceae) versus *Emiliania huxleyi* (Prymnesiophyceae). Limnol Oceanogr 46:899-703

Turner D, Hunter KA and de Baar HJW (2001) Introduction, chapter 1 in: Biogeochemistry of Iron in Seawater, IUPAC Book Series on Analytical and Physical Chemistry of Environmental Systems, Volume 7:1-8

Van den Berg CMG (1982) Determination of copper complexation with natural organic ligands in seawater by equilibration with MnO_2. II. Experimental procedures and application to surface seawater. Mar Chem 11:323-342

Van den Berg CMG (1984) Determination of the complexing capacity and conditional stability constants of complexes of copper(II) with natural or-

ganic ligands in seawater by cathodic stripping voltammetry of copper-catechol complex ions. Mar Chem 15:1-18

Van den Berg CMG (1995) Evidence for organic complexation of iron in seawater. Mar Chem 50:139-157

Van den Berg CMG., Merks AGA and Duursma EK (1987) Organic complexation and its control of the dissolved concentration of copper and zinc in the Scheldt estuary. Estu Coast Shelf Sci 24:785-797

Vasconcelas MTSD and Leal MFC (2001) Antagonistic interactions of Pb and Cd on Cu uptake, growth inhibition and chelator release in the marine algae *Emiliania huxleyi*. Mar Chem 75:123-140

Vega M and van den Berg CMG (1997) Determination of cobalt in seawater by catalytic adsorptive cathodic stripping voltammetry. Anal Chem 69:874-881

Wächtershäuser G. (1992) Groundworks for an evolutionary biochemistry: The iron-sulphur world. Progr Biophys Mol Biol 58:85-201

Waite TD, Szymczak R, Espey QI, Furnas JM (1995) Diel variations in iron speciation in northern Australian shelf waters. Mar Chem 50:79-91

Watson AJ, Bakker DCE, Ridgwell AJ, Boyd PW and Law CS (2000) Effect of iron supply on Southern Ocean CO_2 uptake and implications for glacial atmospheric CO_2. Nature 407:730-733

Wells ML and Bruland KW (1998) An improved method for rapid preconcentration and determination of bioactive trace metals in seawater using solid phase extraction and high resolution inductively coupled mass spectrometry. Mar Chem 63:145-153

Wells ML, Smith GJ and Bruland KW (2000) The distribution of colloidal and particulate bioactive metals in Narragansett Bay, Rhode Island. Mar Chem 71:143-163

Wen LS, Santschi P, Gill G and Paternostro C (1999) Estuarine trace metal distributions in Galveston Bay: Importance of colloidal forms in the speciation of the dissolved phase. Mar Chem 63:185-212

Young RW, Carder KL, Betzer PR, Costello DK, Duce RA, DiTullio GR, Tindale NW, Laws EA, Uematsu M, Merrill JT, Feely RA (1991) Atmospheric iron inputs and primary productivity: phytoplankton responses in the north Pacific. Glob Biogeochem Cycl 5:119-134

Zettler ER, Olson RJ, Binder BJ, Chisholm SW, Fitzwater SE, Gordon RM (1996) Ironennchment bottle experiments in the equatorial Pacific: Responses of individual phytoplankton cells. Deep-Sea Res 43:1017-1030

Zhang Y, Amakawa H and Nozaki Y (2001) Oceanic

profiles of dissolved silver: Precise measurements in the basin of western North Pacific, Sea of Okhotsk, and the Japan Sea. Mar Chem 75:151-163

Zhu XK, O'Nions RK, Guo YL and Reynolds BC (2000a) Secular variation of iron isotopes in North Atlantic Deep Water. Science 287:2000-2002

Zhu X, O'Nions RK, Guo Y, Belshaw NS and Rickard D (2000b) Determination of natural copper isotope variation by plasma source mass spectrometry - implications for use as geochemical tracers. Chem Geol 63:139-149

Ocean-Atmosphere Exchange and Earth-System Biogeochemistry

D.W.R. Wallace[1]*and R. Wanninkhof[2]

[1] *Forschungsbereich Marine Biogeochemie, Institut für Meereskunde an der Universität Kiel, Düsternbrooker Weg 20, 24105 Kiel, Germany*
[2] *Ocean Chemistry Division, NOAA Atlantic Oceanographic and Meteorological Laboratory, 4301 Rickenbacker Causeway, Miami, FL 44149, USA*
*corresponding author (email): dwallace@ifm.uni-kiel.de

Abstract: Atmosphere-ocean chemical exchanges exert a significant control on both the gas-phase and particulate composition of the atmosphere. As a consequence, there are close linkages of these exchanges with atmospheric radiative transfer, climate and atmospheric chemistry. Such exchanges are important to human society both because they are fundamental to predictions of future climate forcing and because of an immediate need for science-based assessment of the effectiveness of proposed emission controls and other mitigation strategies. In this article we present a subjective overview of the current state of knowledge related to the exchange of climate-relevant materials including long-lived gases (CO_2, N_2O), short lived gases including volatile sulphur and halogen-containing compounds, and sea salt. Because of the importance of quantification of air-sea exchanges for short-term control of the concentration of these species in ocean and atmosphere, we also review recent progress and present opportunities related to studies of air-sea gas transfer. Finally we present a range of open questions, promising approaches and major challenges related to the measurement, modelling and understanding of air-sea chemical exchanges. Many of these issues will be addressed within the new global change research program SOLAS (Surface Ocean Lower Atmosphere Study).

Background

The chemical composition of the Earth's atmosphere is determined by biological processes acting in concert with physical and chemical change (Wayne 1991). Atmospheric composition is a key factor determining Earth's climate, due to the controls exerted on radiative transfer via a range of gases and particles. There is interplay between the physical climate system and biogeochemistry which maintains habitable conditions on Earth.

Over the past 200 years, mankind has significantly altered the chemical composition of the atmosphere. Some well-known examples are:

• Atmospheric CO_2 levels have increased by 40% since the industrial revolution and are higher than at any time over the past 20 million years (Prentice et al. 2001).

• Atmospheric CO_2 will likely double in concentration during this century.

• Reactive Cl and Br atoms of human origin have altered the chemistry of the stratosphere.

• Sulfate aerosols from the burning of fossil-fuels have had a reginal net cooling effect on climate.

• Release of fixed-nitrogen species is altering atmospheric chemistry regionally and may be impacting primary production on land and in parts of the ocean.

• Desertification is altering the amount of mineral dust within the atmosphere with potential impacts on climate and ocean productivity.

• Release of CH_4 and other reactive trace gases may be affecting the oxidising capacity of the global atmosphere.

From WEFER G, LAMY F, MANTOURA F (eds), 2003, *Marine Science Frontiers for Europe.* Springer-Verlag Berlin Heidelberg New York Tokyo, pp 107-129

• The concentration of the greenhouse gas N_2O has increased by ~15% since the industrial revolution.

Most of these perturbations are now large enough to significantly affect the Earth's climate-biogeochemistry system. The anthropogenic pertur-bations themselves originate primarily on or over land. However atmosphere-ocean exchanges across the vast surface area of the global ocean play a key role not only in controlling the sinks for anthropogenic species (e.g. CO_2), but also the sources of a wide-variety of radiatively- and chemically-active compounds and particles. Anthropogenic changes to the marine atmosphere and climate (including climate-related changes in ocean circulation), as well as anthropogenic impacts in the coastal oceans (including effects of fishing) may alter ocean productivity and ecology. Such changes in the biology and chemistry of the oceans may, in turn, feedback on climate via shifts in the atmosphere-ocean exchange of gases and particles.

Significance for society

Global change prediction

Scientific understanding of the biogeochemical processes affecting radiatively-active species and particles in the atmosphere is required in order to identify, predict, and plan for future climate change.

In the case of CO_2, it is likely that mitigation efforts will be too little and too late to avoid significant global warming. Climate-related processes within the natural carbon cycle could accelerate or reduce the rate of accumulation of CO_2 in the atmosphere, hence the magnitude of climate change. Such feedbacks should be assessed in order to predict the magnitude of future changes in radiative forcing for a given CO_2 emissions scenario.

It is tempting to assume that the Earth's natural controls of atmospheric composition and climate will not be significantly perturbed as a result of recent anthropogenic forcing. Sometimes such assumptions are made from no firmer basis than the consideration that the Earth's biological and chemical processes are, 'too complicated' to understand or predict. In particular, feedbacks involving biologi-

cal and chemical processes may involve unexpectedly strong non-linear behaviour (i.e. 'surprises'). There are already well-known examples of such global change surprises (e.g. ozone destruction, and particularly the ozone hole). Identification and early warning of such effects generally comes from basic research, including long-term observation.

With respect to CO_2, the relative stability of atmospheric CO_2 (variability <20 µatm: Indermuehle et al. 1999) over 8000 years prior to the Industrial Revolution, and the long timescales associated with the larger glacial-interglacial pCO_2 variations (100 µatm changes over thousands of years) has led some to suggest that the natural carbon cycle is immune to decadal and centennial climate change. The argument might follow that climate variability is unlikely to drive changes in natural carbon cycling that can affect future atmospheric CO_2 by as much as 100 µatm (which is small with respect to uncertainty associated with projected emissions). However such arguments, at least if based on the paleo-record alone, are weak. Present-day climate forcing due to the release of greenhouse gases is of a nature (anthropogenic), magnitude (larger), timescale (faster) and sign (higher than interglacial CO_2 concentration) that is unprecedented at least for the past 420,000 years and likely longer (Prentice et al. 2001). Hence the paleo-record can act as a useful indicator of processes and mechanisms that may be important for this response, but contains no direct analog that can be used to predict the future.

From what we do know about key processes in the natural carbon cycle, a linear response to anthropogenic forcing is unlikely. Most of the known direct feedbacks are strongly non-linear and decrease the uptake capacity of the oceanic and terrestrial reservoirs. For example: CO_2-fertilization of plant growth saturates at high CO_2 levels and the oceanic uptake capacity for anthropogenic CO_2 decreases with increasing ambient CO_2 levels. At least with these reasonably well-understood chemical and physiological processes, the 'buffering' of atmospheric CO_2 levels evident in the low-amplitude variability of atmospheric CO_2 throughout the Holocene will be strongly reduced over the next 100 years. Further, we are likely moving into a significantly warmer and moister climate in which the rate of microbial carbon cycling both on land and in the upper ocean will increase in a non-linear fashion.

In summary: non-linear responses of biogeochemical cycles to global environmental change are to be expected. Associated changes of atmospheric levels of radiatively-active species should therefore be considered carefully when attempting prediction of future climate response for a given emissions scenario.

Mitigation of global environmental change

A second socio-economic imperative for basic research in biogeochemistry has recently become increasingly important:

• Understanding of biogeochemical cycling, and monitoring of biogeochemical processes, are required bases for planning and assessment of global change mitigation measures.

International agreements to mitigate global change are now major societal, economic and political issues (e.g. Montreal Protocol, Kyoto Protocol and successor agreements). Apart from the need to establish reliable climate predictions and evaluate uncertainties, there is a rapidly growing need for basic research upon which to design, implement and evaluate greenhouse gas mitigation measures. The socio-economic requirement for research into the fate of radiatively-active species such as CO_2 is therefore becoming almost independent of the need to make accurate climate predictions. The implementation of complex and potentially costly global agreements to reduce future greenhouse gas levels requires that we have a strong scientific basis to assess the effectiveness of any measures taken. For many of the gases regulated under the Montreal Protocol (e.g. CFCs, CCl_4) such assessment is relatively straightforward using accurate global measurements of the atmospheric levels of the controlled species because the species have no natural sources (see for example Fig. 1).

For CO_2 and other gases that have large natural as well as anthropogenic sources and/or significant non-atmospheric sinks, such assessments will be considerably more complicated. This will require a high level of understanding and measurement of exchanges between the atmosphere, the

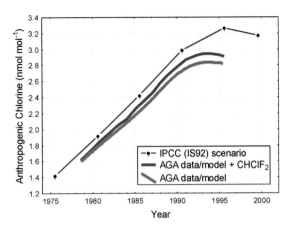

Fig. 1. Anthropogenic chlorine in the lower troposphere based on atmospheric time-series measurements and model predictions for the chlorofluorocarbons, CCl_4 and CH_3CCl_3. This figure is redrawn after Cunnold et al. (1997). AGA data/model refers to measurements made by the Global Atmospheric Gases Experiment from the Advanced GAGE measurement network (Cunnold et al. 1997). The data have been extrapolated and interpolated using a 2-dimension model of the atmosphere. Also plotted is the inventory of anthropogenic chlorine when $CHClF_2$ is included. For comparison, the IS92 emission scenario of the Intergovernmental Panel on Climate Change (IPCC; Prather et al. 1996) is shown. The calculated chlorine loading declined more rapidly than expected based on the emissions estimates derived from industry figures and the IPCC scenario. These types of data relating to Montreal Protocol implementation indicate the value of global studies for evaluating the implementation of international agreements concerning greenhouse gas emissions controls.

ocean and the terrestrial biosphere. An example can be gleaned from the Montreal Protocol itself, which sought initially to control production and release of purely anthropogenic compounds responsible for stratospheric ozone destruction. Subsequent extension to cover CH_3Br, a gas with both natural and anthropogenic sources, raised a scientific and political controversy. Notably, it was unclear to what extent elimination of industrial CH_3Br production would reduce the overall delivery of Br atoms to the stratosphere, because the global budgets of CH_3Br and Br were so poorly characterized (see review by Butler and Rodriguez 1996). In comparison with CO_2, the economic importance of CH_3Br is extremely small, however this example illustrates how scientific understanding of global budgets is required in order to make informed and defensible decisions concerning mitigation.

Atmosphere-ocean biogeochemical interactions

The atmosphere is affected strongly by chemical and physical exchange with both the land and the ocean surface. Similarly the surface ocean is strongly affected by direct interaction with the land. In this chapter the focus is exclusively on ocean-atmosphere chemical interactions, while recognising that the other interfaces are also critical to an understanding of earth systems and their sensitivity to future change. We concentrate on the exchange of gases and sea salt. Many key biogeochemical interactions are mediated by the deposition of dust, and hence bio-limiting trace metals, on the sea surface. This important aspect of air-sea exchange is discussed in the review by deBaar and La Roche (this volume).

Long-lived radiatively active gases

CO_2: The air-sea CO_2 flux is a key exchange between reservoirs and has been perturbed by mankind as a result of the increase of CO_2 levels in the atmosphere. Future changes to this flux are expected due to continued emissions as well as changes in surface ocean chemistry, water temperature, sea-ice extent, wind-speed, ocean circulation, and the microbial and plankton ecology of the upper ocean. The sign of most such changes would be expected to represent positive feedbacks on the accumulation rate of CO_2 in the atmosphere which would accelerate the pace of global climate change (Prentice et al. 2001). On the other hand decreases in biocalcification, driven by a reduction in seawater's carbonate ion content due to uptake of anthropogenic CO_2, would drive a negative feedback via surface water alkalinity changes (e.g. Buitenhuis et al. 1999). Strong decreases in marine biocalcification with increasing pCO_2 have been predicted by Kleypass et al. (1999) for coral reefs, and by Riebesell et al. (2000) for planktonic

calcifiers (coccolithophorids). Further laboratory and mesocosm experiments are required to investigate this response. The effect of reduced bio-calcification may turn out to be more serious for the marine ecosystem itself than for the atmospheric CO_2 level, and considerably more research will be required to assess such effects and assess whether organisms may be able to adapt physiologically to higher pCO_2 levels.

Apart from the need to predict the future, there are present-day reasons to better resolve the air-sea exchange of CO_2. These relate to the role of terrestrial carbon sinks in the contemporary global carbon cycle. Such sinks are controversial in the scientific community and also within the international agreements (e.g. the Kyoto protocol) relating to CO_2 emission controls. The scientific basis for assessing, quantifying and managing such sinks over large scales is weak however. Fluxes of carbon into and out of the terrestrial biosphere are notoriously difficult to measure, and estimates require critical assessment across a variety of scales and approaches if they are to be credible. One particularly promising method to assessing regional- and continental-scale sinks is inverse modeling of atmospheric CO_2 data (e.g. Bousquet et al. 2000).

Such modeling seeks to infer unknown regional and temporal distributions of carbon sources and sinks at the land- and sea-surfaces based on the measured gradients of atmospheric pCO_2 and carbon-related properties (e.g. O_2, ^{13}C) using atmospheric transport models. Geographic and temporal distributions of atmospheric pCO_2 (and other carbon-related properties) arising from known fossil fuel emissions are predicted and the magnitude and regional distribution of ocean and land surface sources and sinks adjusted to match the measured atmospheric distributions. Specification of constraints on regional and/or global air-sea fluxes can greatly improve definition of the location and magnitude of the exceedingly difficult to measure land-atmosphere fluxes (Tans et al. 1990; Fan et al. 1998). Detailed information is required not only on the mean air-sea fluxes but also their temporal variability. Specifically, inverse models of atmospheric CO_2 data would benefit from constraints concerning regional and temporal (seasonal to decades) variability of the air-sea flux of CO_2 and related properties. Recent work in the oceans has also applied inverse models of ocean carbon transport to assess ocean-atmosphere fluxes (e.g. Holfort et al. 1998).

In addition to constraining inverse models, the potential exists to utilize observed variability in surface fluxes, including interannual variability, to diagnose the sensitivity of carbon fluxes to climatic perturbation. The effect of climate variability on atmospheric CO_2 levels is recorded in long time-series of atmospheric pCO_2 and this can be used as a partial test of the reliability of climate-carbon feedbacks in models (e.g. Cox et al. 2000). Presently there remains considerable uncertainty concerning the origins of such variability, including the relative magnitude of the observed variability that is driven by land-atmosphere and ocean-atmosphere fluxes.

Independent time-series estimate of air-sea carbon fluxes would be of considerable use in resolving both of these issues. Presently, such time-series of surface water pCO_2 measurements, which are critical to estimating fluxes, are available for only very limited regions of the ocean: Hawaii (e.g. Winn et al. 1998), Bermuda (Bates et al. 1996), the equatorial Pacific (e.g. Feely et al. 1999), and the Greenland Sea (Skjelvan et al. 1999). The valuable time-series data at Bermuda have shown relatively little interannual variability of surface water pCO_2 (e.g. Bates et al. 1996, 2001) and the equatorial Pacific appears to dominate global variability of the ocean-atmosphere CO_2 flux (e.g. 70% according to model simulations by Le Quéré et al. 2000). However detection of strong CO_2 flux variability at the Hawaii or Bermuda time-series sites might be considered analogous to the discovery of large interannual variability in land-atmosphere CO_2 fluxes in the Gobi Desert (i.e. it is not to be expected because the mean annual air-sea fluxes are small)! Rather, significant interannual flux variability should be sought in regions where the mean annual flux is large (e.g. high latitudes, equatorial regions, coastal upwelling systems). The Equatorial Pacific flux variability has been estimated from measurements and is indeed large, with an amplitude of approx. 0.2 to 1 PgC/yr, driven by ENSO variability in upwelling (Feely et

al. 1999). The only high-latitude region of high net annual CO_2 flux that has been studied (the Greenland Sea, Skjelvan et al. 1999) also apparently is subject to strong inter-annual variability. Surface pCO_2 in the winter of 1996-1997 averaged –55 µatm in comparison to –36 µatm during the winter of 1994-1995. This difference was associated with interannual changes in temperature and ice-cover. Although their global significance is limited due to the small area involved, these results from the Greenland Sea indicate that interannual CO_2 variability also exists in high latitude CO_2 sink regions next to low-latitude CO_2 source regions affected by ENSO.

Variability outside of these regions (e.g. in the vast Southern Ocean) is presently weakly constrained by oceanic measurement, and can only be assessed by models (e.g. Le Quéré et al. 2000) or indirect methods using proxies such as sea surface temperature (Lee et al. 1998). It should be noted that the full expression of within-ocean carbon cycle variability is damped within atmospheric CO_2 time-series by the long gas exchange timescale of CO_2 (~ 1 year). One might therefore expect ocean-forced variability in atmospheric pCO_2 to be expressed at lower frequencies than variability arising from terrestrial carbon cycling. There may be a suggestion of this behaviour in the time-series of fluxes reproduced in Fig. 2 (from Battle et al. 2000). Such low-frequency variability will be very hard to detect in the still short combined time-series of atmospheric pCO_2, ^{13}C and O_2, but may be of considerable importance and relevance to the decadal timescale of anthropogenic climate change.

N_2O: Nitrous oxide (N_2O) acts as a major greenhouse gas and it is also the major source for NO radicals in the stratosphere. Present-day as well as paleo records of tropospheric N_2O show significant long and short term variability which is not yet fully explained (e.g. Flückiger et al. 1999). The present-day atmospheric level (mixing ratio of ~314 x 10^{-9} mol/mol) is ~15% higher than the preindustrial value estimated from ice core data (~270 x 10^{-9}; Ehhalt et al. 2001).

The oceans play a major role in the global budget of atmospheric N_2O (Khalil and Rasmussen 1992; Nevison et al. 1995; Bouwman et al. 1995). Of the present-day sources of N_2O, estimates com-

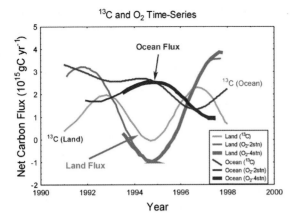

Fig. 2. Temporal variability of land-atmosphere and ocean-atmosphere fluxes inferred from atmospheric time-series of CO_2, ^{13}C and O_2 data (Battle et al. 2000). Positive fluxes refer to net fluxes into the ocean and land biosphere. The two different O_2-based curves for both land and ocean fluxes refer to results from averaging 2 stations (longer time-series of measurements) or 4 stations (shorter time-series). This plot is redrawn after Battle et al. (2000). The time-series indicate that the atmosphere-to-land net flux is smaller and more variable than the atmosphere-to-ocean flux.

piled by Ehhalt et al. (2001) imply anthropogenic and oceanic contributions of 27-45% and 17-24% respectively to a total source strength of between 14.7 and 17.7 TgN yr^{-1}. The oceanic source is likely dominated by estuaries, coastal upwelling regions and continental shelves (Bange et al. 1996; Seitzinger and Kroeze 1998). The overall uncertainty of such global budgets and importance of coastal regions was emphasized by Bange et al. (1996) who estimated a total ocean source of 7-11 TgN yr^{-1} with the majority being emitted from coastal regions. The importance of coastal fluxes may have been seriously underestimated in other budgets, although it should also be noted that the database available for making estimates of fluxes from coastal regions including coastal upwelling systems is very small. Hence estimates of even the long-term mean N_2O source from the ocean are still associated with considerable uncertainty, as is the overall atmospheric budget of N_2O. In modelling studies it is generally assumed that oceanic N_2O fluxes have remained constant (e.g. Rahn and Wahlen 2000) and that anthropogenic sources on

land are responsible for the observed N_2O increase (e.g. industrial releases, use of fertiliser in agriculture, etc). The potential for anthropogenic pertubation of estuarine fluxes is also high however due to large-scale post-industrial changes in riverine nitrogen and organic carbon transports (Navqui et al. 2000).

Sources of N_2O within the ocean include shallow sources associated with bacterial nitrification (Dore and Karl 1996; Dore et al. 1998) as well as (usually) deeper sources associated with nitrification and (in proximity to suboxic waters) denitrification (Codispoti et al. 1992). Karl (1999) has suggested that shallow N_2O production may be climate sensitive, introducing a conceptual framework for an open ocean-atmosphere feedback mediated by the climate-relevant trace gas N_2O. The air-sea flux of N_2O from coastal waters, estuaries and marginal seas may also be climate sensitive, however this possibility has received little study.

A particularly powerful tool in resolving issues concerning the tropospheric budget of N_2O is its stable isotope composition (Kim and Craig 1990), including its 'isotopomeric' composition (i.e. the intramolecular position of ^{15}N in the linear NNO molecule; Yoshida and Toyoda 2000; Roeckmann et al. 2001). Recent measurements show that the isotopic signature of terrestrially-derived N_2O is extremely variable (Perez et al. 2000). This variability has led Rahn and Wahlen (2000) to use the (apparently) less-variable isotopic composition of oceanic and stratospheric N_2O as constraints in their models of the global N_2O budget.

There is potential to use the (apparently) tightly-constrained isotopic composition of near-surface marine N_2O as a constraint in global budgets, in order to resolve mean fluxes of more variable and harder-to-sample processes (e.g. land-atmosphere isotopic composition and fluxes, stratosphere-troposphere exchanges, etc.). This approach is analogous to the proposed use of atmosphere-ocean CO_2 flux data to resolve the magnitude and location of terrestrial carbon sinks presented earlier. The ocean offers the dual advantages of: (1) relatively limited spatial variability (in comparison with terrestrial systems) due to the effects of ocean mixing which makes representative sampling possible, and (2) a reasonably well-tested approach to estimating inter-reservoir fluxes (i.e. the gas-exchange parameterization).

To-date however isotopic measurements of oceanic N_2O are limited and restricted to open ocean regions (e.g. sub-tropical Pacific, Arabian Sea). Notably, little or no isotopic data from coastal, estuarine or continental margin locations were available to be included in the modelling of Rahn and Wahlen (2000) despite their known significance for the ocean-to-atmosphere N_2O flux (e.g. Bange et al. 1996). Hence further measurements may reveal greater variability of near-surface isotopic composition than represented in Fig. 3. Of particular utility would be coupled estimates of ocean-atmosphere N_2O fluxes with isotopic meas-

urements from a wide-variety of regions and seasons in order to define flux-weighted means for oceanic N_2O isotopic composition. In order to attain such a goal, many more measurements of N_2O and its isotopic composition are needed in order to ascertain that the oceanic source is adequately characterized in current models of N_2O sources and sinks.

Short-lived radiatively and chemically active gases

Sulphur Gases: Marine ecosystems are a major source of dimethylsulphide (DMS), the atmospheric oxidation of which plays a significant role in creating non-sea salt sulphate (NSS) aerosol. This aerosol is hypothesized to contribute to

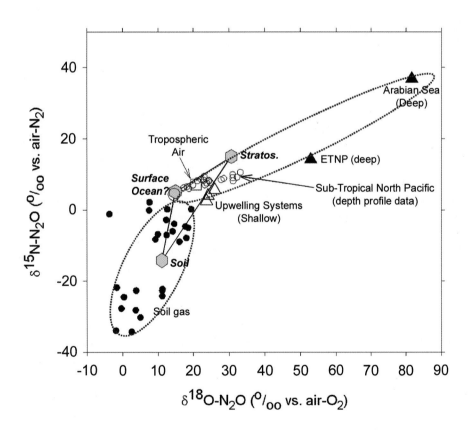

Fig. 3. The delta-^{15}N versus delta-^{18}O of N_2O for oceans (surface and deep), soil-gas, and the atmosphere (troposphere and stratosphere). This plot is based on data and references cited in Rahn and Wahlen (2000) and Dore et al. (1998). The results of individual measurement campaigns in different regions by various investigators are also shown (small symbols). Large-scale averages are denoted by larger symbols, including shaded hexagons which mark the proposed isotopic 'end-members' of Kim and Craig (1990).

increased numbers of cloud condensation nuclei (CCN) over the oceans, thereby affecting the distribution and albedo of clouds, and potentially reflecting more sunlight back to space and cooling the Earth. It has been further hypothesized that the marine production of DMS might be itself climate-sensitive (e.g. via light and temperature effects on phytoplankton productivity) thereby contributing to a self-regulating system (the CLAW hypothesis of Charlson et al. 1987). The complexity and variability of the processes involved makes it difficult to assess the validity of the proposed link between marine ecosystem processes and climate forcing (Andreae and Crutzen 1997).

One major problem has been to understand the factors that control the flux of DMS from the ocean to the atmosphere. Uncertainty in the gas-exchange parameterization itself contributes to this problem. However a more significant and specific problem is our lack-of-understanding of the other factors that regulate the near-surface concentration of this biogenic trace gas.

Three approaches have been utilised to predict DMS concentration (and hence air-sea flux) variability: (1) a correlative approach, relating measured DMS concentrations to proxy variables that can be used for extrapolation of limited DMS data sets over larger spatial and temporal scales; (2) a data compilation approach, in which all existing measurement data are assembled and gridded to produce climatological mean distributions; and (3) a process-level approach, in which the mechanistic and ecosystem factors underlying DMS production, consumption and flux are parameterized. The first and third approaches are not independent as the choice of proxy variables is frequently based on mechanistic understanding.

Initially it was hoped that the proposed phyto-plankton-DMS link in the CLAW hypothesis would be expressed in a simple relationship between phytoplankton biomass and DMS concentration. However the more data that were collected, the clearer it became that most simple correlations would not readily explain the observed variability. The reason for this failure lies clearly with the complexity and variability of the processes underlying DMS production and consumption.

Subsequently, and with the benefit of a collection of ~16000 oceanic DMS measurements world-wide, Kettle et al. (1999) applied a spatial gridding and averaging approach to produce a global climatology of DMS concentrations. This approach largely dispensed with mechanistic understanding, although bigeochemical provinces were utilized in the construction of the climatology. The approach represents essentially a "brute-force" solution to the problem however this approach was fully justified given the lack of success of the other approaches to-date and the large data set that had been compiled. The DMS climatology became the basis for a global estimate of the air-sea flux of DMS (e.g. Kettle and Andreae 2000). However this climatology is determined purely from measurement, mapped on the basis of location and time, and therefore contains little or no information from which to assess how the DMS concentration field might vary in response to long-term changes in forcing.

The apparent failure of the simple correlative approach, implies that a process-level understanding and parameterisation of DMS production and loss pathways is required in order to predict DMS flux response to global environmental change. This is a major challenge however, as almost all components of the marine food chain play a role in DMS cycling (viruses, bacteria, phytoplankton, zooplankton), and several complex abiotic and photochemical loss pathways also play significant roles. One major problem has been to understand the factors that control the production of DMS from its non-volatile precursor, dimethylsulfonio-propion-ate (DMSP). The production of DMSP itself is highly species-specific amongst phytoplankton. Further, the yield of DMS from its precursor is also highly variable. Despite this complexity, understanding is slowly emerging. Notably, the significance of DMSP in satisfying bacterial sulphur and carbon demand, and hence the potential significance of bacterial assimilation of DMSP in controlling DMS yield, is becoming recognised (e.g. Kiene and Linn 2000). Further, correlative approaches of DMS yield with mixed-layer stratification over short-term and seasonal timescales appear promising in that they appear to be based on a rational mechanistic basis that is consistent with much of our present knowledge (Simo and Pedros-Allo 1999; Fig. 4).

Most recently, Anderson et al. (2001) re-examined the data compilation of Kettle et al. (1999) and identified a correlation of higher surface water DMS concentrations with the composite parameter $\log_{10}(CJQ)$, where C is surface chlorophyll concentration, J is the incident light intensity (mean daily short-wave) and Q is a nutrient limitation term based on nitrate. This choice of composite variable was justified on the basis that DMS concentrations are observed to be high in well-lit, nutrient-rich waters. The light-related factor allows for expected seasonality in DMS concentrations, the nutrient limitation factor for variations in growth rate and the chlorophyll factor for the regional variation of biomass. The authors note that despite some 'mechanistic' justification, the specific composite parameter was chosen largely because it worked. That this empirical relationship did not capture all mechanistic factors that underlie DMS conentration variability was also noted by the authors, who emphasised that their relationship was only

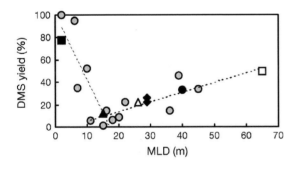

Fig. 4. Yield of DMS from DMSP as a function of mixed-layer depth based on field experiments is various regions by various authors. The DMS yield is defined as the measured DMS production rate divided by the DMSP consumption rate expressed as a percentage. Data points originate from a wide variety of studies cited in Simo and Pedros-Alio (2000) and include data from a wide range of oceanic regimes including the North Atlantic, the Sargasso Sea, the North Sea, the Equatorial and Northeast Pacific, the Gulf of Maine and the Mediterranean.

significant after extensive smoothing of the data set and that it did not explain DMS variability in low-DMS regions (e.g. [DMS] < 2.29 nM).

These developments are however encouraging and indicate that progress is being made. The process-oriented findings imply that a relatively strong climate-DMS concentration link exists within the ocean, with overall sign consistent with the original CLAW hypothesis, but which is based on overall ecosystem response to forcing rather than a simple climate-phytoplankton link as originally proposed. What is now required is rigorous testing of these emerging ideas both in the field and within experimental systems, in which the proposed controlling factors can be independently manipulated.

Once DMS is in the atmosphere, it is subject to multiple pathways and fates with the amount that ultimately contributes to the sulfate aerosol being determined by multiple oxidation and scavenging pathways. The OH radical reacts with DMS with SO_2 and methane sulfonic acid (MSA) being the major products. However the relative yield of these two products is variable and the importance of other potential reactions of DMS is still not well known. The fate of the DMS oxidation products, particularly the proportion that is further oxidised to sulfate and contributes to the sulfate aerosol, is also likely to be variable depending on environmental conditions and requires greatly improved understanding.

Several other volatile sulfur compounds are produced in the ocean and emitted to the atmosphere, including carbonyl sulfide (COS). COS is the most abundant reduced sulphur gas in the atmosphere with a longer lifetime in the troposphere than DMS. It can therefore reach the stratosphere where its oxidation contributes to the formation of stratospheric clouds. It is both produced and consumed in the oceans, with most production arising from the photochemically-mediated reaction of dissolved organic sulfur compounds. The principal loss pathways are hydrolysis, air-sea exchange and downward mixing. Overall however the oceans are a significant source for COS. The dissolved organic sulfur in seawater is rapidly cycled, and its concentration is spatially and temporally variable. Significant progress is being made in modelling the marine production of this gas (Von Hobe et al. submitted) in part because the controlling processes are to some extent simpler and certainly better understood than for DMS.

Volatile organo-halogens: There are three principal sources of reactive halogen species for the atmosphere: (1) volcanic emissions of HCl; (2) the sea-salt aerosol; (3) volatile organohalogen compounds of both natural and anthropogenic origin. Chlorine from sea-salt and HCl generally does not reach high altitudes due to rapid removal processes within the troposphere. Volatile organohalogens on the other hand contribute significantly to the halogen budgets of both the stratosphere and the troposphere. For example, ~12% of the equivalent chlorine of atmospheric ozone-depleting gases is derived from naturally-produced CH_3Cl . Naturally-produced CH_3Br has been estimated to contribute a further 11% of equivalent chlorine ('equivalent chlorine' refers to amounts of various halogens that have been adjusted, according to individual ozone-destroying strength, to an equivalent amount of chlorine). The importance of these organic compounds lies in the fact that their tropospheric lifetimes are sufficiently long that the associated halogen atoms can be transported to the upper troposphere and stratosphere and participate in ozone-destroing reaction pathways there.

Shorter-lived organic and inorganic halogen compounds participate significantly in ozone destruction cycles in the troposphere. These are emitted either directly from the oceans (e.g. CH_2Br_2, $CHBr_3$, CH_3I, CH_2I_2), produced indirectly via photochemical transformation of precursor species, or made by chemical processes involving sea salt aerosol in the marine boundary layer (e.g. HCl, HBr, Br_2, BrCl, Cl_2, BrO, ClO, IO) (Platt and Moortgat 1999). The reactive halogen species (in particular BrO and IO) may play significant roles in free troposphere and boundary layer photochemistry and the significance of reactive iodine for particle production has recently been suggested (O'Dowd et al. 2002). A fundamental difference between the cycling of iodine and bromine/chlorine in the marine atmosphere is that iodine has no significant source from the sea salt aerosol. In contrast. it tends to become strongly enriched above sea-salt ratios within the sea-salt aerosol indicating that the aerosol is a sink for gas-phase iodine. The predominant source of iodine for the marine

boundary layer is the sea-to-air flux of biogenic, iodine-containing gases.

The surface ocean has been postulated as an important source of atmospheric methyl halides since James Lovelock's pioneering measurements in the early 1970's. Most early field studies were limited in geographical and seasonal coverage however, and were subject to analytical problems (Butler and Rodriguez 1996). Previously for example, the open ocean was considered to be a major source of CH_3Br, however measurements in the open ocean (e.g. Lobert et al. 1995; Groszko and Moore 1998) and global assessments (Kurylo et al. 1999) suggest it is a net sink. Nevertheless occasionally high supersaturations of CH_3Br are observed in surface waters, particularly in coastal regions, implying that oceanic production pathways exist. Similarly an oceanic source of CH_3Cl was previously considered to dominate global sources of CH_3Cl (cf. review in Warneck 1999) but recent studies suggest more widespread natural sources of CH_3Cl including salt marshes and other coastal sources (Rhew et al. 2000; Yokouchi et al. 2000). For these two compounds at least, the message seems clear and similar to the message concerning N_2O sources: the oceanic supply of trace gases must be assessed for the entire marine system including open ocean, coastal and brackish domains.

For the other principal methyl halide, CH_3I, the situation is different. Here widespread supersaturation exists throughout much of the the surface ocean, including low productivity regimes such as the tropical Atlantic. Supersaturations of several hundred to several thousand percent are measured in warm surface waters. The factors underlying the source remain unclear, with both biotic and primarily photochemical sources having been postulated (e.g. Manley and Dastoor 1988; Moore and Zafiriou 1994; Happell and Wallace 1996). Strong correlations of CH_3I supersaturation with temperature exist (Groszko 1999) but strong correlations with incident radiation have also been identified (e.g. Happell and Wallace 1996) and these might be more 'fundamental' than the correlation with temperature.

Whereas most attention in the literature has focussed on the oceanic production of methyl halides, other volatile halocarbons are also produced in significant quantities and may also play a significant role in the atmosphere. Bromine-containing gases deserves special attention here because, atom-for-atom, bromine is 50 times more effective for ozone destruction than chlorine. In addition there is synergy between the ozone-destroying power of Br and Cl which arises from the reaction cycle:

$$BrO + ClO \rightarrow Br + Cl + O_2$$
$$Br + O_3 \rightarrow BrO + O_2$$
$$Cl + O_3 \rightarrow ClO + O_2$$

Net: $O_3 + O_3 \rightarrow 3O_2$

This implies that the recent anthropogenic increase in stratospheric Cl will have amplified the ozone-destroying effectiveness of naturally-derived Br. Further, in contrast to Cl, which exists primarily in the form of compounds with long residence times and large atmospheric reservoirs, Br is transported mainly by short-lived gases. With anthropogenic Cl, even major and sudden changes in emissions (such as those arising from the Montreal Protocol) take decades to have a significant effect on ozone depletion, due to the long residence times of the gases involved. In contrast there is potential for short-term variations in the natural production of short-lived gases to significantly and rapidly impact the delivery of reactive Br to the stratosphere. This, combined with the increased sensitivity arising from the elevated reactive-Cl levels, suggests that ozone destruction driven by Br might be sensitive to short-term variability of bromine-containing gas production rates.

In contrast to methyl bromide, the significance of bromoform ($CHBr_3$) for stratospheric, or even tropospheric, ozone chemistry has tended to be overlooked until very recently. For example the excellent and comprehensive review by von Glasow and Crutzen (in press) focuses exclusively on CH_3Br as a potential organic precursor of free troposphere BrO. The significance of bromoform may have been overlooked due to its low background mixing ratio of $\sim 1 \times 10^{-12}$ mol/mol in the marine boundary layer and its short photolytic tropospheric life time of 2-4 weeks (Moortgat et al. 1993; Barrie et al. 1988; Penkett et al. 1985). However there is a growing awareness (Dvortsov et al. 1999; Sturges et al. 2000; Quack and Wallace in press) that bromoform might supply, either directly or indirectly, the majority of the BrO_x found in troposphere and even the lower stratosphere of the midlatitudes and tropics. Intriguingly, this is where most of the recent decreases in ozone have been identified (WMO 1999).

An interesting and presently unexplained signal in atmospheric data is the strong maximum in tropospheric bromoform concentrations in the tropics over both the Pacific and Atlantic Oceans (Fig. 5; Schauffler et al. 1999; Atlas et al. 1993; Class and Ballschmiter 1988). Quack and Suess (1999) and Yokouchi et al. (1997) also detected elevated concentrations of bromoform over regions of the tropical western Pacific. Elevated atmospheric bromoform concentrations are also regularly observed in coastal regions (Carpenter et al. 1999; Quack 1994). The role of such regional 'hot-spots' of bromoform may also have been under-estimated in global assessments of atmospheric halogen chemistry that have tended to focus more on the background concentrations measured far from source regions.

Bromoform has a strong biogenic source in the ocean, with macroalgae being a particularly important source (1.6 (0.4-2.7) Gmol Br yr^{-1} according to Carpenter and Liss 2000). Open ocean sources may also exist, although very few direct measurements of production have been made with the exception of Tokarczyk and Moore's (1994) documentation of bromoform production by cold water diatoms. Important sinks for $CHBr_3$ include substitution by Cl (Class and Ballschmiter 1988).

Quack and Wallace (in Press) assembled all published data on $CHBr_3$ concentrations in surface water and the lower atmosphere. The resulting global sea-to-air flux of bromoform was equivalent to approximately 9-13 Gmol Br yr^{-1} which is about a factor of 3 higher than recent estimates based on tropospheric $CHBr_3$ sink calculations (e.g. Dvortsov et al. 1999). This difference may reflect uncertainty in tropospheric sink calculations due to undersampling of regional 'hot-spots' with high sea-to-air fluxes and high tropospheric bromoform concentrations. On the other hand the air-sea flux estimate is limited by a very meagre database (no more than ~200 mean values) which can be com-

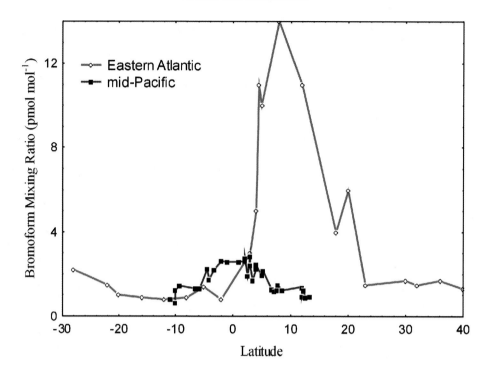

Fig. 5. The tropical maximum of tropospheric bromoform in the marine boundary layer of the Atlantic and Pacific Oceans [data from (1) Class and Ballschmiter, 1988; (2) Schauffler et al. 1999; (3) Atlas et al. 1993]. The origin of this maximum is poorly understood.

pared with the DMS database (16000 measurements), and the 940,000 surface pCO_2 data points available. Bromoform, together with other short-lived organohalogens, has highly variable concentrations in both the air and the gas phases and there are almost no published studies where both phases have been measured simultaneously. (Oceanographers have tended to measure $CHBr_3$ only in the water, whereas atmospheric chemists have measured $CHBr_3$ only in the atmosphere, usually on separate expeditions). Here is a clear example of the need for improved coordination between the atmospheric and marine science communities.

Oxygenated organics: Singh et al. (2001) have reported measurements of a wide array of oxygenated organic compounds (e.g. formaldehyde, acetone, methanol, acetaldehyde, methyl hydroperoxide) in the remote marine troposphere of both hemispheres. The overall abundance of these rarely-measured oxygenated compounds was 3-4 times higher than the abundance of non-methane hydrocarbons (NMHC) and little difference was found between the hemispheres (in contrast to the NMHCs). The level and widespread distribution of some of these compounds was unexpected, and implies large and diffuse sources that are largely unknown. The sources could include progressive oxidation of NMHC through a succession of relatively long-lived intermediates that have not yet been identified and measured. Alternatively, there may be a strong marine source for some of the compounds (e.g. photochemical transformation of dissolved organic carbon in the sea-surface microlayer). The high levels of these compounds (summed mixing ratio of the order 2000 x 10^{-12} mol/mol) implies an important role as a source of free radicals in the troposphere, as well as a potentially important role in contributing to the organic fraction of the marine aerosol. Measurements of these compounds are extremely limited, and experimental and air-sea flux measurements virtually non-existent. This research area appears to be an important issue for the future.

NH₃: Ammonia (NH_3) also has a marine source (Quinn et al. 1987). Within the atmosphere, scavenging of NH_3 (and methylamines) and subsequent reaction with H_2SO_4 to form salts affects aerosol optical properties. Moreover the pH of aerosol, cloudwater, and rainwater as well as SO_2 oxidation rates in these condensed phases is also affected. While ammonia may also control new particle nucleation rates, its atmospheric concentrations, sources and sinks are largely unknown. This is particularly true in coastal and shelf waters where its concentration and sea to air flux could be elevated due to higher algal and bacterial activity. On the other hand, proximity to continental NH_3 sources could suppress or even reverse this flux. The balance between these effects determines the extent of recycling of N inputs with distance from land. As a consequence further measurements are required before its role in new particle nucleation rates can be assessed.

Sea-salt aerosol

Sea-salt particles are a major reactive medium in the marine boundary layer as well as a significant source of atmospheric alkalinity, organic material and precursors for volatile reactive halogens. The production of several classes of compounds, particularly reactive halogen species, as well as the chemical processing, scavenging and deposition of important sulphur and nitrogen species are directly tied to the cycling of sea-salt.

These particles also play an important climatic role. Sea-salt can be responsible for the majority of light scattering by aerosols in the remote marine atmosphere, and sea-salt also comprises a significant fraction of inferred cloud nuclei (e.g. Murphy et al. 1998; Mason 2001).

As noted earlier, shorter-lived inorganic halogen species (e.g. HCl, HBr, Br_2, BrCl, Cl_2, BrO, ClO, IO; Platt and Moortgat 1999) can be produced by auto-catalytic chemical processes involving sea salt aerosol in the marine boundary layer (see review by von Glasow and Crutzen in press). These species can contribute to O_3 destruction in the troposphere and impact the atmospheric oxidation of a wide-range of organic compounds and pollutants. The impact of reactive halogen species such as atomic Br and BrO on tropospheric ozone is exemplified by the polar 'Tropospheric Ozone Holes' which occur during springtime in the Arctic and Antarctic boundary layers (e.g. Barrie and Platt 1997). In addition, potentially important interactions between polluted air masses and the sea-salt aerosol in affecting the oxidizing capacity of the remote marine atmosphere have been identified (e.g. Sander and Crutzen 1996). The chemistry of such interactions is complex as indicated by even the highly simplified scheme of the heterogeneous halogen chemistry shown in Figure 6. A more recent schematic of the major reactions of bromine, chlorine and sulphur, presented by von Glasow and Crutzen (in press), depicts considerably more detail concerning halogen transfers within the aqueous phase and the role of nitrogen oxides in releasing halogens from sea salt.

Exchange mechanisms

Generation of the sea salt aerosol

For determining the chemical and climatic effects of sea-salt, both the sea-salt aerosol content of the atmosphere, and its size distribution are crucial parameters. These are strongly wind-speed and sea-state dependant quantities and also dependent on meteorological conditions within the marine boundary layer. They are surprisingly poorly quantified. Spray droplets are the source of the local marine aerosol. Sea spray forms from droplet-formation associated with bubble bursting (jet and film drop formation), the action of droplets (e.g. rain) striking the water surface, and spume formation (i.e. tearing off of wave crests by the wind). The characteristics of the droplets formed (e.g. number, size and surface area, chemical composition) depends on many factors including wind-speed, turbulence, wave characteristics, and likely the organic composition of seawater. The residence time of the droplets depends on droplet microphysics and the role of turbulent dispersion in the wave boundary layer. Droplets tend to fall

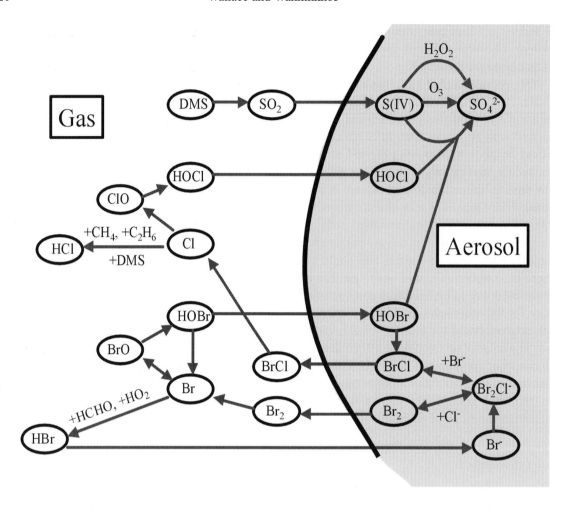

Fig. 6. A simplified scheme for heterogenous and homogeneous reactive halogen cycling in the marine boundary layer. Figure is redrawn after Vogt et al. (1996). Transformations within the shaded area occur in the deliquesced sea-salt aerosol. Additional reactions are important and a more complex and up-to-date representation can be found in von Gladow and Crutzen (in press).

back to the surface, but evaporation reduces their radius while turbulence can keep them suspended.

Overall the generation of sea-salt aerosol (e.g. as a function of meteorological conditions) is complex and our ability to model its generation with sufficient accuracy for inclusion in chemical and radiative transfer models is questionable. For example, Andreas (1998) noted that published sea spray generation functions, for a given wind speed, differ by up to 6 order of magnitude in predicting the volume flux of droplets of a given radius.

Improved parameterisation of aerosol generation appears to be an area of research that will be required for both the climate and chemical studies of the future. Because the production of sea-salt particles from the ocean is strongly related to wind velocity, climatically driven changes in global wind fields can be expected to alter concentrations and lifetimes of atmospheric sea salt. This in turn could alter the associated reactant and product species, and the effects of sea-salt aerosol on radiative transfer.

Air-sea gas exchange – background

As the database for surface ocean and atmospheric trace gas concentrations increases, the need for a solid quantitative representation of air-sea gas exchange rates grows. Such a representation is critical for assessing the variability of air-sea fluxes: including both natural variability and the future variability associated with global environmental change. We therefore briefly review the state-of-understanding in this field, highlighting areas of active research.

The flux of a gas between ocean and atmosphere is controlled by its partial pressure, or more correctly, fugacity difference between surface water and air (ΔpC), its Ostwald solubility (s), the chemical enhancement (α), and a gas transfer velocity (k) according to:

$$F = k\,(\alpha\,s)\,\Delta pC \qquad (1)$$

The fugacity difference is affected by physical, chemical and biological processes operating within the surface ocean and/or atmosphere. Transport of gases across the interface is limited by resistance within the air-side and/or water-side molecular boundary layers located right at the interface. The gas transfer velocity is therefore ultimately regulated by molecular diffusion and near-surface turbulent processes. For slightly soluble gases (s <10) the transfer is controlled by processes operating within the liquid boundary layer, whereas soluble gases experience major resistance to transfer in the air boundary layer. Gases with high reactivity with water, such as SO_2 and O_3 are also subject to air boundary layer resistance. The reaction of gases with water or other species in the boundary layer is referred to as chemical enhancement, in which the rate-limiting step of molecular transfer through the liquid boundary layer is effectively bypassed. For gases with (αs) ~ 30 the resistance of gas transfer is equally distributed between the air and water boundary layers. If (α s) is much less than this, transfer across the water boundary layer will be the rate limiting step. Liss and Merlivat (1986) and Liss and Slater (1974) provide an overview of the basics of gas transfer dynamics while McGillis et al. (2001) provide a detailed account of gas-phase versus liquid-phase resistance using DMS exchange as an example.

In the absence of bubbles, gas transfer velocities of various slightly soluble gases (e.g. gases a and b) over wavy surfaces can be interchanged using the Schmidt number (Sc) according to:

$$k_a = k_b\,(Sc_a/\,Sc_b)^{-1/2} \qquad (2)$$

where Sc is the kinematic viscosity of seawater divided by the gas diffusion coefficient. For soluble gases the transfer velocities are inter-related through their molecular weight according to:

$$k_a = k_b\,(M_a/\,M_b)^{-1/2} \qquad (3)$$

Gas transfer parameterisation

Many of the gases of environmental interest are slightly soluble, and this is where much of the work that relates gas transfer to environmental forcing has been focussed. Further, most regional and global estimates of gas fluxes have been made using parameterisations of the gas transfer velocity that are based on wind speed. These relationships have been derived using a variety of empirical approaches and theoretical considerations, including experiments in wind-wave tanks, compilations of field-based estimates, and (often) combinations thereof (Liss and Merlivat 1986; Nightingale et al. 2000; Wanninkhof 1992). These different parameterisations yield regional and global flux estimates that differ by a factor of two or more, even when the same ΔpC data are used. Most of the published relationships point towards a non-linear dependence on wind-speed that has been expressed variously as a quadratic, cubic, or polynomial relationship. The limited work for soluble gases suggests a linear dependence with wind speed with gas transfer velocities that are ~30 fold higher than for the slightly soluble gases (Liss and Slater 1974).

The differences in gas exchange-wind speed relationships are caused, in part, by the non-unique nature of a parameterisation based on wind-speed. Boundary layer stability effects (Erickson III 1993), variability in surfactant loading (Frew 1997), and gas-specific bubble enhanced gas exchange (Asher et al. 1996) are amongst the possible causes for 'simple' relationships with wind speed to break down. In addition, the non-linear dependence of k on wind-speed, and the possible cross correlation between k and ΔpC, implies that the bulk flux relationship given in equation (1) has to be expanded when averaged quantities are used:

$$F = k \, s \, \Delta pC = [(ks)_{av} + (ks)'] \, \Delta pC + (ks)' \, \Delta pC' \tag{4}$$

where the subscript "av" refers to a quantity that has been averaged over some (specified) time scale. The effect of the non-linearity of the gas exchange-wind speed relationships is expressed in the $[(ks)_{av} + (ks)']$ term. Nonlinearity in the dependence of the gas transfer velocity on wind speed can lead to flux bias of up to 100 %, depending on the parameterisation used, if monthly averaged wind speeds are used rather than hourly data. The cross correlation term $(ks)' \, \Delta pC'$ potentially accounts for the correlation between the two quantities but its magnitude is usually poorly known. For most gases, an increase in k will drive ΔpC towards equilibrium on timescales of days to weeks. This is somewhat different for CO_2 which is strongly buffered in seawater and which will approach equilibrium on a timescale of order 1 year. However, additional effects can contribute to co-variance. For examples, a cross correlation can result because high wind speeds can drive both increased gas transfer velocities but also mixed layer deepening. For many gases, mixed-layer deepening drives entrainment of water with different dissolved gas concentrations into the mixed layer, thereby altering ΔpC.

Gas flux measurements and scaling

Recent years have seen several important advances in methods used to quantify gas fluxes as well as an improved understanding of the environmental factors responsible for driving gas transfer. Direct flux measurements using micrometeorologi-cal techniques, along with accurate measurements of concentration gradients and/or temporal fluctuations in the lower atmosphere, now make it possible to directly determine fluxes over time scales of < 1 hour. Previous methods of determining gas transfer velocities relied on changes in the water column that could only be detected over periods of days to weeks. Four direct flux methods for gas transfer are now in use: the co-variance method, the gradient method, the (relaxed) eddy accumulation method, and the inertial dissipation method. Each has different requirements with respect to basic assumptions of boundary layer dynamics, similarity theories, and the frequency of gas concentration and micrometeorological measurements. An overview of these methods and tradeoffs can be found in Fairall et al. (2000). Since the gradient of slightly soluble gases within the atmospheric boundary layer is usually small, methods that rely on such measurements can be extremely difficult. For soluble gases the fluxes and concentration gradients in air are frequently larger and hence easier to measure. Direct flux methods are, for example, frequently used to determine water vapour fluxes. Direct flux measurements have also recently been used successfully to quantify both CO_2 and DMS fluxes over the ocean (McGillis et al. 2001). This method can, however, only be applied at the upper range of CO_2 fluxes that are encountered within the open-ocean environment. The particular value of these approaches is that the field measurements of gas fluxes can be performed at the same time and space scales as fluctuations in physical forcing.

The simultaneous measurement of gas fluxes and turbulence offers the opportunity to more fully investigate controls on fluxes in the natural environment and develop an improved parameterisation of gas transfer. Proxies of gas transfer such as heat fluxes and surface thermal structure also have potential to both quantify gas transfer and elucidate the processes that control exchange at the surface (Jessup et al. 1997; Zappa et al. 2002).

Remote sensing provides a means to scale up local findings and measurements to regional and global scales. High quality, daily-resolution wind products are now available globally from active radiometers (scatterometers) such as ERS II and QuickScat. Passive radiometers that yield surface brightness temperatures are used to derive wind products at greater temporal resolution (up to five times per day) but with lower precision. The return from passive radiometers are also sensitive to whitecap coverage and thereby offer an alternative means to estimate global gas transfer rates using algorithms that relate whitecap coverage and gas transfer (e.g. Asher et al. 1996). The full utility of scatterometers will be realized when the return from satellite sensors can be directly related to gas transfer, thereby bypassing problems involved in relating gas transfer to wind speed. For example, the backscatter signal is strongly related to Bragg scattering from wave slopes of waves of particular wavelengths (1-10^2 cm depending on the frequency of the scatterometer). In principle, backscatter should be more closely related to gas transfer than wind speed, as suggested by several wind-wave tank studies. Initial estimates of global gas transfer velocities using wave slopes derived from altimeter data show promise, although issues of characterising high wind regimes remain (Frew et al. 1999).

To summarise: in order to characterise present-day air-sea fluxes and also understand the future response of the atmosphere and ocean to global environmental change, an improved understanding of the chemical and biological processes affecting gas concentrations in the ocean and atmosphere is required. However, our understanding of the physical exchange process is also limited as shown by the disagreements between various gas-exchange parameterisations discussed above. This fundamentally limits our ability to make quantitative estimates of air-sea fluxes. The non-linearity of gas transfer with wind-speed implies that physical exchange processes must be better defined if the future response of gas fluxes to climate change (including shifting wind regimes) is to be assessed. Recently, direct flux techniques have shown significant potential to improve our parameterisation of gas transfer. Such process scale measurements of fluxes together with measures of near surface turbulence will be of particular value if they can be related to forcing at larger spatial and temporal scales through the optimal use of remote sensing.

Challenges and open questions

It has been determined that a surprisingly wide range of trace species (as well as sea salt) that originate in the ocean are transferred to the atmos-phere. These species participate in complex chemical and radiative transfer pathways within the atmosphere. In the context of global environmental change, including future changes to marine ecosystem structure, surface temperature, ice-cover, wind stress, and circulation, it cannot be expected that the production and flux of these species will stay constant.

The history of atmospheric composition recorded in ice cores (e.g. Petit et al. 1999) implies that it has varied roughly in phase with changes in the Earth's climate. This implies feedbacks between climate change and biogeochemistry which must include ocean biogeochemistry (at least for CO_2 and DMS). To date, in addition to a climate feedback involving DMS (Charlson et al. 1987), several mechanisms underlying CO_2-climate feedbacks have been proposed and examined. It should be noted again however that the magnitude and rate of change of greenhouse gas forcing that will be encountered during this present century is likely unprecedented over the past 20 million years. The effect of such forcing on, for example, future oceanic CO_2 uptake, or the flux of N_2O, organohalo-gens and sea salt across the air-sea interface is therefore largely unknown.

The expected warming trend (in contrast to the transition to cooler, glacial climates) has some unique consequences. For example, a warmer world implies decreased oxygen solubility in ocean surface waters. For many coastal seas, the wintertime equilibrium oxygen concentrations define an upper bound for the amount of organic material that can be aerobically decomposed in deeper waters during the stratified season. Hence a warmer climate might imply increased frequency of hypoxic or anoxic conditions in coastal waters. Hypoxic conditions represents a major "switch point" for biogeochemical reactions. The effect of increased hypoxia on ocean biogeochemistry in general, and air-sea fluxes in particular (e.g. on the large coastal fluxes of N_2O) is presently impossible to estimate but could contribute an additional feedback on climate change.

Alternatively, an altered climate could be expected to result in changes in the production of various short-lived bromine-containing gases in tropical surface waters. Altered wind-speeds or atmospheric convection might alter the flux across the air-sea interface and consequently the supply across the tropopause. Such flux changes, while speculative, would be expected to affect strato-spheric and tropospheric ozone cycling rapidly. Our ability to predict such changes is seriously limited by our understanding. For example, with-out knowing the reason(s) for extremely high CH_3I supersaturation in tropical waters (is it tempera-ture-controlled? or light-controlled?) it is impossible to predict the future cycling of this species and its effects on tropospheric chemistry.

The effect of global environmental change on air-sea carbon cycling involves the upper ocean climate, air-sea gas exchange, the upper ocean ecosystem as well as the circulation and biogeo-chemical processes within the ocean interior. The prediction of the future magnitude and distribution of the air-sea flux of CO_2 therefore requires sophis-ticated models of the entire ocean and an under-standing of biogeochemical processes within the ocean interior. This is also true for many of the other chemicals and properties that have been dis-cussed above, although the complexity and extent of the model is dependent on the complexity of the production/loss pathways and where they operate. On the other hand there are also pressing scientific, economic and political reasons to be able to deter-mine the present-day air-sea flux of CO_2 to high accuracy.

In order to assess the significance of air-sea exchanges for our planetary environment, we need to understand them, measure them, and ideally rep-resent them in models. A suitable set of field meas-urements is often the requirement for improving understanding and modelling. For the long-lived greenhouse gases (N_2O and CO_2), the determina-tion of large-scale terrestrial sources and sinks di-rectly (e.g. from inventory or flux studies) appears to be difficult due to the extreme heterogeneity of the terrestrial biosphere. The potential exists to utilise constraints on ocean-atmosphere exchanges in the context of inverse modelling or budgeting of atmospheric observations to assist with the reso-lution of representative mean values for terrestrial sinks. This implies that for these long-lived gases, the ocean-atmosphere-terrestrial exchanges must be dealt with as a single system in a coordinated

manner. In order to resolve atmosphere-ocean fluxes of N_2O, CO_2 and DMS adequately, a large-scale observation network is required (e.g. cover-ing ocean basin scales). Such networks must cover coastal as well as open ocean regions if the impor-tant fluxes are not to be missed. A suitable vehicle for such large-scale measurements would be a combination of a limited number of instrumented time-series sites (to resolve temporal variability) together with volunteer observing ship measure-ments (to resolve geographical variability). For air-sea exchange issues, commercial shipping vessels have many advantages over dedicated research vessels and moorings, and have been seriously underutilised by the marine biogeochemistry com-munity to-date.

The establishment of a global measurement network suited to investigating and measuring air-sea exchanges is a major challenge, but one that must ultimately be faced if we are to understand the response of the atmosphere and ocean to an-thropogenic forcing. Such networks also provide the opportunity to make assessments of the effec-tiveness of any climate-change mitigation efforts that are implemented in the future.

Such air-sea exchange measurement networks should be designed in the context of the new satel-lite sensor missions that are being proposed to monitor chemical species in the atmosphere (e.g. BrO, CO, sulphate aerosol, even CO_2). They must also be designed with the growing strengths of large-scale atmospheric chemical modelling in mind. The combination of coordinated satellite observations of trace species, air-sea flux estimates based on ship-of-opportunity data, and 3-D chemical modelling of the atmosphere will be powerful. Most importantly, atmospheric and oceanic campaigns and measurements should be coordinated. Based on past experience, this will be a major challenge because the atmospheric and ocean science communities have traditionally worked separately, and funding opportunities are generally uncoordinated. The forthcoming international SOLAS (Surface Ocean Lower Atmosphere Study) project of the International Geosphere-Biosphere Program repre-sents an important attempt to bridge this gap.

In addition to measurement and modelling ef-forts, there is also an urgent need for detailed and carefully controlled process and experimental studies. For example, in addition to the chemical measurement networks proposed, the physical process of gas exchange and aerosol formation must be better understood, parameterised and related to remote sensing data.

Acquiring understanding of processes and controlling factors on the basis of field measurements alone is frequently frustrating because in the ocean and atmosphere many of the candidate controlling factors are themselves internally correlated. A simple example for CH_3I production are the strong correlations between light intensity, water temperature and (sometimes) photosynthesis. It is not immediately obvious on the basis of field measurements alone what the prime controlling factor underlying a trace gas flux is. The history of DMS research confirms this, and shows that there is a great need for careful manipulative experimental work with suitable controls. It is worth noting that a broad range of scales of experimental manipulations are now possible, ranging from laboratory culture and incubation studies, through fixed-location mesocosm studies, to deliberate tracer-addition, whole-ecosystem experiments conducted in the open ocean (e.g. the recent iron-fertilization experiments) and beyond to experiments that make use of the natural forcing of ecosystem and chemical processes. Going from the small-scale laboratory tests to the whole-ecosystem experiments there is an overall decrease in the ability to implement multiple controls and adequate replication (de Baar and Boyd 2000). Hence interpretation of results from large-scale experiments can be problematic and ambiguous. On the other hand the experiments at larger scales are more "realistic" and less prone to artifacts than laboratory studies. The challenge for biogeochemists will be to design suitable experiments and work across a hierarchy of scales in order to find consistency of results and findings (Lalli 1991; Lalli and Parsons 1993).

Acknowledgements

Some of the material in this chapter is derived directly or indirectly from the SOLAS Science Plan. The authors therefore gratefully acknowledge the contribution and insight of the SOLAS Editorial Team and reviewers who were responsible for preparing the SOLAS Science Plan (see http://www.uea.ac.uk/env/solas/science.htm). D.W. also acknowledges input from members of the Chemical Oceanography group at the Institut für Meereskunde in Kiel, particularly input from Birgit Quack, Hermann Bange and Arne Körtzinger.

References

Anderson TR, Spall SA, Yool A, Cipollini P, Challenor PG and Fasham MJR (2001) Global fields of sea surface dimethylsulfide predicted from chlorophyll, nutrients and light. J Mar Syst 30:1-20

Andreae MO and Crutzen PJ (1997) Atmospheric aerosols: Biogeochemical sources and role in atmospheric chemistry. Science 276(5315):1052-1056

Andreas EL (1998) A new sea spray generation function for wind speeds up to 32 ms[-1]. J Phys Oceanogr 28:2175-2184

Asher WE, Farley PJ, Higgins BJ, Karle LM, Monahan EC and Leifer IS (1996) The influence of bubble plumes on air/seawater gas transfer velocities. J Geophys Res 101:12027-12041

Atlas E, Pollock W, Greenberg J, Heidt L (1993) Alkyl nitrates, nonmethane hydrocarbons, and halocar-bon gases over the equatorial Pacific Ocean during Saga 3. J Geophys Res 98:933-947

Bange HW, Rapsomanikis S, and Andreae MO (1996) Nitrous oxide in coastal waters. Glob Biogeochem Cycl 10:197-207

Barrie LA and Platt U (1997) Arctic tropospheric chemistry: Overview to Tellus special issue. 49B:450-454

Barrie LA, Bottenheim RC, Crutzen PJ, and Rasmussen RA (1988) Ozone destruction and the chemical reactions at polar sunrise in the lower Arctic atmosphere. Nature 334:138-141

Bates NR (2001) Interannual variability of oceanic CO_2 and biogeochemical properties in the Western North Atlantic subtropical gyre. Deep-Sea Res II 48:1507-1528

Bates NR, Michaels AF and Knap AH (1996) Seasonal and interannual variability of oceanic carbon dioxide species at the US JGOFS Bermuda Atlantic Time-series Study (BATS) site. Deep-Sea Res 43:347-383

Battle M, Bender M, Tans PP, White JWC, Ellis JT, Conway T and Francey RJ (2000) Global carbon sinks and their variability, inferred from atmospheric O_2 and $\delta^{13}C$. Science 287:2467-2470

Bousquet P, Peylin P, Ciais P, LeQuere C, Friedlingstein P and Tans PP (2000) Regional changes in carbon dioxide fluxes of land and oceans since 1980. Science 290:1342-1346

Bouwman AF, Van der Hoek KW and Olivier JGJ (1995) Uncertainties in the global source distribution of nitrous oxide. J Geophys Res 100:2785-2800

Buitenhuis ET, de Baar HJW and Veldhuis MJW (1999) Photosynthesis and calcification of *Emiliania huxleyi* (Prymnesiophyceae) as a function of inorganic carbon species. J Phycol 35:949-959

Butler JH and Rodriguez JM (1996) Methyl bromide in the atmosphere. In: Bell CH, Price N and Chakrabarti B (eds) J Wiley and Sons, New York pp 2-90

Carpenter LJ and Liss PS (2000) On temperate sources of bromoform and other reactive organic bromine gases. J Geophys Res 105:539-547

Carpenter LJ, Sturges WT, Penkett SA, Liss PS, Alicke B, Hebestreit K and Platt U (1999) Short-lived alkyl iodides and bromides at Mace Head, Ireland: Links to biogenic sources and halogen oxide production. J Geophys Res 104:679-689

Charlson RJ, Lovelock JE, Andreae MO and Warren SG (1987) Oceanic phytoplankton, atmospheric sulphur, cloud albedo, and climate. Nature 326:655-661

Class T and Ballschmiter K (1988) Chemistry of organic traces in air, VIII, Sources and distribution of bromo- and bromochloromethanes in marine air and surface water of the Atlantic Ocean. J Atmos Chem 6:35-46

Codispoti LA, Elkins JW, Yoshinari T, Friederich GE, Sakamoto CM and Packard TT (1992) On the nitrous oxide flux from productive regions that contain low oxygen waters. In: Desai BN (ed) Oceanography of the Indian Ocean. Oxford Publishing, New Dehli pp 271-284

Cox P, Betts R, Jones C, Spall S and Totterdell I (2000) Acceleration of global warming due to carbon cycle feedbacks in a coupled climate-carbon model. Nature 408:184-187

Cunnold DM et al. (1997) GAGE/AGAGE measurements indicating reductions in global emissions of CCl_3F and CCl_2F_2 in 1992-1994. J Geophys Res 102:1259-1269

De Baar HJW, Boyd PM (2000) The Role of Iron in Plankton Ecology and Carbon Dioxide Transfer of the Global Oceans. In: Hanson RB, Ducklow HW and Field JG (eds) The Dynamic Ocean Carbon Cycle: A Midterm Synthesis of the Joint Global Ocean Flux Study. International Geosphere Biosphere Programme Book Series, Vol. 5, Cambridge University Press (ISBN 0 521 65603 6) pp 61-140

DeBaar HJW, LaRoche J (2003) Trace metals in the Ocean: Evolution, biology and global change. In: Wefer G, Lamy F, Mantoura F (eds) Marine Science Frontiers for Europe. Springer, Berlin pp 79-104

Dore JE and Karl DM (1996) Nitrification in the euphotic zone as a source for nitrite, nitrate, and nitrous oxide at station ALOHA. Limnol Oceanogr 41:1619-1628

Dore JE, Popp BN, Karl DM and Sansone FJ (1998) A large source of atmospheric nitrous oxide from subtropical North Pacific surface waters. Nature 396:63-66

Dvortsov VL, Geller MA, Solomon S, Schauffler SM, Atlas EL and Blake DR (1999) Rethinking reactive halogen budgets in the midlatitude lower stratosphere. Geophys Res Lett 26:1699-1702

Ehhalt D, Prather M, Dentener F, Derwent R, Dlugokencky E, Holland E, Isaksen I, Katima J, Kirchhoff V, Matson P, Midgley P and Wang M (2001) Atmospheric Chemistry and Greenhouse Gases. In: Houghton JT, Ding Y, Griggs DJ, Noguer M, van der Linden PJ, Dai X, Maskell K, Johnson CA (eds) Climate Change 2001: The Scientific Basis. Contributions of Working Group I to the Third Assessment Report of the Intergovernmental Panel on Climate Change. Cambridge University Press, Cambridge, UK and New York USA pp 881

Erickson III DJ (1993) A stability-dependent theory for air-sea gas exchange. J Geophys Res 98:8471-8488

Fairall CW, Hare JE, Edson JB and McGillis W (2000) Parameterization and micrometeorological measurement of air-sea gas transfer. Boundary-Layer Meteorol 96:63-105

Fan S, Gloor M, Mahlman J, Pacala S, Sarmiento J, Takahashi T and Tans P (1998) A large terrestrial carbon sink in North America implied by atmospheric and oceanic carbon dioxide data and models. Science 282:442-446

Feely RA, Wanninkhof R, Takahashi T and Tans P (1999) Influence of El Niño on the equatorial Pacific contribution to atmospheric CO_2 accumulation. Nature 398:597-601

Flückiger J, Dällenbach A, Blunier T, Stauffer B, Stocker TF, Raynaud D and Barnola J-M (1999) Variations in atmospheric N_2O concentration during abrupt climatic changes. Science 285:227-230

Frew NM (1997) The role of organic films in air-sea gas exchange. In: Liss PS, Duce RA (eds) The Sea Surface and Global Change. Cambridge University Press, Cambridge pp 121-163

Frew NM, Glover DM, Bock EJ, Goyet C, McCue SJ and Healy RJ (1999) Estimation of Global Air-Sea Transfer of CO_2 Using TOPEX/Poseidon Dual-Frequency Backscatter. IUGG abstracts, Birmingham UK

Groszko W (1999) An estimate of the global air sea flux of methyl chloride, methyl bromide, and methyl iodide. PhD Thesis, Dalhousie University, Halifax, Canada

Groszko W and Moore RM (1998) Ocean atmosphere exchange of methyl bromide: NW Atlantic and Pacific Ocean studies. J Geophys Res 103:16737-16741

Happell JD and Wallace DWR (1996) Methyl iodide in the Greenland/Norwegian Seas and the tropical Atlantic Ocean: Evidence for photochemical production. Geophys Res Lett 23:2105-2108

Holfort J, Johnson KM, Schneider B, Siedler G and Wallace DWR (1998) The meridional transport of dissolved inorganic carbon in the South Atlantic Ocean. Glob Biogeochem Cycl 12:479-499

Indermuehle A, Stocker TF, Joss F, Fischer H, Smith HJ, Wahlen M, Deck B, Mastroianni D, Tschumi J, Blunier T, Meyer R and Stauffer B (1999) Holocene carbon-cycle dynamics based on CO_2 trapped in ice at Taylor Dome, Antarctica. Nature 398:121-126

Jessup AT, Zappa CJ, Loewen MR and Hesany V (1997) Infrared remote sensing of breaking waves. Nature 385:52-55

Karl DM (1999) A sea of change: Biogeochemical variability in the North Pacific Subtropical Gyre. Ecosystems 2:181-214

Kettle AJ and Andreae MO (2000) Flux of dimethylsulfide from the oceans: A comparison of updated data sets and flux models. J Geophys Res 105:793-808

Kettle AJ, Andreae MO, Amouroux D, Andreae TW, Bates TS, Berresheim H, Bingemer H, Boniforti R, Curran MAJ, DiTullio GR, Helas G, Jones GB, Keller MD (1999) A global database of sea surface dimethyl-sulfide (DMS) measurements and a simple model to predict sea surface DMS as a function of latitude, longitude and month. Glob Biogeochem Cycl 13(2): 399-444

Khalil MAK, Rasmussen RA (1992) The global sources of nitrous oxide. J Geophys Res 97(D13):14651-14660

Khalil MAK and Rasmussen RA (1999) Atmospheric chloroform. Atmos Environ 33:1151-1158

Kiene RP and Linn LJ (2000) Distribution and turnover of dissolved DMSP and its relationship with bacterial production and dimethlsulphide in the Gulf of Mexico. Limnol Oceanogr 45:849-861

Kim K-R and Craig H (1990) Two isotope characterization of N_2O in the Pacific Ocean and constraints on its origin in deep water. Nature 347:58-61

Kleypas JA, Buddemeier RW, Archer D, Gattuso J-P, Langdon C and Opdyke BN (1999) Geochemical consequences of increased atmospheric CO_2 on coral reefs. Science 284:118-120

Kurylo MJ, Rodriguez JM, Andreae MO, Atlas EL,

Blake DR, Butler JH, Lal S, Lary DJ, Midgely PM, Montzka SA, Novelli PC, Reeves CE, Simmonds PG, Steele JP, Sturges WT, Weiss RF and Yokouchi Y (1999) Short-lived ozone-related compounds. In: Ennis CA (ed) Scientific Assessment of Ozone Depletion: 1998. Global Ozone Research and Monitoring Project, Report 44, pp 2.1-2.56 World Meteorol. Org, Geneva, Switzerland

Lalli CM (1991). Enclosed Experimental Marine Ecosystems; A review and recommendations; a contribution of SCOR Working Group. Springer Verlag, New York 218p

Lalli CM and Parsons TR (1993) Biological Oceanography, an Introduction. Oxford: Pergamon Press

Lee K, Wanninkhof R, Takahashi T, Doney SC and Feely R (1998) Low interannual variability in recent oceanic uptake of atmospheric carbon dioxide. Nature 396:155-159

Le Quéré C, Orr JC, Monfray P, Aumont O and Madec G (2000) Interannual variability of the oceanic sink of CO_2 from 1979 through 1997. Glob Biogeochem Cycl 14:1247-1265

Liss PS and Slater PG (1974) Flux of gases across the air-sea interface. Nature 247:181-238

Liss PS and Merlivat L (1986) Air-sea gas exchange rates: Introduction and synthesis. In: Buat-Menard P (ed) The Role of Air-Sea Exchange in Geochemical Cycling. Reidel, Boston pp 113-129

Lobert J, Butler JH, Montzka SA, Geller LS, Myers RC and Elkins JW (1995) A net sink for atmospheric CH_3Br in the East Pacific Ocean. Science 267:1002-1005

Manley SL and Dastoor MN (1988) Methyl iodide production by kelp and associated microbes. Mar Biol 98:477-482

Mason BJ (2001) The role of sea-salt particles as cloud condensation nuclei over the remote oceans. QJR Meteorol Soc 127:2023-2032

McGillis WR, Edson JB, Ware JD, Dacey JWH, Hare JE, Fairall CW and Wanninkhof R (2001) Carbon dioxide flux techniques performed during GasEx 98. Mar Chem 75:267-280

Moore RM and Zafiriou OC (1994) Photochemical production of methyl iodide in seawater. J Geophys Res 99:415-420

Moortgat GK, Meller R and Schneider W (1993) Temperature dependence (256K-296K) of the absorption cross-sections of bromoform in the wave-length range 285-360nm. In: Niki H and Becker RH (eds) The Tropospheric Chemistry of Ozone in the Polar Regions. Springer-Verlag, New York pp 359-370

Murphy DM, Anderson JR, Quinn PK, Mcinnes LM,

Brechtel FJ, Kreidenweis SM, Middlebrook AM, Pósfai M, Thomson DS and Buseck PR (1998) Influence of sea-salt on aerosol radiative properties in the Southern Ocean marine boundary layer. Nature 392:62-65

Naqvi SWA, Jayakumar DA, Narvekar PV, Naik H, Sarma VVSS, D'Souza W, Joseph S and George MD (2000) Increased marine production of N_2O due to intensifying anoxia on the Indian continental shelf. Nature 408:346-349

Nevison CD, Weiss RF and Erickson III DE (1995) Global oceanic emissions of nitrous oxide. J Geophys Res 100:15809-15820

Nightingale PD, Malin G, Law CS, Watson AJ, Liss PS, Liddicoat MI, Boutin J and Upstill-Goddard RC (2000) *In situ* evaluation of air-sea gas exchange parameterizations using novel conservative and volatile tracers. Glob Biogeochem Cycl 14:373-387

O'Dowd CD, Jimenez JL, Bahreini R, Flagan RC, Seinfeld JH, Hameri K, Pirjola L, Kulmala M, Jennings SG and Hoffmann T (2002) Marine aerosol formation from biogenic iodine emissions. Nature 417(6889):632-636

Penkett SA, Jones BMR, Rycroft MJ and Symmons DA (1985) An interhemispheric comparison of the concentrations of bromine compounds in the atmosphere. Nature 318:550-553

Perez T, Trumbore SE, Tyler SC, Davidson EA, Kellar M and DeCamargo PB (2000) Isotopic variability of N_2O emissions from tropical forest soils. Glob Biogeo-chem Cycl 14(2):525-535

Petit JR, Jouzel J, Raynaud D, Barkov NL, Barnola JM, Basile I, Bender M, Chappellaz J, Davis M, Delaygue G, Delmotte M, Kotlyakov VM, Legrand M, Lipenkov VY, Lorius C, Pepin L, Ritz C, Saltzman E and Stievenard M (1999) Climate and atmospheric history of the past 420,000 years from the Vostok ice core, Antarctica. Nature 399(6735):429-436

Platt U and Moortgat GK (1999) Heterogeneous and homogeneous chemistry of reactive halogen compounds in the lower troposphere. J Atmos Chem 34:1-8

Prather M, Derwent R, Ehhalt D, Fraser P, Sanhueza E and Zhou X (1996) Other trace gases and atmospheric chemistry, in IPCC Second Assessment of Climate Change. Cambridge University Press, New York section 2.2

Prentice I, Fasham MJR, Goulden ML, Heimann M, Jaramillo VJ, Kheshgi HS, LeQuere C, Scholes RJ and Wallace DWR (2001) The Carbon Cycle and Atmospheric CO_2. In: Houghton JT, Ding Y, Griggs DJ, Noguer M, van der Linden PJ, Dai X, Maskell K and Johnson CA (eds) Climate Change 2001: The Scientific Basis. Contributions of Working Group I to the Third Assessment Report of the Intergovernmental Panel on Climate Change. Cambridge University Press, Cambridge, UK and New York USA pp 881

Quack B (1994) Volatile halogenated hydrocarbons in the marine atmosphere: Inventory, sources and mass balances over North Sea and Baltic. Ph D thesis, IfM, Univ Kiel, Germany

Quack B and Suess E (1999) Volatile halogenated hydrocarbons over the western Pacific between 43°N and 4°N. J Geophys Res 104:663-678

Quack B and Wallace DWR (in press) Rates and controls of the air-sea flux of bromoform ($CHBr_3$): A review and synthesis. Glob Biogeochem Cycl

Quinn PK, Charlson RJ and Zoller WH (1987) Ammonia, the dominant base in the remote marine troposphere: A review. Tellus 39B:413-425

Rahn T and Wahlen M (2000) A reassessment of the global isotopic budget of atmospheric nitrous oxide. Glob Biogeochem Cycl 14:537-543

Rhew RC, Miller BR and Weiss RF (2000) Natural methyl bromide and methyl chloride emissions from coastal salt marshes. Nature 403:292-295

Riebesell U, Zondervan I, Rost B, Tortell PD, Zeebe RE and Morel FMM (2000) Reduced calcification of marine plankton in response to increased atmospheric CO_2. Nature 407:364-367

Roeckmann T, Kaiser J, Crowley JN, Brenninkmeijer CAM and Crutzen PJ (2001) The origin of the anomalous or "mass-independent" oxygen isotope frac-tionation in tropospheric N_2O. Geophys Res Lett 28:503-506

Sander R and Crutzen PJ (1996) Model study indicating halogen activation and ozone destruction in polluted air masses transported to the sea. J Geophys Res 101D:9121-9138

Schauffler SM, Atlas EL, Blake DR, Flocke F, Lueb RA, Lee-Taylor JM, Stroud V and Travnicek W (1999) Distributions of brominated organic compounds in the troposphere and lower stratosphere. J Geophys Res 104:513-535

Seitzinger SP and Kroeze C (1998) Global distribution of nitrous oxide production and N inputs in freshwater and coastal marine ecosystems. Glob Biogeochem Cycl 12:93-113

Simó R and Pedros-Alló C (1999) Role of vertical mixing in controlling the oceanic production of dimethyl sulphide. Nature 402:396-399

Singh H, Chen Y, Staudt A, Jacob D, Blake D, Heikes B and Snow J (2001) Evidence from the Pacific troposphere for large global sources of oxygenated organic compounds. Nature 410:1078-1081

Skjelvan I, Johannessen T and Miller LA (1999) Inter-annual variability of fCO$_2$ in the Greenland and Norwegian Seas. Tellus 51B:477-489

Sturges WT, Oram DE, Carpenter LJ and Penkett SA (2000) Bromoform as a source of stratospheric bromine. Geophys Res Lett 27:2081-2084

Tans PP, Fung IY and Takahashi T (1990) Observational constraints on the global atmospheric CO$_2$ budget. Science 247:1431-1438

Tokarczyk R and Moore RM (1994) Production of volatile organohalogens by phytoplankton cultures. Geophys Res Lett 21:285-288

Vogt R, Crutzen PJ, Sander R (1996) A mechanism for halogen release from sea-salt aerosol in the remote marine boundary layer. Nature 382:327-330

Von Glasow R and Crutzen PJ (in press) Tropospheric halogen chemistry. In: Keeling R (ed) Treatise on Geochemistry, Volume 4: The Atmosphere. Elsevier Science

Von Hobe M, Najjar R, Kettle AJ and Andreae MO (submitted) Photochemical and physical modeling of carbonyl sulfide in the ocean. J Geophys Res

Wanninkhof R (1992) Relationship between gas exchange and wind speed over the ocean. J Geophys Res 97:7373-7381

Warneck P (1999) Chemistry of the Natural Atmosphere (2nd ed). Academic Press 923p

Wayne RP (1991) Chemistry of Atmospheres (2nd ed) Oxford Science Publications, Oxford, UK 447 p

Wennberg PO (1999) Bromine explosion. Nature 397:299-301

Winn CD, Li YH, Mackenzie FT and Karl DM (1998) Rising surface ocean dissolved inorganic carbon at the Hawaii Ocean Time-series Site. Mar Chem 60:33-47

WMO Scientific Assessment of Ozone Depletion (1998) Global Ozone Research and Monitoring Project, Report 44, CA Ennis (ed) 2.1-2.56 World Meteorol. Org, Geneva, Switzerland

Yokouchi Y, Mukai H, Yamamoto H, Otsuki A, Saitoh C and Nojiri Y (1997) Distribution of methyl iodide, ethyl iodide, bromoform, and dibromomethane over the ocean (east and southeast Asian seas and the western Pacific). J Geophys Res 102:805-809

Yokouchi Y, Noijiri Y, Barrie LA, Toom-Sauntry D, Machida T, Inuzuka Y, Akimoto H, Li HJ, Fujinuma Y, Aoki S (2000) A strong source of methyl chloride to the atmosphere from tropical coastal land. Nature 403(6767):295-298

Yoshida N and Toyoda S (2000) Constraining the atmospheric N$_2$O budget from intramolecular site preference in N$_2$O isotopomers. Nature 405(6784):330-334

Zappa CJ, Asher WE, Jessup AT, Klinke J and Long SR (2002) Effect of microscale wave breaking on air-water gas transfer. In: Donelan M, Drennen W, Saltzman E and Wanninkhof R (eds) Gas Transfer at Water Surfaces. Geophys Monogr 127, American Geophysical Union, Washington DC

What Controls the Sequestration of Phytogenic Carbon in the Ocean?

E. Epping

Royal Netherlands Institute for Sea Research, P.O. Box 59, 1790 AB, Den Burg, NL
corresponding author (email): epping@nioz.nl

Abstract: A major fraction of phytogenic organic carbon, produced in the surface ocean, is processed by the heterotrophic biota in the water column, whereas only a minor fraction is buried in marine sediments. The depth distribution of organic carbon mineralisation in the water column determines the time scales for carbon dioxide withdrawal from the atmosphere. The transfer of organic carbon from the surface to the deep ocean through gravitational settling and downward advection are therefore key processes in determining the efficiency of the biological pump. Sediment trap studies indicate the downward flux of particulate organic carbon to decrease exponentially with increasing water depth. The vertical export flux from the euphotic zone correlates positively with primary production up to productivities of ~200 g C m^{-2} y^{-1} and attains a constant value at higher production rates. The amount and quality of organic carbon exported from the surface mixed layer strongly depends on the food web structure in the upper ocean. Some of the current perspectives on the role of grazing by micro- and mesozooplankton on retention versus export of organic carbon in the mixed layer are presented. The transformation through solubilization, respiration and fragmentation of aggregates and fecal pellets by particle associated food webs diminishes the downward flux of particulate organic carbon with increasing water depth. Calculated and measured metabolic activities of these food webs suggest turn-over times of the aggregate organic carbon in the order of weeks. The hydrodynamics of sinking particles appears crucial for the mass exchange between the particle and its surrounding water and regulates the activity of the associated community. In addition, while sinking through the water column the particle associated communities are exposed to changes in physico-chemical conditions such as temperature and oxygen concentration which, by controlling the activity of the community, may regulate the downward flux of particulate organic carbon in deeper waters. Our current understanding of biological processes, especially in the upper water column, is insufficient to predict the composition and downward fluxes of organic matter.

Introduction

In recent decades, the estimate of annual net primary production in the world oceans has been revised upward to 45-50*10^{15} g C (e.g. Field et al. 1998). A major fraction of the phytogenic organic carbon, produced by the photosynthetic reduction of carbon dioxide in the surface ocean, is processed by the heterotrophic biota in the water column. This conversion of phytogenic carbon results in the formation of new heterotrophic biomass and in organic waste products, regenerated nutrients and dissolved inorganic carbon. Only a small fraction of the primary production will not be susceptible to remineralisation and will ultimately be buried in the sediment. The annual burial of organic carbon is estimated at 0.16-0.25*10^{15} g C (Romankevich et al. 1999), which amounts to only ~0.4% of the total primary production in the ocean. Hence, the majority of the phytogenic carbon production is remineralised in the water column and in surface sediments, which is a prerequisite to sustain primary production in the ocean.

The vertical distribution of organic carbon (OC) remineralisation has a number of biological and geochemical implications. First, the downward flux of organic carbon supplies the heterotrophic biota thriving in deeper compartments of the ocean with

From WEFER G, LAMY F, MANTOURA F (eds), 2003, *Marine Science Frontiers for Europe.* Springer-Verlag Berlin Heidelberg New York Tokyo, pp 131-146

food. Remineralisation of OC in the water column above will preferentially remove the relatively easily degradable OC fraction. Remineralisation will therefore not only reduce the amount but also the quality of the delivered food and will set an upper limit to the metabolism of deeper communities. Second, the depth distribution of remineralisation determines the vertical gradients in geochemical parameters and thereby affects the exchange of the ocean with the atmosphere and the geosphere. In the surface ocean, primary production lowers the saturation state of carbon dioxide and thereby drives the ocean's uptake of carbon dioxide from the atmosphere. Part of the produced organic matter is re-mineralised in the euphotic zone, whereas another part is lost due to export to deeper compartments of the ocean. Remineralisation in the upper ocean releases carbon dioxide which can be exchanged with the atmosphere on seasonal time scales or less. With increasing water depth below the winter mixed layer, the time scales for atmospheric equilibration increase dramatically to hundreds or thousands of years, depending on the hydrography of the basin. Carbon dioxide released during the remineralisation of organic carbon in deeper compartments, will therefore be removed from the atmosphere for a longer period of time. The most efficient withdrawal of carbon dioxide and nutrients, i.e. involving longest time scales, results from the burial of OC in marine sediments. Thus, the transfer of organic carbon from the surface ocean to deeper compartments is a key process in determining the efficiency of the biological pump in withdrawing and storing carbon dioxide from the atmosphere.

This contribution aims to present an overview of the processes and factors that are currently believed to control the rate of OC remineralisation and its vertical distribution in the ocean.

Organic carbon pools

Primary production by phytoplankton in the euphotic zone of the ocean is the most important source of organic carbon in the ocean. The reservoir of phytogenic carbon comprises a continuum of molecular weights, ranging from small molecules such as amino acids, organic acids, sugars, nucleotides, to high molecular weight polymers. This reservoir supplies carbon, energy and nutrients to the heterotrophic biota in the ocean. The pool of organic carbon is, therefore, not merely composed of primary phytogenic carbon, but also of metabolites and structural matter originating from higher trophic levels. The production by phytoplankton, the transformation by heterotrophic biota, and transport processes determine the distribution and the chemical composition of total organic carbon (TOC).

Total organic carbon is divided into a dissolved fraction and a particulate fraction, which are operationally differentiated by the poresize of the filters used for separation. The organic fraction retained on the filter, often a Whatman GF/F or silver filter having nominal pore sizes of 0.5-0.7 μm, is considered to be particulate organic carbon (POC), whereas the fraction passing through the filter is assumed to be dissolved organic carbon (DOC). This division, however, is arbitrary since a continuum exists from dissolved molecules through colloids to macroscopic particles. Many organic particles such as bacteria and viruses, which are believed to make up a substantial fraction of the organic carbon in the ocean, may pass these filters and are therefore considered to be in the dissolved phase. Conversely, dissolved and colloidal carbon may be retained on the filter by sorption either to the filter or to the particulate phase accumulating on it.

Dissolved Organic Carbon (DOC)

DOC is often considered the largest reservoir of organic carbon in the ocean. The concentration of DOC in sea water has been a matter of considerable debate in the early 1990's, since the commonly used techniques yielded diverging results during intercalibrations. The currently accepted concentration for deep oceanic water is ~40 μM. In productive surface waters, the concentration may reach values of up to several hundreds of micromolar and may show a seasonal amplitude of ~30 μM. The average DOC concentration for the entire ocean amounts to 100 μM, which is equivalent to a total amount of $1600*10^{15}$ g C (Romankevich et al. 1999). For comparison, this total mass of DOC is approximately 30 times the annual primary production in the ocean.

Particulate Organic Carbon (POC)

The determination of POC in sea water is straight-forward, and is commonly performed by dry combustion of the particulate fraction (after acidification to remove particulate inorganic carbon) and subsequent measurement of the CO_2 produced. POC is composed of a variety of organic particles, including phytoplankton cells and debris, fecal pellets from grazers, gelatinous feeding nets and houses produced by zooplankton, crustacean exoskeleton molds, and amorphous organic particles. These particles can be found freely suspended in the water column, or in aggregated form, mixed with inorganic particles and cemented by sticky substances. In studies focusing on organic carbon fluxes in the ocean, POC is frequently classified as suspended or sinking. The suspended particles dominate the average pool size of POC in the water column, whereas the sinking particles determine the vertical exchange between surface waters and the deeper compartments of the ocean.

The concentration of POC in sea water is much less than the concentration of DOC and range from 0.4-0.8 µM in deep water up to several tens of micromolar in surface waters. The average concentration is about 3 µM, which amounts to $55*10^{15}$ g C for the entire ocean (Romankevich et al. 1999). Thus, the stock of POC in the oceanic water column is approximately equal to the annual primary production in the ocean.

Vertical transport of organic carbon

Modes of organic carbon transport

The most important downward vector for organic carbon is the gravitational settling of relatively rare but large particles. Most of these particles appear to be fecal pellets from grazing organisms and mixed organic aggregates with a diameter >100 µm (Fowler and Knauer 1986). The magnitude of this organic particle flux is determined by the concentration and sinking rates of the particles. The sinking rate of a particle depends on its size, shape and density, and on the viscosity and density of the surrounding medium. Laboratory measurements indicate aggregate sinking rates of 50-150 m d[-1] although considerably higher rates (300-1020 m d[-1])

were measured for phytodetrital aggregates collected from the seafloor at 2700 m water depth (Lampitt 1985). A possible explanation for the higher values is that these aggregates may have incorporated relatively heavy inorganic material during deposition-resuspension cycles in the benthic boundary layer. A growing number of *in situ* camera observations of sinking aggregates suggest a range of sinking velocities of 1-500 m d[-1] (Diercks and Asper 1997), which is in close agreement with estimates from the time-lags between export signals recorded in sediment traps moored at different depths. Most of the POC, however, appears to settle at only several meters/day, whereas a minute fraction, e.g. fecal pellets, may be sinking at considerably higher rates of up to 2000 m d[-1] (Angel 1989). The latter fraction, however, may represent a vehicle for the downward transport of small slowly sinking particles. Due to the varying composition of POC, its wide range of settling rates and facilitated transport of smaller particles, a weighted average settling velocity for bulk POC is hard to obtain.

For DOC and non-sinking POC, diffusion and advection are probably the most important vertical transport mechanisms. The diffusive and advective exchange between the surface waters and the deeper ocean is strongly impeded by density stratification of the water column. Density stratification retains non-sinking particulate and dissolved organic carbon and associated nutrients in surface waters, which upon remineralisation may sustain part of the primary production. The input of turbulent kinetic energy due to wind stress and tides and internal waves may temporarily reduce stratification and enhance the diffusive and advective downward transport of DOC. This transport, however, does not extend to below the depth of maximal ventilation and does therefore not contribute to a sequestration flux on climatically relevant time scales. In temporarily stratified areas, the downward advection during winter may even be quantitatively more important than the flux of POC (Carlson et al. 1994). Subduction of surface water, e.g. due to Ekman transport or deep water formation in polar regions, may be regionally important mechanisms for the downward transport of DOC and may represent the sole means of DOC sequestration.

Apart from these physical modes, OC can be transported downwards by the vertical migration of biota (Longhurst et al. 1990). Zooplankton and necton may feed at night near the surface and release fecal material and DOC at depth during the day. The contribution of these processes to the total downward flux has not been quantified yet, and may show considerable regional differences.

Export flux measurements

The export flux of OC is defined as the vertical flux of OC at a specified depth in the water column. Many studies on export fluxes of OC only take into account the settling of POC since this is considered the dominant pathway. This flux is generally determined by collecting settling material using drifting or fixed sediment traps. Traps can be positioned at different depths in a vertical array and can be deployed for a prolonged period of time. Samples, collected during defined time intervals are used to reconstruct the quantity and composition of the sinking matter. A second approach to assess the downward flux of POC relies on the determination of radio-isotope disequilibria in the water column, e.g. ^{234}Th and ^{238}U. ^{238}U is relatively soluble and is present in a constant ratio to salinity. ^{234}Th, a daughter isotope of ^{238}U, is quite insoluble and readily adsorbs to particles. As these particles sink, the ^{234}Th is removed from the water column. This removal can be calculated from the deficiency of ^{234}Th in the water column, relative to the amount that should be present from the decay of ^{238}U.

Empirical relationships

The spatial coverage of export flux estimates using sediment traps and the method of radio-isotope disequilibria is relatively low compared to that of primary production measurements although the seasonal coverage at single sites is considerably better than that of productivity measurements. A number of empirical relationships has been established which relate export fluxes to water depth and the mean annual rate of primary production. Many of these relationships were actually built on the basis of instantaneous (daily, weekly, seasonal) flux and productivity measurements. These relationships are used to predict the delivery of particulate organic

carbon to a given water depth, regionally as well as globally from primary production rates obtained by less elaborate techniques, such as remote sensing or ^{14}C incorporation measurements. In his evaluation, Bishop (1989) demonstrated that none of the equations fitted all data sets particularly well, and that the particle flux to deep waters could be better predicted if the flux at 100 m would be known, by using the relationship of Martin et al. (1987):

$$J(z) = J_{100}^{OC} \times (z/100)^{-0.858} \qquad (1)$$

where z is the water depth (m) and J_{100}^{OC} is the flux of organic carbon at 100 m water depth.

With respect to the amount of OC that can be exported from the euphotic layer it is important to distinguish between 'new' and 'regenerated' primary production. For a one-dimensional steady state scenario, the export production at the base of the euphotic zone is approximately equal to the primary production which is supported by the allochthonous input of nutrients, at least if integrated over a time span of several years. The primary production that is supported by the supply of new nutrients is defined as 'new production' (Dugdale and Goering 1967). The supply of new nutrients to the euphotic zone, e.g. from waters below the euphotic zone, is largely dependent on turbulence generated by winds and tides. Therefore, new production is relatively important in coastal seas and areas of intense upwelling. Regenerated production is the primary production that is sustained by pelagic recycling of nutrients within the mixed layer. This recycling is relatively important in the oligotrophic ocean and during summer stratification, when the input of new nutrients is limited. The relative contribution of new and regenerated production is indexed as the f-ratio, which is the ratio of new to total primary production. A comparison of new production with total production for different deep oceanic regions showed a positive correlation of the f-ratio with primary production values up to ~200 g C m^{-2} y^{-1} and a more or less constant ratio at higher productivities (Eppley and Peterson 1979). The f-ratio ranged from values of 0.1-0.3 for oligotrophic gyres and open oceans to 0.5 for coastal upwelling areas. At present this positive correlation is a matter of debate. Lohrenz et al. (1992) suggest a negative correlation by plotting their spot measurements of export fluxes versus primary production in the Sargasso Sea. However, their approach assumes steady state on a daily timescale between primary production and export fluxes at 200 m water depth. A literature survey on long-term simultaneous measurements of total primary production and export production in boreal coastal zones of the North Atlantic showed the export production to be positively and nonlinearly related to total primary production. For a total production range from 60 to 230 g C m^{-2} y^{-1}, the relationship was well described by:

$$P_E = 0.049(P_T)^{1.41} \qquad (2)$$

where P_E denotes the export production and P_T the total primary production. Thus, the f-ratio is given by the following equation:

$$f = 0.049 P_T^{0.41} \qquad (3)$$

It was stressed that such clearcut relationships may only be obtained if the appropriate time scales are considered for the system of interest. A compilation of OC fluxes throughout many areas of the open ocean (Lampitt and Antia 1997) demonstrates, in line with earlier suggestions made by Eppley and Peterson (1979), that the export flux, normalised to 2000 m depth, increases linearly with increasing rates of primary production up to a rate of 200 g C m^{-2} y^{-1}. Above this value, the export flux stabilizes, indicating a negative correlation between the f-ratio and primary production. Similar results were obtained by and by Palmer and Totterdell (2001) in a theoretical study using the Hadley Centre Ocean Carbon Cycle model, showing a stabilization of export fluxes at a rate of 50 g C m^{-2} y^{-1}.

The complex relationship between primary production and export flux from the euphotic zone is to some extent controlled by the structure of the foodweb in the euphotic zone. The general view is that the phototrophic community in areas of low and recycled production is dominated by picoplankton. Neither the picoplankton nor the microzooplankton grazing on them contribute significantly to the export of OC. In contrast, new

production scenarios are dominated by larger phytoplankton, e.g. diatoms, which may contribute directly to the export flux as single cells or in aggregates or are grazed by meso-zooplankton that produce larger fecal pellets. Berger et al. (1989) developed the following non-linear relationship between new production and primary production based on the ^{15}N uptake measurements of Eppley and Peterson (1979):

$$P_{new} = PP^2/400 - PP^3/340000 \qquad (4)$$

where P_{new} denotes the rate of new production and PP denotes the rate of primary production. This relationship shows a slightly sigmoidal behaviour but with an inflection point at a primary production rate of ~250 g C m^{-2} y^{-1}. Moreover, it predicts an f-ratio which is positively correlated with the rate of primary production for a range of 75 to 400 g C m^{-2} y^{-1}, and a range of values that is much higher (0.17 to 0.53) than observed by Palmer and Totterdell (2001) (0.13 to 0.17). This equation does neither support the findings of Lampitt and Antia (1997) nor those of Palmer and Totterdel (2001). Assuming that the export fluxes as compiled by Lampitt and Antia (1997) obey the export function of Martin et al. (1987), the export flux of OC at 100 m water depth can be described by combining the empirical relationship of Lampitt and Antia (1997) and the export relationship of Martin et al. (1987):

$$J_{100}^{OC} = P_{new} = 21.331 \times \tanh(2.87(PP - 132.7)/$$
$$132.7) + 22.48$$
$$\qquad (5)$$

This equation predicts the export flux to increase with increasing productivity up to a maximum and constant export of ~45 g C m^{-2} y^{-1} at primary production rates >200 g C m^{-2} y^{-1}. The predicted f-ratio increases exponentially with increasing primary production to a maximum of 0.22 at a productivity of ~180 g C m^{-2} y^{-1}, and subsequently declines with a further increase in productivity. The relationship between primary production and f-ratio corresponds well with the theoretical data of Palmer and Totterdell (2001). The predicted f-ratio is 0.17 at a productivity of 134 g C m^{-2} y^{-1}, which is the average productivity for the entire ocean, calculated from the ocean surface area, $362*10^6$ km^2, and the productivity of the ocean, 48.5 Gt y^{-1} (Field et al. 1998). This *f*-ratio suggests 8.35 Gt C y^{-1} to be exported from the euphotic layer globally, which closely resembles the values reported for the POC fluxes estimated by using global circulation models, e.g. Yamanaka and Tajika (1997).

Only a limited number of studies have quantified the export of OC through other mechanisms. Several authors have suggested that the downward transport of OC may be underestimated by measuring the POC flux, since this approach does not consider the release of DOC from these particles while collected by sediment traps. Using ^{14}C-labeled fecal pellets from large calanoid copepods, Urban-Rich (1999) studied the initial release of DOC present in fecal pellets at the time of their production. While sinking through a simulated water column, 39% of the fecal pellet ^{14}C was released as DO^{14}C after 5.5 days, the majority being released during the first 48-72 hours. Based on their estimated settling velocity of these pellets of 50 m d^{-1}, this carbon would have been released while settling over a distance of 100-150 m which would be sufficient to transport DOC out of the euphotic zone and to supply the waters below with DOC.

Due to the complexity of pathways, the total downward transport of DOC from the euphotic layer is extremely difficult to assess experimentally. As an alternative, the transport of DOC has been evaluated by applying general circulation models (GCM's). Earlier exercises, in an attempt to reproduce the erroneous high DOC concentrations of Sugimura and Suzuki (1988), suggested that the export of DOC at the base of the euphotic zone should be twice the export of POC. Yamanaka and Tajika (1997), using recent estimates of DOC, concluded that the global annual downward flux of OC at a water depth of 100 m amounted to 11 Gt, partitioned in 8 Gt POC and 3 Gt DOC. The downward flux of DOC, accomplished by convection, decreased strongly with depth and below 400 m, the export of OC was exclusively due to sinking POC. Assuming a global annual production of

48.5 Gt (Field et al. 1998), Yamanaka and Tajika's analyses suggest that 22% of the primary production is exported from the euphotic zone and that the contribution of DOC to this export amounts to 27%.

For open ocean settings, i.e. distant from the continental margins which may be a source of laterally advected particles, export fluxes of particulate organic carbon decrease exponentially with increasing water depths below the mixed layer. From equations (1) and (5) it is predicted that at the global average rate of primary production, 17.2, 4.3%, 2.4%, 1.3%, and 0.73% of the primary production arrives at a depth of 100, 500, 1000, 2000, and 4000m, respectively. This shows that the export flux from the euphotic layer is reduced already by 75% at a depth of 500 m. However, the 25% residual flux to deeper compartments is still of major importance for the biogeochemistry of the ocean and the efficiency of the biological pump.

The transformation of organic carbon

The decrease in export flux with increasing water depth is the net result of a number of physical and biological processes. Bacteria, protozoa, zooplankton, and viruses, free living or associated with organic particles remineralise part of the OC and will transform another part into new biomass and excretion products. Secondly, biological and physical processes may result in the formation as well as in the fragmentation of aggregates, thereby affecting their sinking velocities. These transformations are quantitatively important in the upper ocean and may regulate the downward flux of OC and thus the relationship of this flux with primary production.

The role of bacteria in POC-DOC interconversion

Particulate organic carbon cannot be taken up by bacteria directly. Extracellular and membrane-bound hydrolytic enzymes, produced by certain strains of bacteria, are required for the transformation of biopolymers and macromolecular organic material into soluble compounds. This hydrolysate is subsequently taken up by the microbial consortium.

A number of studies have indicated that ~50% of the daily primary production is required to support bacterial production in the upper ocean (Ducklow and Carlson 1992). Since bacteria are restricted to the uptake of dissolved molecules, half of the daily primary production should go through a dissolved state and thus contribute to the poolsize and dynamics of DOC. Besides the bacteria-mediated enzymatic hydrolysis of POC, three more sources of DOC have been identified.

• During photosynthesis, a wealth of organic compounds including sugars, amino acids, lipids, and fatty acids is released by phytoplankters. The release of DOC is often normalised to the rate of photosynthesis, a quantity known as percent extracellular release (PER). Culture studies indicate that the PER for healthy cells, growing exponentially under optimal conditions, is generally <10%. Suboptimal conditions with respect to nutrient supply, temperature, and irradiance may result in a significant increase in PER. Values of up to 80% have been reported for field studies, however, a data compilation by Baines and Pace (1991) showed that extracellular release is linearly related to the rate of primary production and has an average value of 13% across various aquatic systems.

• Phytoplankton and bacteria are substantially grazed by zooplankton and protozoa. Sloppy feeding, excretion and release from fecal pellets all contribute to the production of DOC (Jumars et al. 1989; Urban-Rich 1999), although their relative contributions are far from clear. Although the number of studies quantifying the release of DOC from grazers is limited, it is estimated that 10 to 20% of the ingested prey is converted to DOC (Strom et al. 1997). Consequently, if primary production is balanced by grazing, then 10 to 20% of the production is converted to DOC through grazing.

• Viruses, omnipresent in seawater, exploit the metabolism of their host cell for reproduction. Upon reproduction, the host cell bursts, releasing the reproduced viruses and cell contents. In productive nearshore surface waters the typical virus abundance is 10^{10} dm^{-3} decreasing to 10^9 dm^{-3} in less productive offshore waters (Cochlan et al. 1993). Their abundance is typically one order of magnitude higher than the abundance of hetero-trophic bacteria. Over a decade ago it became apparent that grazing on bacteria alone was insufficient to balance bacterial production. Viral infection and subsequent cell lysis proved to be an additional control mechanism on bacterial abundance. On average, about 20% of the marine heterotrophic bacteria appears to be infected by viruses and 10-20% of the cells are lysed on a daily basis (Suttle 1994). A number of studies demonstrate that virus induced lysis and microzooplankton grazing contribute equally to bacterial mortality (Fuhrman and Noble 1995; Weinbauer and Höfle 1998). Consequently, a significant fraction of bacterial carbon is not being transferred to higher trophic levels, but retained in a bacteria-viruses-DOC loop (viral loop), which may regulate bacterial production (Bratbak et al. 1992).

A large fraction of the DOC released through these processes is readily respired and assimilated by the heterotrophic bacterial population on time scales in the order of days or less. As a result of its short turn-over time, this labile fraction probably contributes less than 1% to the total reservoir of DOC. Conventionally, labile DOC is believed to consist of low-molecular weight material, whereas the more refractory pool would be composed of high-molecular weight compounds formed by abiotic condensation of organic molecules. Instead, Amon and Benner (1994) demonstrated that bacterial growth on the high-molecular weight fraction was three times greater than on the low-molecular weight fraction, suggesting the low-molecular weight fraction to be less available to the bacterial community. This remarkable result seems consistent with the observation that 65-80% of the DOC reservoir is composed of low-molecular weight compounds. Obviously, the differentiation of DOC in low- and high-molecular weight fractions is not informative on bacterial bioavailability and degradability.

Of the released DOC, only a small fraction is less susceptible, or rendered less susceptible to enzymatic degradation. This fraction persists for a considerable period of time in the water column and constitutes the bulk of the oceanic DOC reservoir. ^{14}C dating on DOC in surface waters of the central North Pacific Ocean suggested an average

age of ~1300 years (Williams and Druffel 1987). Assuming an ocean surface area of $362*10^6$ km^2, a mixed layer depth of 100 m, and an average concentration of DOC in the mixed layer of 80 μM (Williams and Druffel 1987) yields a total of $3.48*10^{16}$ g C as DOC. By further assuming steady state conditions suggests that each year, an approximate fraction of 1/1300 of this pool is replaced, which is equivalent to $2.7*10^{13}$ g C y^{-1}. Adopting a global annual primary production of $50*10^{15}$ g C indicates that only 0.05% of the primary production is required to be or to become refractory in order to maintain a DOC concentration of 80 μM with an average age of 1300 years. Since the chemical nature of this refractory DOC is largely unknown, it is impossible to reconstruct whether it is produced directly by the phytoplankton or rendered refractory through microbial mediation. Recent studies, however, have indicated that bacterial remains such as peptidoglycans, a class of refractory compounds associated with bacterial cell walls, contribute to this pool of DOC, suggesting that at least part of the refractory pool passes the bacterial shunt. The majority of this pool, 65-80% however, is composed of low-molecular weight compounds of unknown chemical signature and cannot, therefore, be traced to any source.

To further improve the understanding of the biogeochemical cycling of DOC in the ocean requires techniques for the identification and quantification of individual organic compounds, techniques that allow the determination of compound-specific turn-over times, and information on *in situ* substrate versatility of the microbial population.

The role of zooplankton in organic carbon transformation

Traditionally, the biomass of phytoplankton in the ocean has been considered to be regulated by resources, especially irradiance and nutrients. More recently, especially on the recognition that small phytoplankton contributes significantly to overall production, it has been demonstrated that the regulation is more complex, involving not only bottom-up but top-down mechanisms such as grazing and viral lysis as well. One of the leading concepts in studies on trophodynamics is that of size selective feeding, according to which a single species cannot feed on the entire size range of available prey. Feeding on a specific size class results in selective mortality and by affecting the competitive balance among prey, size selective feeding adds to chiseling the trophic architecture of the ecosystem. Accordingly, large phytoplankton would be grazed by the mesozooplankton (size range 200 μm - 2 mm), whereas the small phytoplankton and bacterioplankton would be grazed by the microzooplankton (size range 20 – 200 μm). However, phytoplankton and zooplankton are not simply large or small but comprise a continuum of sizes. Moreover, the zooplankton is taxonomically and above all functionally a highly diverse group which is not exclusively herbivorous, but may be bacterivorous, omnivorous, carnivorous, or detrivorous. It might be questioned whether a classification according to size in trophodynamic studies is appropriate when functional responses are of interest. This trophic versatility in combination with regionally specific physical controls such as irradiance, temperature and mixing, gives rise to spatially and temporally variable trophic scenarios.

It is argued by Banse (1995) that the solubilization of organic carbon through grazing is the principle process that leads from physically controlled new production to total production over much of the ocean area as it permits the DOC limited heterotrophic bacteria to remineralise nutrients for regenerative production. At present, microzooplankton, often dominated by ciliates and heterotrophic nanoflagellates, are recognized as the most important grazers in the open ocean (Burkill et al. 1993; Sherr and Sherr 1994; Landry et al. 1997; Stelfox-Widdicombe et al. 2000) and a growing body of evidence suggests that they may also regulate the phytoplankton biomass in eutrophic coastal areas at different latitudes (Burkill et al. 1995; Strom and Strom 1996, Archer et al. 2000, Strom et al. 2001). During nutrient depleted conditions, the phytoplankton typically consists of small cells and the temporal variation in biomass is generally small. Of the loss factors controlling the stock of phytoplankton, grazing by microzooplankton is considered most important. The maximum specific growth rates of the microzooplankton have been shown to decrease with increasing cell volume

(Fenchel and Finlay 1983), and to match those of the small phytoplankton. These high growth rates enables the microzooplankton population to keep pace with changes in population density of the phytoplankton. Preventing major accumulations of phytoplankton biomass by efficient grazing limits aggregate formation and vertical export of phytogenic carbon. Moreover, the fecal pellets produced by the microzooplankton are small and generally slowly sinking. This tight coupling between primary production and grazing by microzooplank-ton under nutrient depleted conditions limits vertical sinking losses and retains nutrients and carbon in the productive surface layer. The microzoo-plankton in turn may serve as food for the mesozoo-plankton and several studies suggest that predation rather than food availability regulates the biomass of the microzooplankton (Smetacek 1981; Nielsen and Kiørboe 1994). Thus, the grazing of microzoo-plankton on phyto-and bacterioplankton (Putland 2000) and predation of mesozooplankton on the microzooplankton may link microbial processes to higher trophic levels.

The mesozooplankton are generally considered to be important grazers in the more productive ecosystems where diatoms contribute substantially to primary production. A short linear food chain of diatom-copepod-fish results in a high trophic efficiency and may support a high production of pelagic fish in these ecosystems. A prerequisite for this scenario is a close matching between primary production, grazing and carnivorous predation, a condition which is not always satisfied. For example, in Nordic fjords, the springtime vertical export of POC is usually dominated by fecal pellets, indicating extensive grazing of copepods and a tight coupling with primary production similar to the situation in subtropical upwelling areas. In some years, however, the vertical POC flux is dominated by diatoms, indicating a mismatch between the diatom production and mesozooplankton grazing. The interannual differences in the contribution of diatoms to the vertical POC export has been suggested to result from interannual differences in oceanic-coastal coupling which is crucial for the advective import of mesozooplankton from the shelf into the fjords and determines the degree of matching between primary production and grazing (Reigstad

et al. 2000). A low initial stock of meso-zooplankton has also been suggested the cause for the open ocean spring bloom in the North Atlantic. Deep mixing during the winter season severely limits primary production and results in a strong reduction in the stock of mesozooplankton. The reduction in grazing pressure and the shallowing of mixing in early spring may result in high net rates of diatom production and the buildup of diatom biomass. Extensive diatom biomass formation during the spring bloom in temperate coastal areas also indicates a loose coupling between primary production and grazing, and suggests that the population of grazers cannot keep pace. For Danish coastal waters, the seasonal stock of mesozooplankton closely followed the trend in temperature, and showed an increase in spring, a single maximum in summer and a 10-fold reduction during winter. The chlorophyll concentration peaked in April and coincided with the maximum in copepod productivity, suggesting that the production of copepod biomass was not food limited and that the biomass of diatoms was not regulated by grazing. The autumn bloom in phytoplankton also resulted in substantial copepod productivity but was, in contrast to spring, accompanied by a reduction in copepod biomass, indicating that the stock of mesozooplankton was primarily controlled by mortality (Kiørboe and Nielsen 1994). In general, the production of diatom biomass in temperate coastal waters is ultimately limited by the availability of silicic acid which is an essential nutrient for the production of diatom frustules. The depletion of silicic acid stock often preludes the cessation of the diatom bloom, the subsequent aggregation and vertical export of diatoms.

These scenarios may suggest an important role for mesozooplankton in phytoplankton carbon removal in productive ecosystems and a minor role in unproductive waters. A cross-system survey of primary production and mesozooplankton grazing revealed a significant increase in mesozooplankton grazing with increasing primary production (Calbet 2001). However, the relationship between grazing and primary production was non-linear, with the relative proportion of primary production removed by mesozooplankton grazing declining from unproductive to productive communities. The daily grazed percentage of primary production averaged 40% for low productive communities (< 250 mg C m^{-2} d^{-1}), 22% for moderate (250-1000 mg C m^{-2} d^{-1}), and 10% for highly productive (> 1000 mg C m^{-2} d^{-1}) communities. The amount of phytoplankton carbon ingested per unit mesozooplankton biomass appeared to be lower in the low productive than in the moderate and highly productive communities, suggesting that alternative food sources such as microzooplankton contribute to the mesozooplankton diet in low productive communities. In contrast with the common view, this analysis indicates that the grazing pressure of meso-zooplankton on oceanic primary production is comparable to that of microzooplankton, whereas in more productive ecosystems the microzooplankton would exert a higher pressure than the meso-zooplankton.

Clearly, grazers play an eminent role with respect to retention versus vertical export of POC. Grazing by microzooplankton would result in an efficient recycling of carbon and nutrients, however, the feeding mode and the biomass of mesozoo-plankton may be crucial in determining the recycling efficiency as illustrated by Wassman (1998) in a number of theoretical scenarios. A low meso-zooplankton grazing pressure reduces the aggregation and settling of phytogenic carbon and yields a somewhat increased sedimentation of larger fecal pellets resulting in a slightly increased retention of carbon and nutrients in the upper ocean. Changing to a omnivorous or carnivorous mode by additionally or exclusively feeding on the microzooplankton may sustain a higher standing stock of mesozoo-plankton and increase the export of fecal pellets. The consequent reduction in grazing pressure of the microzooplankton on small phytoplankton may increase the aggregation and vertical loss of aggregates from the mixed layer, resulting in a reduced retention of carbon and nutrients and a shift to an export food web. As stated by the author, these scenarios represent only a very general comprehension and a more thorough understanding of the dynamics of processes controlling retention and export is urgently needed to predict the quantity and composition of the vertically transported POC.

Aggregates

Vertical POC transport in the ocean happens primarily through the settling of aggregates and fecal pellets. Aggregates evolve through the process of coagulation, the repeated collision of particles due to turbulent shear or differential particle sinking rates and subsequent sticking of smaller ones (Jackson 1990). Several studies, instigated by the discovery of high concentrations of sticky transparent exopolymer particles (TEP) (Alldredge et al. 1993), suggest that TEP form the matrix of marine snow aggregates and facilitate the coagulation during diatom blooms (Logan et al. 1995; Passow et al. 1994). The formation of large aggregates is considered to enhance the downward flux of OC, since an increase in particle size may increase the sinking velocity. Aggregates, however, are extremely fragile structures that may easily be disrupted, resulting in a reduction in settling velocity and therefore in a longer residence time in the water column. In surface waters, the break-up of large aggregates has been observed to coincide with increased windspeed resulting in enhanced turbulent mixing (Riebesell 1992). However, fluid shear rates in the deeper ocean are considered insufficient to break-up marine snow (Alldredge et al. 1990).

Biological processes, however, such as enzymatic hydrolysis, grazing by zooplankton and shear stress generation during swimming activities of zooplankton (Dilling and Alldredge 2000) result in the fragmentation of marine snow as well. Aggregates represent hot spots of organic carbon and nutrients in a relatively depleted water column, and are therefore attractive sites for heterotrophic microbiota and mesozooplankton. Aggregates harbour complex foodwebs comprising bacteria, amoebae, ciliates, flagellates, and larger organisms like mesozooplankton and polychaete larvae. Quantitative studies on the abundance of zooplankton in aggregates are scant, however, recent studies suggest that zooplankton may be concentrated on aggregates relative to the bulk water (Green and Dagg 1997). A recent compilation on the abundance of aggregate associated zooplankters illustrates the potential importance of zooplankton in organic carbon turn-over in the euphotic zone (Kiørboe 2000). The abundance of zooplankters associated with aggregates, when normalised to their abundance in the surrounding sea water, scaled with the aggregate radius raised to the power 2.27. From the zooplankton abundance in aggregates, the estimates of their metabolic and grazing rates, and the organic carbon content of aggregates, it was calculated that the turnover rate of aggregates in the size range of 0.1-1.0 cm would be in the order of days to a week. This is similar to or faster than the turnover rates due to the activity of bacteria and implies that 20-70% of the aggregate carbon would be degraded by associated zooplankters before the aggregate leaves a 50 m deep upper mixed layer. Hence, zooplankters colonizing sinking aggregates may effectively retain nutrients and dissolved inorganic carbon within the euphotic zone. Kiørboe's suggestion that aggregation of smaller particles into larger, rapidly sinking aggregates may encourage the remineralisation through enhancing the abundance of zooplankters, thereby sustaining primary production in the upper ocean, competes with the viewpoint that aggregation enhances the export of nutrients from the euphotic zone and certainly merits further attention. The net removal of POC and the decrease in vertical POC flux with water depth are regulated by the metabolic activities and carbon conversion efficiencies of the heterotrophic communities inhabiting these aggregates. Metabolic activities are strongly affected by the prevailing physico-chemical conditions, some of which will be discussed below.

Hydrodynamic control on solute exchange

Bacteria which are freely suspended in sea water are faced with an environment that is dominated by viscous forces. As a result, mass exchange between a bacterium and its surrounding water is controlled by molecular diffusion. The relative importance of inertial forces acting on a particle increases with increasing particle diameter, and therefore, the contribution of turbulence to mass exchange increases (Kiørboe et al. 2001). Bacteria and microzooplankton may benefit from this turbulent contribution for larger particles by attachment to these particles, thereby overcoming diffusional limitation on external substrate supply. Remarkably, only a limited number of studies has considered the potential control of hydrodynamics on the activity of bacteria and microzooplankton associated with aggregates. Studies on the effect of fluid shear on microbial activity have indeed shown that particle attached bacteria may benefit nutritionally if attached to sufficiently large particles.

Ploug and co-workers showed the importance of applying the appropriate hydrodynamic conditions, i.e. those mimicking a sinking aggregate through the water column, in studies on the mass transfer between settling aggregates and the surrounding water (e.g. Ploug and Jørgensen 1999). Using oxygen microelectrodes, they showed that the distribution of oxygen in individual aggregates changed dramatically upon transfer from sinking conditions to stagnant conditions while lying onto a solid surface. The change in hydrodynamic condition and the concomitant increase in diffusion distance clearly affected the distribution of oxygen in the aggregate. The total oxygen flux towards the aggregate, i.e. the rate of total oxygen consumption by the associated microbial consortium, however, was not altered. It was speculated by the authors that the increase in diffusion distance to the surrounding water by concentrating aggregates or by incubating a single aggregate while sitting on a solid surface in theory may affect the metabolic activities of aggregate-attached bacteria.

Ploug and Grossart (1999) examined the effect of pooling on the exchange of oxygen and on bacterial production measurements in aggregates. They showed that the mean diffusion distance for solutes increased and that the rate of thymidine and leucine incorporation per aggregate decreased with increasing numbers of aggregates in a vial under stagnant conditions. Pooling 5 aggregates in a single vial resulted in a ~5-fold reduction in incorporation rates as compared to aggregates that were incubated individually and kept suspended. Concentrating the aggregates is common practice for production measurements of aggregate-attached bacteria in order to obtain a detectable response and a representative measurement. Previous studies on the organic carbon turn-over rate in aggregates suggested that the carbon demand of attached bacteria, as estimated from the rates of bacterial production, was very small, which would imply extremely long turn-over times. Nevertheless, aggregate solubilisation through enzymatic hydrolysis of the organic particulate matter appeared to be fast and efficient (Smith et al. 1992). These observations suggested that the enzymatic hydrolysis of particulate organic carbon was not tightly coupled to the uptake of the hydrolysate and bacterial production, resulting in a diffusive loss of hydrolysate from the aggregate. Contrary to these earlier views, the calculated specific growth rates and growth efficiencies of attached bacteria on individually suspended aggregates (Ploug and Grossart 1999) suggested a rapid and efficient conversion of particulate carbon into biomass of attached bacteria and carbon dioxide. The dramatic effect of hydrodynamic conditions on solute distribution and therefore on the activity of the microbial consortium suggest that earlier estimates of the activity of attached bacteria obtained by pooling aggregates should be carefully reconsidered.

Oxygen availability and the remineralisation of organic carbon

The effect of oxygen availability on the rate and the extent of OC degradation continues to be a subject of debate. The effect has been studied in the laboratory using different experimental setups with a variety of organic substrates. At first glance, these studies yielded seemingly conflicting results. Some studies suggested OC to degrade faster under oxic conditions than under anoxic conditions, whereas other studies demonstrated the opposite to be the case. It was concluded that all evidence for carbon oxidation to be inherently slower via anaerobic pathways was ambiguous, at least for easily degradable substrates (Blackburn 1991). This suggests that anoxia during transport of OC through the water column would not affect the export flux of OC. Evidence against this suggestion was presented by Harvey et al. (1995), who studied the decomposition rates for a diatom and a cyanobacterium under oxic and anoxic conditions in a laboratory flow-through system mimicking settling through the water column. The concentrations of POC, PON, lipids, proteins, and carbohydrates, classes of compounds with different reactivities, were monitored during the course of the experiment. Their results demonstrated that even for extremely fresh phytogenic carbon the absence of oxygen resulted in a slower rate of decay for all cellular components of phytoplankton. In addition to decreasing decomposition rates, anoxia also resulted in a higher concentration of residual organic carbon. Contrasting the common view, these results imply that the oxygen availability for carbon remineralisation during transport through the water column could control the flux of organic carbon to the sea floor.

Sinking through a deep anoxic water column may increase the export flux, not only due to a reduction in bacterial degradation rates, but also because of the reduced contribution of zooplankton to remineralisation. A recent sediment trap study of OC fluxes in the Cariaco Basin, one of the world's largest marine anoxic basins (maximum water depth ~1400 m), however, yielded contrasting results (Thunell et al. 2000). These authors observed a reduction of 60% in the OC flux between the shallow trap at the oxic-anoxic interface (275 m) and the deepest trap (1255 m), which indicates a substantial degradation of OC during transport through the anoxic water column. The observed changes in OC flux with water depth were similar to those reported for oxic waters of the open ocean. These results accord with the view that the rates of aerobic and anaerobic remineralisation rates of fresh OC are not significantly different.

An (partially) anoxic water column, however, is not a prerequisite for anaerobic OC degradation in the water column. Anoxia may occur in somewhat larger, fresh aggregates even under hydrodynamic conditions mimicking their sinking through the water column (Ploug et al. 1997). However, the efficient supply of oxygen to the aggregate due to its 3-dimensional diffusion geometry requires extremely high rates of oxygen consumption to create anoxic conditions within the aggregate. High rates of carbon remineralisation, however, will result in a rapid turnover of OC. Theoretical calculations showed that carbon limitation of heterotrophic processes within the aggregate would limit anoxia to a few hours only, depending on the size of the aggregate. This short-term anoxia in sinking aggregates will probably not seriously affect the export flux of OC.

The effect of temperature

Temperature is an important physical parameter controlling the activity and growth of bacteria, micro- and mesozooplankton. Temperature coefficients, Q_{10}, are generally between 2 and 3.5, indicating a 2 to 3.5-fold reduction in activity for each 10°C lowering in temperature below the optimal value. Much of the seasonal variation in bacterial production and growth rates in temperate areas can be explained by water temperature. Pomeroy and Deibel (1986) suggested that bacterial metabolism and growth at low sea water temperatures were more depressed than those of phytoplankton. This would result in a reduced microbial loop activity and a higher availability of phytoplankton carbon for grazers. Low temperatures, however, do not a priori imply low biological activities. In perennially cold surface waters of the Arctic Ocean, heterotrophic bacterial activity and production were high (Rich et al. 1997), and the clearance rates of dinoflag-ellates and ciliates were similar to the values reported for herbivorous protists in temperate waters (Sherr et al 1997). In seasonally cold Newfoundland coastal waters, the microbial food web remained active throughout the year and substrate production by microzooplankton grazing probably supported active bacterial growth during the spring bloom (Putland 2000). Bacterial communities from deep waters (~4700 m) in the eastern Atlantic showed cell specific activity levels at *in situ* conditions comparable to those of bacteria from a depth of 150 m (Patching and Eardly 1997). Thus, it appears that bacterial communities and microbial food webs can be perfectly active in cold waters. Based on a literature compilation Rivkin et al. (1996) proposed that bacteria-based food webs and microbial trophic pathways are as important in overall energy and material cycling in high latitude oceans as they are at lower latitudes.

Sinking aggregates and associated communities however, may experience a change in temperature on a much shorter than a seasonal time scale.

The average sea surface temperature for all oceans outside the upwelling areas range from >26°C at the equator to < 0°C in the polar regions. Bottom water temperatures, however, do not show a strong latitudinal gradient and amount to 1-2°C. Hence, the bacteria and zooplankton colonizing aggregates and fecal pellets in the upper mixed layer in tropical regions are faced with a considerable decrease in temperature upon downward transport. The temperature coefficients suggests that the activity of the attached and exported communities in tropical areas would be reduced by a factor of 8 to 40 in the deeper waters relative to the surface mixed layer. The strongest decrease in activity, however, would occur in the upper 200 m of the water column where the strongest gradient in temperature is observed. In contrast, communities colonizing aggregates in the upper mixed layer in polar regions are exposed to more or less constant temperatures during downward transport. Studies on the effect of decreasing temperatures in the upper water column on the remineralisation of natural aggregates and fecal pellets are scarce. Are the communities which colonised particles in warmer surface layers outcompeted by newly scavenged, better adapted or acclimated communities residing in deeper waters or are they able to acclimate to reduced temperatures? How does it affect the quantity and quality of OC delivered to the sea floor in tropical regions as compared to temperate and polar regions? These are still open questions which need to be addressed in order to understand the regional controls of water temperature on POC export fluxes.

Conclusions

The biological mechanisms controlling the export flux from the euphotic layer are presently not fully understood, which limits the use of satellite imagery of primary production to predict regional and temporal patterns in export fluxes.

To improve the understanding of the biogeochemical cycling of POC and DOC requires techniques for the identification and quantification of individual organic compounds and methods that allow the determination of compound specific turn over times.

Hydrodynamics appear to exert a strong control on the remineralisation of aggregates and future activity measurements of particle associated biota should be performed under hydrodynamic conditions mimicking *in situ* hydrodynamics.

Most of our information on POC remineralisation and the role of biota and physical conditions originates from studies in the upper ocean. Studies on the remineralisation of POC in midwaters and in the benthic boundary layer are scant, which may be the consequence of the extreme difficulties to sample highly fragile aggregates at these depths.

Acknowledgement

The author greatly appreciated the thoughtful and constructive comments of Avan Antia and Paul Wassman for improving the original manuscript.

References

Alldredge AL, Granata TC, Gotschalk GC, Dickey TD (1990) The physical strength of marine snow and its implications for particle disaggregation in the ocean. Limnol Oceanogr 35:1415-1428

Alldredge AL, Passow U, Logan BE (1993) The abundance and significance of a class of large, transparent organic particles in the ocean. Deep-Sea Res I 40:1131-1140

Amon RMW, Benner R (1994) Rapid cycling of high-molecular-weight dissolved organic matter in the ocean. Nature 369:549-552

Angel MV (1989) Does mesopelagic biology affect the vertical flux? In: Berger WH, Smetacek VS, Wefer G (eds) Productivity of the Ocean: Present and Past. J Wiley &Sons, Chichester pp 155-173

Archer SD, Verity PG, Stefels J (2000) Impact of micro-zooplankton on the progression and fate of the spring bloom in fjords of northern Norway. Aquat Microb Ecol 22:27-41

Baines SB, Pace ML (1991) The production of dissolved organic matter by phytoplankton and its importance to bacteria: Patterns across marine and freshwater systems. Limnol Oceanogr 36:1078-1090

Banse K (1995) Zooplankton: Pivotal role in the control of ocean production. ICES J Mar Sci 52:265-277

Berger WH, Smetacek VS, Wefer G (1989) Ocean productivity and paleoproductivity - An overview. In: Berger WH, Smetacek VS, Wefer G (eds) Productivity of the Ocean: Present and Past. J Wiley &Sons, Chichester pp 1-34

Bishop JKB (1989) Regional extremes in particulate matter composition and flux: Effects on the chemistry of the ocean interior. In: Berger WH, Smetacek VS, Wefer G (eds) Productivity of the Ocean: Present and Past. J Wiley & Sons, Chichester pp 117-137

Blackburn TH (1991) Accumulation and regeneration: Processes at the benthic boundary layer. In: Mantoura RFC, Martin J-M, Wollast R (eds) Ocean Margin Processes in Global Change. J Wiley & Sons, Chichester pp 181-195

Bratbak G, Heldal M, Thingstad TF, Riemann B, Haslund OH (1992) Incorporation of viruses into the budget of microbial C-transfer. A first approach. Mar Ecol Prog Ser 83:273-280

Burkill PH, Edwards ES, John AWG, Sleigh MA (1993) Microzooplankton and their herbivorous activity in the north-east Atlantic Ocean. Deep-Sea Res II 40: 479-494

Burkill PH, Edwards ES, Sleigh MA (1995) Microzooplankton and their role in controlling phytoplankton growth in the marginal ice zone of the Bellingshausen Sea. Deep-Sea Res II 42:1277-1290

Calbet A (2001) Mesozooplankton grazing effect on primary production: A global comparative analysis in marine ecosystems. Limnol Oceanogr 46: 1824-1830

Carlson CA, Ducklow HW, Michaels AF (1994) Annual flux of dissolved organic carbon from the euphotic zone in the northwestern Sargasso Sea. Nature 371: 405-408

Cochlan WP, Wikner J, Steward GF, Smith DC, Azam F (1993) Spatial distribution of viruses, bacteria and chlorophyll a in neritic, oceanic and estuarine environments. Mar Ecol Prog Ser 92: 77-87

Diercks A-R, Asper VL (1997) *In situ* settling speeds of marine snow aggregates below the mixed layer: Black Sea and Gulf of Mexico. Deep-Sea Res I 44:385-398

Dilling L, Alldredge AL (2000) Fragmentation of marine snow by swimming macrozooplankton: A new process impacting carbon cycling in the sea. Deep-Sea Res I 47: 1227-1245

Ducklow HW, Carlson CA (1992) Oceanic bacterial production. Adv Microb Ecol 12:113-181

Dugdale RC, Goering JJ (1967) Uptake of new and regenerated forms of nitrogen in primary production. Limnol Oceanogr 12: 196-206

Eppley RW, Peterson BJ (1979) Particulate organic matter flux and planktonic new production in the deep ocean. Nature 282: 677-680

Fenchel T, Finlay BJ (1983) Respiration rates in heterotrophic, free-living protozoa. Microb Ecol 9:99-122

Field CB, Behrenfeld MJ, Randerson JT, Falkowski P (1998) Primary production of the biosphere: Integrating terrestrial and oceanic components. Science 281: 237-240

Fowler SW, Knauer GA (1986) Role of large particles in the transport of elements and organic compounds through the oceanic water column. Prog Oceanogr 16:147-194

Fuhrman JA, Noble RT (1995) Viruses and protists cause similar bacterial mortality in coastal seawater. Limnol Oceanogr 40:1236-1242

Green EP, Dagg MJ (1997) Mesozooplankton associations with medium to large marine snow aggregates in the northern Gulf of Mexico. J Plankton Res 19: 435-447

Harvey HR, Tuttle JH, Bell JT (1995) Kinetics of phytoplankton decay during simulated sedimentation: Changes in biochemical composition and microbial activity under oxic and anoxic conditions. Geochim Cosmochim Acta 59:3367-3377

Jackson GA (1990) A model of the formation of marine algal flocs by physical coagulation processes. Deep-Sea Res 37:1197-1211

Jumars PA, Penry DL, Baross JA, Perry MJ, Frost BW (1989) Closing the microbial loop: Dissolved carbon pathway to heterotrophic bacteria from incomplete ingestion, digestion and absorption in animals. Deep-Sea Res 36:483-495

Kiørboe T (2000) Colonization of marine snow aggregates by invertebrate zooplankton: Abundance, scaling, and possible role. Limnol Oceanogr 45:479-484

Kiørboe T, Nielsen TG (1994) Regulation of zooplankton biomass and production in a temperate, coastal ecosystem. 1. Copepods. Limnol Oceanogr 39:493-507

Kiørboe T, Ploug H, Thygesen UH (2001) Fluid motion and solute distribution around sinking aggregates. I. Small-scale fluxes and heterogeneity of nutrients in the pelagic environment. Mar Ecol Prog Ser 211:1-13

Lampitt RS (1985) Evidence for the seasonal deposition of detritus to the deep-sea floor and its subsequent resuspension. Deep-Sea Res 32:885-897

Lampitt RS, Antia AN (1997) Particle flux in deep seas: regional characteristics and temporal variability. Deep-Sea Res I 44:1377-1403

Landry MR, Barber RT, Bidigare R, Chai F, Coale KH, Dam HG, Lewis MR, Lindley ST, McCarthy JJ, Roman MR, Stoecker DK, Verity PG, White JR (1997) Iron and grazing constraints on primary production in the central equatorial Pacific: An EqPac synthesis. Limnol Oceanogr 42:405-418

Logan BE, Passow U, Alldredge AL, Grossart H-P, Simon M (1995) Rapid formation and sedimentation of large aggregates is predictable from coagulation

rates (half-lives) of transparent exopolymer particles (TEP). Deep-Sea Res II 42:203-214

Lohrenz SE, Knauer GA, Asper VL, Tuel M, Michaels AF, Knap AH (1992) Seasonal variability in primary production and particle flux in the northwestern Sargasso Sea: US JGOFS Bermuda Atlantic Time-series study. Deep-Sea Res 39:1373-1391

Longhurst AR, Bedo AW, Harrison WG, Head EJH, Sameoto DD (1990) Vertical flux of respiratory carbon by oceanic diel migrant biota. Deep-Sea Res 37: 685-694

Martin JH, Knauer GA, Karl DM, Broenkow WW (1987) VERTEX: Carbon cycling in the northeast Pacific. Deep-Sea Res 34:267-285

Nielsen TG, Kiørboe T (1994) Regulation of zooplankton biomass and production in a temperate, coastal ecosystem. 2. Ciliates. Limnol Oceanogr 39:508-519

Palmer JR, Totterdell IJ (2001) Production and export in a global ocean ecosystem model. Deep-Sea Res 48: 1169-1198

Passow U, Alldredge AL, Logan BE (1994) The role of particulate carbohydrate exudates in the flocculation of diatom blooms. Deep-Sea Res I 41:335-357

Patching JW, Eardly D (1997) Bacterial biomass and activity in the deep waters of the eastern Atlantic - evidence of a barophilic community. Deep-Sea Res 44:1655-1670

Ploug H, Grossart H-P (1999) Bacterial production and respiration in suspended aggregates- a mattter of the incubation method. Aquat Microb Ecol 20:21-29

Ploug H, Jørgensen BB (1999) A net-jet flow system for mass transfer and microsensor studies of sinking aggregates. Mar Ecol Prog Ser 176:279-290

Ploug H, Kühl M, Buchholz-Cleven B, Jørgensen BB (1997) Anoxic aggregates-an ephemeral phenomenon in the pelagic environment? Aquat Microb Ecol 13: 285-294

Pomeroy LR, Deibel D (1986) Temperature regulation of bacterial activity during the spring bloom in Newfoundland coastal waters. Science 233:359-361

Putland JN (2000) Microzooplankton herbivory and bacterivory in Newfoundland coastal waters during spring, summer and winter. J Plankton Res 22:253-277

Reigstad M, Wassman P, Ratkova T, Arashkevich E, Pasternak A, Øygarden S (2000) Comparison of the springtime vertical export of biogenic matter in three northern Norwegian fjords Mar Ecol Prog Ser 201: 73-89

Rich J, Gosselin M, Sherr E, Sherr B, Kirchman DL (1997) High bacterial production, uptake and concentrations of dissolved organic matter in the Central

Arctic Ocean. Deep-Sea Res 44:1645-1663

Riebesell U (1992) The formation of large marine snow and its sustained residence in surface waters. Limnol Oceanogr 37: 63-76

Rivkin RB, Anderson MR, Lajzerowicz C (1996) Microbial processes in cold oceans. I. Relationship between temperature and bacterial growth rate. Aquat Microb Ecol 10:243-254

Romankevich EA, Vetrov AA, Korneeva GA (1999) Geochemistry of organic carbon in the ocean. In: Gray JS, Ambrose W, Szaniawska A (eds) Biogeochemical Cycling and Sediment Ecology. Kluwer Academic Publishers, Dordrecht, pp 1-27

Sherr EB, Sherr BF (1994) Bacterivory and herbivory: Key roles of phagotrophic protists in pelagic food webs. Microb Ecol 28:223-235

Sherr EB, Sherr BF, Fessenden L (1997) Heterotrophic protists in the Central Arctic Ocean. Deep-Sea Res 44:1665-1682

Smetacek V (1981) The annual cycle of proto-zooplankton in the Kiel Bight. Mar Biol 63:1-11

Smith DC, Simon M, Alldredge AL, Azam F (1992) Intense hydrolytic enzyme activity on marine aggregates and implications for rapid particle dissolution. Nature 359:139-142

Stelfox-Widdicombe CE, Edwards ES, Burkill PH, Sleigh MA (2000) Microzooplankton grazing activity in the temperate and sub-tropical NE Atlantic: Summer 1996. Mar Ecol Prog Ser 208:1-12

Strom SL, Strom MW (1996) Microplankton growth, grazing and community structure in the northern Gulf of Mexico. Mar Ecol Prog Ser 130:229-240

Strom SL, Benner R, Ziegler S, Dagg MJ (1997) Planktonic grazers are a potentially important source of marine dissolved organic carbon. Limnol Oceanogr 33:1217-1220

Strom SL, Brainard MA, Holmes JL, Olson MB (2001) Phytoplankton blooms are strongly impacted by microzooplankton grazing in coastal North Pacific waters. Mar Biol 138:355-368

Sugimura Y, Suzuki Y (1988) A high-temperature catalytic oxidation method for the determination of nonvolatile dissolved organic carbon in sewater by direct injection of a liquid sample. Mar Chem 24:105-131

Suttle CA (1994) The significance of viruses to mortality in aquatic microbial communities. Microb Ecol 28: 237-243

Thunell RC, Varela R, Llano M, Collister J, Müller-Karger F, Bohrer R (2000) Organic carbon fluxes, degradation, and accumulation in an anoxic basin: Sediment trap results from the Cariaco Basin. Limnol Oceanogr 45:300-308

Urban-Rich J (1999) Release of dissolved organic carbon from copepod fecal pellets in the Greenland Sea. J Exp Mar Biol Ecol 232:107-124

Wassman P (1990) Relationship between primary and export production in the boreal coastal zone of the North Atlantic. Limnol Oceanogr 35:464-471

Wassman P (1998) Retention versus export food chains: Processes controlling sinking loss from marine pelagic systems. Hydrobiologia 363:29-57

Weinbauer MG, Höfle MG (1998) Significance of viral lysis and flagellate grazing as factors controlling bacterioplankton production in a eutrophic lake. Appl Environ Microbiol 64:431-438

Williams PM, Druffel ERM (1987) Radiocarbon in dissolved organic matter in the Central North Pacific Ocean. Nature 330:246-248

Yamanaka Y, Tajika E (1997) The role of dissolved organic matter in the marine biogeochemical cycle: studies using an ocean biogeochemical general circulation model. Glob Biogeochem Cycl 11:599-612

Coupled Biogeochemical Cycling and Controlling Factors

A.N. Antia[1*], P.H. Burkill[2], W. Balzer[3], H.J.W. de Baar[4], R.F.C. Mantoura[5],
R. Simó[6], D. Wallace[1]

[1]*Institut für Meereskunde, FB 2 - Marine Biogeochemie, Düsternbrooker Weg 20,
24105 Kiel, Germany*
[2]*Plymouth Marine Laboratory, Prospect Place, Plymouth PL1 3DH, UK*
[3]*Universität Bremen, FB 2 - Meereschemie, Klagenfurter Straße, 28359 Bremen, Germany*
[4]*Netherlands Institute of Sea Research, PO Box 59, 1790 AB Den Burg, The Netherlands*
[5]*Centre for Coastal and Marine Sciences, Prospect Place,
The Hoe, Plymouth PL1 3HD, UK*
[6]*Institut de Ciències del Mar (ICM-CSIC), Pg. Joan de Borbó s/n, 08039 Barcelona, Spain*
* *corresponding author (e-mail): aantia@ifm.uni-kiel.de*

Abstract: The changing climate of the planet is closely linked to biogeochemical processes in the oceans with important feedbacks between oceanic, atmospheric and terrestrial components of the earth system. This chapter identifies key processes that mediate the response of marine ecosystems to a changing environment and recommends implementation strategies for future studies. Technological and methodological advances such as the use of new biochemical and molecular techniques have led to the discovery of unknown metabolic pathways and identification of genetic diversity in marine systems. Ecosystem changes, reflected in shifts in dominant plankton groups are likely to have a large global but also regional impact in the European context. In terms of marine biogeochemical cycling, key processes that respond to a changing climate include photosynthesis (and its modulation by trace metal availability and nitrogen fixation), calcification and the production and release of a suite of volatile, climate-reactive gasses. Implementation of future research strategies should focus on the ability to monitor key variables from stationary platforms and ships of opportunity with sufficient stability and accuracy to resolve natural and anthropogenic signals. Large-scale *in situ* manipulation experiments and mesocosm studies are also recommended as well as the application of molecular and genetic techniques that are a powerful means to investigate physiological and biogeochemical transformations that drive the oceans's response to climate change.

The response of ocean biogeochemistry to global change

Recent confirmation of global warming of the planet (IPCC 2001) will lead to changes in the oceans in coming decades to centuries. Increased thermal stratification as well as changes of the deep thermohaline circulation will change the supply of major nutrients (nitrate, phosphate, silicate) and trace nutrients (iron, cobalt, manganese, zinc, copper) from deep to surface waters. This will cause shifts in the biological diversity of the plankton ecosystems in various ocean regions. For example the relative dominance of major phytoplankton groups (diatoms, *Phaeocystis* spp., calcifiers, nitrogen-fix-

From WEFER G, LAMY F, MANTOURA F (eds), 2003, *Marine Science Frontiers for Europe.* Springer-Verlag Berlin Heidelberg New York Tokyo, pp 147-162

ers and picoplankton) is known to depend on limited availability of the suite of major and trace nutrients. It is now commonly realized that growth and abundance of plankton species are regulated by co-limitations of major and trace nutrient elements, next to regulation by the light spectrum (visible and UV), temperature and grazing pressure.

The considerable (~31 %) increase of atmospheric CO_2 concentrations due to human activity is well known, and various other greenhouse gases are also changing. The exchanges of these gases between air and sea are strongly linked with the seasonality and diversity of the plankton community. Hence forecasting of atmospheric trace gas levels and concomitant warming of the planet depends strongly on our understanding and ability to predict changes in marine ecosystems. For the major trace gases carbon dioxide, methane and nitrous oxide, estuaries and coastal seas are significant sources and sinks compared to the atmosphere. There are other natural biogenic trace gases, such as DMS and halocarbons, for which production pathways, major source regions and fate in the atmosphere are only beginning to be realized. Moreover, levels of metals have increased both in the open ocean (Hg, Pb), coastal seas (Zn, Cu, Cd) and estuaries due to pollution inputs during previous decades and centuries. This has lead to shifts in plankton diversity. Furthermore, the increasing human population, land use change and terrestrial climate shifts alter the inputs of organic matter, fertilizer nutrients and iron-rich dust into coastal seas and oceans.

Large scale Ocean Biogeochemical Climate Models (OBCM) can now predict future changes in the oceanic carbon cycle and CO_2 exchanges with the atmosphere. Such models link to ocean circulation and analyse changes with simplified descriptions of biological production fuelled by a single nutrient (e.g. phosphorus). Currently efforts are underway to improve these models by including a more realistic plankton ecosystem involving five major bloom-forming groups and including co-limitation by several major and trace nutrients (EU IRONAGES, de Baar and Croot 2001). Modelling the shift from a recycling, retention food web to productive blooms with subsequent export of matter into the deep-sea is clearly a key step. Improved realistic OBCM's are required for predicting global change not only due to atmospheric warming and increasing CO_2, but also for other changes such as iron-rich dust input and pollutants other than fossil fuel CO_2.

Decadal to centennial changes are superimposed on natural variations of the global marine biosphere over short intervals of days (e.g. weather) and seasons, but also over longer timescales up to the 20,000-100,000 year periodicities of the glacial-interglacial cycles (Petit et al. 1999). All these timescales of natural global change require much attention in order to adequately distinguish anthropogenic factors from natural changes in climate.

Climate models predict significant changes to surface properties of the oceans during the 21st century. In the IPCC scenario, surface water pCO_2 concentrations will continue to increase from present values to 750 µAtm, while the pH will drop from 8.2 to 7.8. At the same time, the abundance of carbonate ions (CO_3^{-2}) will drop to half the pre-industrial value. There is less certainty for other factors that influence climate including atmospheric temperature, which for Europe will depend critically on the patterns of thermohaline circulation. Irrespective of whether thermohaline circulation in the North Atlantic will partly or completely shut down, vertical stratification in the surface ocean is likely to increase as the planet warms. Aeolian input to the ocean may also change. In the short term, predicted increasing wind strengths may generate greater capability to deposit dust into the ocean. However, the sources of this dust may alter as land use changes. This may result in biological stabilisa-tion of present deserts, thus decreasing dust transport as well as the formation of new deserts in other regions.

Superimposed on such global issues, more local and regional influences will be important in the European context. Europe is characterised both by a long coastline and by a large number of regional seas, and the effects of global change are likely to be more strongly felt in these regions. One particular concern in coastal waters is that ambient O_2 levels are predicted to decrease in a warmer climate. This will have important consequences for mankind directly (via water quality and health) and indirectly (via fisheries). Climatic warming in the

Arctic and Baltic seas will result in the reduction of sea-ice with concomitant effects on stratification and vertical mixing. Sea-level rises will also have local effects on water column stratification and fronts, and on the erosion versus deposition patterns in coastal waters.

The role that biology plays in this scenario is complex. Biological communities both drive biogeochemistry, and respond to environmental changes that occur. It is the capacity for ocean biogeochemistry, mediated by the plankton community composition, to respond to global changes that lies at the heart of this challenge. Given the fundamental role played by biological communities in biogeochemical cycles, it is important to understand how communities respond to the predicted global change scenarios. Current predictions are based on the fundamental assumption that biogeochemical functions and activities are distributed according to the biogeography of pelagic communities and that they respond in a predictable manner to environmental change. There is little evidence that this is necessarily true. There may be major geographical shifts in communities with consequential changes in biogeochemical cycling at the regional level. Also, fast reproducing microbial populations may be able to adapt genetically to new environmental conditions, thereby further contributing to the re-distribution of biogeochemical functions and activities.

The concern is that we do not know how major biogeochemical changes are mediated and in some cases even the sign of change under projected climatic scenarios is unclear. An example is the estimation of primary production, a fundamental biogeochemical process. On the one hand, increased stratification will enhance availability of light energy for photosynthesis, in for example the Southern Ocean, a key region for ocean/atmosphere distribution of CO_2 and DMS. On the other hand, stratification will simultaneously reduce deep-water nutrient supply to the euphotic zone, and so reduce net production. However, reduced nutrient supply may be compensated for or even exceeded in localities receiving significant local inputs of terrigenic material from estuarine or aeolian sources. The net effect of these multiple controls on primary production remains unknown. It is unclear whether such shifts as well as the type of nutrient supply will alter phytoplankton community structure and so alter biogeochemical cycling.

Another fundamental biogeochemical process associated with plankton production is calcification (Buitenhuis et al. 1999). This is very sensitive to pH effects and the predicted drop to pH 7.8 could reduce bio-calcification in surface waters by 20-30% (Riebesell et al. 2000). This is due to the decrease in carbonate ion (CO_3^{2-}) concentrations that are required for formation of calcium carbonate ($CaCO_3$). Calcification is of global significance but is restricted to only a few biological groups. The inevitable shifts in their community composition due to global CO_2 increase would not only affect calcification itself, but also influence the net uptake (i.e. net sequestration) of CO_2 by the oceans (Antia et al. 2001).

Although carbon remains the element of reference in climatic change scenarios due to the large anthropogenic effects on atmospheric CO_2, it is now abundantly clear that the cycles of other greenhouse gasses as well as major and trace elements are inextricably linked and form feedback and control mechanisms that interact with and drive global climate change. Thus, changes in upper ocean conditions may affect the chemical speciation and availability of bio-essential trace metals such as iron, zinc, cobalt and manganese. Such metals are of central importance to the biogeochemical function of the oceanic biota with respect to CO_2 uptake. In summary, we are only beginning to realize the mechanisms and feedback effects with biogeochemical consequences associated with IPCC-predicted change scenarios. Drawing on position papers available at the start of the workshop we synthesize below our knowledge and views into three major themes each concluded by some key issues, as well as recommendations for implementation. This group report has relied heavily on the available position papers by Wollast (this volume), Wallace (this volume), Wassmann et al (this volume), and de Baar et al. (this volume).

Key issues and approaches

To address future global change scenarios and to achieve satisfactory representation of biogeochemical cycles, we need to better understand the responses of marine plankton communities to those factors that are predicted to change.

To achieve this will require experimental studies to determine the consequences of changing climate on key functional groups of organisms, and studies on biogeochemical processes within natural planktonic communities. These will involve manipulation experiments in which environmental variables are changed singly and in concert. This type of manipulation of natural systems has been carried out successfully on marine iron fertilisation in which European scientists have played a major role. One goal might be to generate marine "ecotrons" that are equivalent to those used so successfully by terrestrial ecologists to address the biological drivers and consequences of global change scenarios.

Plankton metabolism and biogeochemical cycling

Biogeochemical cycling is mediated by the physiological response of organisms to environmental conditions, and occurs primarily at the molecular level. So far, however, biogeochemical processes have been considered as chemical (enzymatic) processes without much knowledge of the taxa involved. Only recently, the application of molecular biological techniques has resulted in the discovery of novel organism groups and turnover processes. Examples of this include methane reducing bacterial consortia (Boetius et al. 2000), pelagic autotrophic bacteria (Kolber et al. 2000) as well as Archaea (Karner et al. 2001). These discoveries raise fundamental new questions on the diversity and function of pelagic systems and their capacity to respond to environmental change. We are only now perceiving the immense potential and diversity of microbial organisms whose mere existence, let alone function, was previously unknown.

Succession alters biogeochemical ecosystem function

Although we recognise that alterations in species composition have a major impact on marine elemental cycling, we understand little about the forces that drive succession within plankton eco-

systems. 1. For example, despite the concerted efforts of the Global *Emiliania Huxleyi* Modelling (GEM) project (Westbroek et al. 1994), there is still no clear understanding of or ability to predict why, where and when blooms of the coccolithophore *Emiliania huxleyi* occur. We understand that diatoms require silicate for their opal frustules, but only recently it was shown that silification is sensitive to Fe enrichment (de Baar and Boyd 2000; Timmermans et al. 2001). We also do not understand the controls on major bloom-forming groups such as diatoms, *Phaeocystis* or coccolithophorids. Major functional groups of plankton such as calcifiers, diatoms, and nitrogen fixers, have distinct biogeochemical functions, notably in cycling of chemical elements (e.g. C, N, Si, Fe) and molecules (e.g. DMS, N_2, N_2O, CH_3Br). Changes in the occurrence and function of these groups take place in response to changes in their environment. Yet the triggers and dependencies between these groups and the essential elements remain largely unidentified. Information in long-term changes in species composition are rare, although a decadal shift in the relative abundance of plankton groups has been recorded in the Sargasso Sea (Deuser et al. 1995).

Selectivity, that drives succession, is a response to altered conditions including predation pressure and other top-down controls. Microbial communities with their high turnover rates are recognised as key biogeochemical drivers. However, the roles of behavioural adaptation and grazing protection as a means of survival optimisation (Smetacek 1999) in these communities are often neglected. Larger organisms that operate at higher trophic levels are also key players. Large zooplankton can be crucial for the function of the oceans' biological pump and operate as gateways in driving the downward fluxes of particulate matter. These larger organisms can shift the balance from "retention" to "export" systems with concomitant CO_2 replenishment from the atmosphere.

Molecular techniques, despite their many advantages, serve to study bottom-up physiological controls that may not always provide clues to understanding succesion and function of ecosystems. Europe has a long tradition of organism-based ecology and this is strongly needed to understand patterns of change in ecosystem structure. In a global change scenario, the diversity and succession of organisms and the shifts between functional groups are expected to both react to, and drive, alterations in the oceans' biogeochemistry.

Key issues:
- Understanding the patterns of succession of taxa that drive biogeochemical cycles.
- Identifying the environmental cues that drive changes in the relative abundance of bloom-forming organisms including top-down and bottom-up control.
- Understanding long-term changes in organism diversity, through analysis of existing time-series data.

Functional diversity: a process-based approach

A process-based view of microbial diversity is required to assess the role of genetic diversity on biogeochemical cycling. In the future, detailed studies will be required to determine what species or consortia of species drive the cycling of elements and nutrients. One focus should be on identifying the environmental cues that control the growth and abundance of the main species and functional groups. Whereas some key fluxes such as photosynthesis are driven by a wide diversity of organisms others, such as N_2 fixation, are mediated by particular species or groups. For marine sulphur emissions, there is still uncertainty as to the key organisms and processes that mediate DMS emission in natural communities. A regrouping of organismal diversity based on biogeochemical function may be required for biogeochemical models. Molecular probes are a powerful means of identifying biogeochemical function through the identification of genetic potential (for example the Nif gene for N_2-fixation), physiological expression of function (for example methane oxidation) and environmental control of elemental cycling (such as iron limitation of diatom growth). Development of these tools further should advance our understanding of the regulation of biogeochemical fluxes at the organism and molecular levels. It should also identify the microniches in which these fluxes are driven. Ultimately, it is these effects that drive the biological response and feedbacks to global climate change.

There is increasing evidence that some functional groups of organisms drive the cycling of multiple elements. Such multitasking of some organisms link different biogeochemical cycles and should form a focus of further investigations. For example, the coccolithophore *Emiliania huxleyi* participates simultaneously in photosynthesis, calcification and DMS release, thereby linking different aspects of the carbon and sulphur cycles.

Key issues:
- Regroup plankton organisms into functional groups based on key roles in elemental cycling.
- Refocus on the molecular/physiological level of interaction between elemental cycles.
- Assess the role of genetic biodiversity on biogeochemical cycling.

The genetic basis of functional diversity

Molecular techniques based on a genomic and proteomic approach, provide excellent opportunities for studying organismal biogeochemical function. At the genome level, identification of key genes and their detection in natural populations can determine the genetic potential for function, as for example the presence of the Nif gene for nitrogen fixation. Although gene expression, i.e. biochemical cell function, can be determined, the rates of protein synthesis cannot yet be quantitatively assessed. There is a need for development of quantitative probes coupled with rapid analytic techniques such as flow cytometry for biogeochemical studies. Once gene sequences for specific function are identified, their expression can be used to conduct systematic investigations of multiple environmental cues that switch gene expression on or off. This provides a powerful means to determine the physiological response to a number of simultaneous stresses (e.g. iron, zinc, light, nitrogen availability). Through development of labelled probes or extraction of genes from natural populations, the species-specific and community physiological response can be determined. An example of this is flavodoxin production induced by iron stress in diatoms (La Roche et al. 1996).

Phytoplankton growth is seen increasingly to be limited by trace elements such as iron as well as major elements such as nitrogen. Transcription and translation of genetic information is intimately linked with availability of zinc fingers; once enzymes are formed they need a suite of metals as cofactors and for structural stability for quaternary protein folding. Additionally the roles of trace metals such as Cu and Cd that are central to cell function are potentially important; multiple limitations of growth through elemental interactions at the cellular level have been shown (zinc/bicarbonate or iron/nitrogen/carbon cycles), and need to be investigated through basic physiological studies. In biogeochem-ical studies, molecular probes could also unravel the compositional complexity of biopolymeric components of dissolved organic matter.

Key issues:
• determination of genetic and functional diversity in natural communities.
• environmental regulation of gene expression including multiple cues.
• development of quantitative methods for translation of gene expression to rates of protein/enzyme synthesis.
• development of quantitative molecular probes for *in situ* studies.

Microhabitats

Many biogeochemical transformation processes occur in microenvironments that are chemically distinct from the surrounding environment. Such microenvironments allow these processes to occur in bulk habitats where they would not be expected. For example, anaerobic or sub-aerobic processes such as methane production occur in the suboxic microenvironments of faecal pellets and aggregates within the oxygenated water column. Organisms may also produce optimal microenvironments for themselves (e.g. colony formation by *Trichodesmium*) or for others (Ploug et al. 1999). Generally, bulk whole-sample approaches have been used to measure biogeochemical fluxes and transformations, yet such bulk approaches are unable to represent the spatial heterogeneity of microhabitats. New methodologies are needed to sample, identify and analyse these habitats and the turnover processes they mediate. Microsensors and genetic/ immunological markers are powerful tools for investigating the colonisation of microhabitats

by multiple organisms as well as their biogeochemical function.

Key issues:
• the recognition and quantification of microhabitats in bulk environments.
• identifying processes within microhabitats and their impact on biogeochemical cycling in sub-optimal surroundings.

The production of trace gasses

Trace biogenic gasses are important components of biogeochemical cycles in the context of the response of oceans to global change. Although many of the organisms involved in the cycling of trace gasses remain unknown, for gases such as DMS, some of the players are recognised. However, even with DMS, there is a fundamental uncertainty in the processes that control the production and destruction of DMSP and DMS. These processes vary widely between taxa and in response to environmental cues such as nitrogen availability and grazing pressure. Degradation of DMSP to DMS occurs primarily through the enzyme DMSP-lyase, yet its production and activity are largely unchar-acterised. It is thus currently impossible to assess the impact of marine biogeochemistry on atmospheric trace gasses. We cannot therefore predict feedback effects under climatic change. For other volatile and climatically reactive biogenic gasses (e.g. halocarbons, N_2O) still less is known as to what organisms /enzymes are involved in their production and release; yet these are potentially important players in an ocean-climate feedback system.

Key issues:
• Identification of the biogenic pathways responsible for production and destruction of climatically reactive volatile compounds
• Identification of the organisms and environmental cues controlling the production and flux of marine volatile biogenic compounds

Biogeochemical linkages between the ocean, atmosphere and climate

The composition of the atmosphere helps regulate the habitability of Earth, including its climate. Biogeochemical processes within the ocean exert strong and varied influences on atmospheric composition, through ocean-atmosphere material exchanges. These exchanges are subject to change in response to environmental perturbation, which can potentially drive feedbacks between climate change and ocean biogeochemistry.

Carbon dioxide exchanges

About 90 GT C yr^{-1} of CO_2 are exchanged annually between the atmosphere and the ocean, and ocean waters contain approximately 50 times more carbon than the atmosphere. Hence ocean biogeochemical processes exert a profound control on atmospheric CO_2. Direct positive feedbacks on future increases in atmospheric CO_2 involving ocean temperature and buffer capacity changes are understood, however a wide range of additional biogeochemical feedbacks are poorly understood (e.g. effects of changes of ocean calcification, stratification, circulation and nutrient supply, trace element fluxes and changes in nitrogen fixation). In addition, the regional, and particularly, the temporal (sub-decadal) variability of air-sea CO_2 fluxes are not presently resolved from measurements, yet are critical to determining the location and magnitude of terrestrial carbon sinks based on the interpretation of atmospheric CO_2 data.

Key issues:
• effect of altered vertical stratification and nutrient inputs from the subsurface ocean (including trace elements) on ecosystem structure and vertical carbon fluxes.
• effect of altered upper ocean conditions and altered nutrient and aolian inputs on oceanic nitrogen fixation.
• magnitude and temporal (seasonal, sub-decadal) variability of regional air-sea carbon fluxes and the factors driving such variability.
• effect of increased CO_2 levels and temperature changes on calcification.

Volatile sulphur compounds

The biogenic production of dimethyl sulphide (DMS) and its flux into the atmosphere is responsible for about 40% of the non-sea-salt sulphate aerosols. Other naturally produced compounds, e.g. carbonylsulfide (COS), play an important role in the ozone chemistry of the stratosphere. A self-regulating feedback involving phytoplankton production, DMS production and flux, oxidation of DMS, the formation of cloud condensation nucleii and cloud albedo has been proposed (Charlson et al. 1987). Evidence for a link between atmospheric DMS concentrations and the numbers of cloud condensation nucleii has emerged from long time-series observations at Cape Grim (Ayers et al. 2000), and from short-term observations in the tropical Pacific and the tropical Atlantic (Andreae et al. 1995; Clarke et al. 1998). In both cases, the observed correlations have been inferred to result from a causal relationship.

Key issues:
• development of deterministic models for the concentration of DMS and COS in the upper ocean; controlling factors include photochemistry, vertical mixing, composition of dissolved organic material, biological production and consumption reactions.
• assessment of the present-day magnitude of air-sea fluxes, particularly their regional and seasonal variability.
• development of models and experimental approaches designed to evaluate the response of volatile sulphur emissions to climate-driven changes in temperature, vertical mixing, light regimes and nutrient inputs.

Nitrous oxide

Nitrous oxide (N_2O) is a major greenhouse gas and plays a key role in stratospheric ozone cycling. Its atmospheric concentration is currently increasing, and ice-core records show that atmospheric N_2O levels are climate-sensitive. N_2O has a strong oceanic source, particularly in coastal regions. Isotopic measurements in the troposphere suggest an oceanic source associated with near-surface nitrification. The sensitivity of the coastal N_2O sources to climate change is unknown.

Key issues:
• improved characterisation of the magnitude of the ocean-atmosphere N_2O flux and its isotopic composition.

• identification of processes responsible for coastal N_2O sources. Assessment of their sensitivity to climate change and more direct human impacts including eutrophication.

Volatile halocarbons

Volatile halocarbons, both anthropogenic and natural, (e.g. CFC's, Halogens, Methybromide, Chloroform) are integrated into natural biological cycles. Notably the ocean acts as a source (or sink) for a variety of volatile halogenated organic compounds. Breakdown of these gases in the atmosphere releases reactive halogen species that are important in tropospheric and stratospheric chemical reactions, including ozone destruction cycles. Although most reactive Cl in the stratosphere is of anthropogenic origin, the natural background level is significant (e.g. ~20% derived from CH_3Cl). Recent measurements and modelling of the atmosphere has suggested that reactive Br derived from short-lived trace gases produced in the oceans (e.g. $CHBr_3$) may play a greater-than-expected role in stratospheric chemical reactions involving ozone.

Key issues:
• identify the main regions for air-sea transfers of natural and anthropogenic halocarbons.
• identify the balance between coastal vs. open ocean for volatile halocarbons.
• determine the seasonal cycle for volatile halocarbons.
• identify key biotic and abiotic pathways responsible for the production and degradation of these compounds.
• identify the key enzymes involved and their rates and controlling factors.

Fixed nitrogen deposition

Significant quantities of fixed nitrogen are delivered from land to the ocean via the atmosphere. Atmospheric wet and dry deposition includes nitrate, nitrite and ammonium. Much of this is from anthropogenic sources (combustion of fuels, utilization of fertilizers, animal and human waste, etc.). Delivery of atmospheric nitrogen to coastal regions in Europe and North America is estimated to have increased four-fold during the past 50 years (Paerl and Fogel 1994), leading to considerable enhance-ment of the coastal eutrophication that was caused originally by riverine outflow, direct wastewater discharge and storm water runoff. Atmospheric deposition is a major term for the nitrogen budgets of all European marginal seas. There is also concern over the increasing input of human-derived nitrogen to the open ocean where nitrogen is also often a limiting nutrient. Although quantitatively smaller than coastal inputs, atmospheric inputs to the open ocean are highly episodic and concentrated, and may drive short-term biogeochemical changes in excess of what might be expected from their long term average input. Deposition of organic forms of nitrogen may be equal to or perhaps greater than the inorganic nitrogen flux in the pelagic ocean (Cornell et al. 1995).

Key Issues:
• improve characterization of the chemical forms of nitrogen deposited at the air-sea interface and their biological availability.
• quantify the deposition of nitrogen in a range of marine ecosystems.
• determine the response of marine ecosystems to episodic N-deposition (e.g. productivity and species composition changes).

Impact of trace metal and mineral deposition on marine processes

The impact of trace metals on the biology and chemistry of the upper ocean has several facets. First, recent research has revealed that certain metals (notably Fe, Zn, Mn, Cu) play important roles in marine productivity and there is no doubt today that iron acts as a limiting trace nutrient in 40 % of the oceans, the High Nutrient Low Chlorophyll (HNLC) regions (de Baar and Boyd 2000). In addition, there is also evidence that N_2-fixation in oligotrophic sea regions may be enhanced by iron which is mainly supplied by dust deposition. Secondly, the exponential increase in mining and industrial application of metals (such as Pb, Cu, Zn, Hg, Sn) and their emission to the atmosphere have produced a global impact in the surface ocean where these metals naturally used to occur only at ultra-trace levels. Lead with its already decreasing levels in the open Atlantic due to the phasing-out of leaded gasoline is a good example for ocean-wide pollution via long-range atmospheric transport. Yet it is also most encouraging with respect to societal action due to the now almost complete switch to unleaded gasoline. However for other pollutant metals (e.g. Hg, Zn, Cu) emissions (legal and illegal) remain high or are still increasing. Apart from the aeolian input route affecting all of the northern hemisphere, there is also substantial input via rivers affecting mostly the regional estuaries and coastal seas.

In spite of their seasonally independent input routes, there is long-term evidence for a joint sedimentation of aluminosilicates and organic matter. Thus, a further impact on the carbon cycle may be exerted by aggregation of the refractory dust grains with organic matter thus influencing the relation between retention and sinking out of particulate matter.

Key issues
• identify the differences between the bioavailabilities of trace metal in the surface ocean and the requirements by marine organisms.
• examine the relationship between N_2-fixation in the open Atlantic and atmospheric deposition of iron
• examine the relationship between the availability of dissolved iron in the surface ocean and its increased solubilization due to pH decrease in the atmosphere and in the surface ocean due to anthropogenic emissions.
• examine the role of mineral particles as ballast in the sinking of organic material and its contribution to export fluxes.

Biogeochemical reactivity of the land-ocean interface

Estuaries, shelf seas and ocean margins are dynamically coupled and biogeochemically reactive highways for the transportation, transformation and export of terrestrial and anthropogenic products into the ocean.

Europe's estuaries span a wide range of sizes (10-1000 km, rias, fjords, lagoons, inland seas), flushing times (<10 day to > 1000 days) and biogeochemical reactivities (turbid, stratified, anoxic, eutrophic, acidic). These sizes and time constants have profound consequences on the cycling of carbon, nitrogen, trace elements and pollutants. It

is becoming increasingly possible to reliably estimate river fluxes of elements and pollutants from modelling of catchment-based industrial, domestic and natural weathering fluxes. However the downstream transformation of these fluxes and fate of substances in estuaries and shelf seas is generally poorly understood. Estuaries are biogeochemical reactors. For example (1) intense bacterial degradation of organic matter not only depletes oxygen but may also result in outgassing of vast amounts of CO_2, which for some European estuaries could amount to 5-10 % of the regional emissions from fossil fuel burning (Frankignoulle et al. 1998), (2) macrotidal estuaries exhibit steep gradients in pH, redox, ionic strength and turbidity which trigger numerous non-conservative flocculation and sorption reactions affecting trace metals and hydrophobic organic pollutants and (3) the high nitrate and phosphate levels in European estuaries often stimulate nuisance blooms, with or without toxicity, in coastal marine waters.

Biogeochemical transformation of nitrogen compounds (e.g. N-assimilation, nitrification, denitrification, N_2 fixation) in estuaries is complex, non-conservative and varies between different estuaries and shelf waters. However, in an overview of Atlantic estuaries by Nixon et al. (1996) a clear pattern was evident between the percentage of NO_3 lost by denitrification and the water residence times. Recent studies on the NW European Margin indicate that > 80 % of shelf wide phytoplankton production is supported by cross shelf inflow of oceanic NO_3. However within the coastal and estuarine zones, primary production is almost entirely fuelled by regional inputs of riverine nutrients.

On a global level, we still do not know the molecular composition, reactivity and fate of the massive inputs (0.5-1.0 GT C/year) of terrestrial dissolved and particulate organic carbon (DOC, POC) and nitrogen (DON, PON) to the marine environment. Recent mapping of pCO_2 suggests that tha North Sea is net heterotrophic (i.e. C degradation exceeds production) indicating that an unknown proportion of terrestrial DOC and POC is degradable within the shelf flushing time. Photodegradation also appears to be an important sink for riverine DOC in coastal and offshore waters.

Towards a shelf typology

On the gradient from estuaries to the ocean margins, European continental margins exhibit a sequence of several distinct biogeochemical sub-provinces bounded within hydrodynamically distinct regimes. For example, in the case of a transect from the Southern Bight of the North Sea westwards to the Atlantic Ocean, both remote sensing and models show a clearly delineated sequence of (1) estuarine plumes driven by episodic flushing of estuarine sediments, (2) tidally-mixed and light-limited coastal regimes, (3) tidal fronts, (4) seasonally stratified deeper waters, and (5) shelf edge up-welling zones. The planktonic foodweb, nutrients and trace metal biogeochemistry and the degree of bentho-pelagic coupling are unique and reproducible within each shelf province. In microtidal and narrow shelf regimes (e.g. Mediterranean) a simpler system of biogeochemical sub-provinces may occur. This 'biogeochemical typology' proposed for shelf seas could form the basis of a new functional ordination scheme which could be nested within hydrodynamic modelling, and which would find applications to each of the European systems (estuarine to shelf seas to margins). Provided we have critical observations of biogeochemical concentrations, rates and parameters, then societally-important scenarios (e.g. eutrophication, pollution inputs, climate change drivers and ecosystem restructuring) should be addressable in a consistent way.

   Key issues
• delineation of shelf and coastal typology according to biogeochemical function..
• identification of regional and functional "hot-spots" of relevance to European society and their susceptibility to change under projected climatic scenarios.

Margins

Recent biogeochemical and oceanographic flux studies at three stations along the N E Atlantic margins identified mid-water degradation as the primary sink of organic matter. Unlike the W Atlantic margins, cross slope fluxes were weak and the slopes were not acting as 'depocentres' for carbon. On the other hand, margins characterised by upwelling (e.g. Iberian) or cascading (Norwe-gian trench) or deltaic (Rhone) regimes showed significant sedimentation and burial of organic matter. The fundamental question of how much, and where, shelf and slope carbon is degraded or buried remains unanswered; yet it is critical in closing the oceanic budget of carbon. The recent discovery of methane oxidising microbial mats along the Oregon slopes rich in methane clathrates and seeps may be a more widespread phenomenon which could significantly contribute to benthic productivity. Future research must focus on (1) unravelling the biogeochemical and physical processes (flocculation, burial, advection, microbial or photo-degradation) controlling the fate of terrestrial, oceanic and anthropogenic matter, and (2) applying novel techniques (e.g. LC-IR-MS, isotopic and molecular probes) to characterise the molecular complexity and reactivity of organic matter.

   Key issues
• determine the potential for atmospheric CO_2 sequestration at the European continental margins.
• identify key margin types including canyons and upwelling systems and characterisation of their biogeochemical functioning.

Trace metals

Recent trace metal studies have revealed for the first time that many of the bio-essential trace metals (e.g. Fe, Zn) exist at ultra low levels (< 1 nM) in the oceans and therefore limit marine productivity. However, at modest levels (10-100 nM) found in coastal waters some trace metals may be toxic (e.g. Cu, Cd). This leads us to hypothesise that present day estuarine and shelf ecosystems which are chronically exposed to modest to high levels of trace metals (10^2-10^4 nM) operate under long term trace metal controls, originating from the massive increase in riverine and atmospheric inputs of trace metal since the industrial revolution. Well before the ecology of coastal seas and estuaries was studied systematically their ecosystem may already have been altered by human inputs of metals.

   Key issues.
• unravel the link between speciation chemistry and microbial availability of trace metals.

• bioessential trace metals should now be routinely co-profiled (using ultraclean rapid shipboard FI-CL techniques) with major nutrients in tracking the biogeochemical status and controls on marine productivity and carbon fluxes in estuaries, coastal seas and the open ocean.

Implementation

To address the various key issues above, a wide variety of existing approaches are required. These should be supplemented with the development of new methods tailored to the issues involved. Rather than listing all of these, an effort is here made to focus on those selected approaches which have an optimal benefit/cost ratio. Sometimes this happens to be by advocating continuation of existing strong efforts or expertise within Europe, at other times by proposing new lines of research and development.

The key issues identified above can be addressed by a common set of approaches. These include:

Experimental studies at all scales

To identify ecosystem controlling factors and determine rates of key processes, a strong emphasis on experimental studies is required. Relevant experimental approaches range from laboratory studies (e.g. determination of rate constants for biotic reactions using culture studies) through mesocosm experiments with multiple communities to open-ocean, whole-ecosystem manipulation experiments as recently applied with great success to examine the effect of iron enrichment in marine ecosystems. European scientists have been pioneers (Watson et al. 1991) in the development and application of mesocosm and whole-ecosystem manipulation techniques in marine biogeochemistry (Cooper et al. 1996; Watson et al. 2000).

Recommendations for implementation: Europe has a rich tradition in marine mesocosm studies. The new strategy of biogeochemical perturbation experiments (e.g. SOIREE, EU-CARUSO, EU-CYCLOPS) within an SF_6 labelled Lagrangian patch of open seawater is a powerful approach for tracking whole ecosystem responses to the perturbation. We recommend that European marine science:

i) maintains and expands existing coastal experimental facilities such as the Mesocosm facility at Bergen. Such facilities are essential for internationally coordinated experimental studies of the inherent complexity of plankton ecosystems

ii) develops new capability for experiments on open ocean plankton. This would be a natural extension to the Bergen mesocosm concept to open water communities. Such experimental facilities would require manipulation of environmental variables that can be changed singly and in concert. This type of manipulation on natural systems has been carried out successfully with *in situ* iron fertilisation in which European scientists have played a major role. Marine scientists would benefit from working with terrestrial ecologists who have used "ecotrons" so successfully to address the drivers and consequences of global change scenarios.

Time-series and moorings

Critical to several of the key issues identified above are issues relating to temporal variability, including seasonal variability. Studies at fixed-location, time-series sites have already provided a wealth of information and important new ideas concerning the sensitivity of key biogeochemical processes and rates to environmental perturbations. Time-series sites, such as the European-supported time-series site at ESTOC, offer the major benefit of conducting intensive and complex studies of biogeochemical processes in the context of a continuous time-series of oceanographic conditions. Such sites should be at minimum be maintained and preferably extended, for example through the establishment of high-quality, biogeochemical time-series in a high-latitude region and/or a European marginal sea. In order to support research into atmosphere-ocean linkages, wherever possible combined time-series observations of both the atmosphere and the ocean should be established and maintained. It is important to state clearly that, by their nature, the funding of time series stations must be long term, and this is at variance with contemporary European policy.

Recommendations for implementation: For long-term recordings of major variables in remote areas with good temporal resolution moored systems are indispensable. Such platforms are equipped with sensors, automated measuring systems, sampling and possibly profiling systems from the ocean surface to the deep-sea. Such moored systems at strategic locations should form platforms for the deployment of new sensors and conducting process studies. They would also generate data on the range of natural variability. The philosophy should be to:

i) maintain existing time-series stations and add measurements of new key indicators of change. Here the focus would be to identify key processes and functional groups of organisms prior to measuring them at time-series stations.

ii) initiate a northern European time series station at high latitude (ca 72°N) associated with deep-water formation. This would capitalise on the expertise of the Norwegian scientific community. Combine this with trans-Atlantic VOS measurements to generate time series of key biogeochemical measurements to monitor how the Atlantic is changing.

iii) Include existing historical time series. Lessons can be learnt from the past using for instance high-resolution analysis of sedimentary archives. These have the advantage that they can cover wider spatial areas of interest with less long-term investment.

Volunteer observing ships (VOS)

Biogeochemical research in the oceans has been conducted primarily from specialised research vessels, and more recently, from fixed-point time-series and moored sensor systems. Research vessels offer many advantages but are limited in terms of their temporal and spatial coverage, and therefore frequently miss episodic but important biogeochemical events. Moored systems have restricted spatial coverage and more important only a limited number of biogeochemical measurements can be made. Morever, the prospects of developing mooring-based systems for many of the key, modern biogeochemical tools and measurement approaches are technologically remote (e.g. mass spectrometry, molecular biological techniques, flow cytometry).

A previously under-utilised platform for biogeochemical research is the ship-of-opportunity or volunteer-observing-ship (VOS). In the North Atlantic Ocean and the European marginal seas, commercial shipping activity is intense, and covers large spatial scales (e.g. Europe to North America) with high frequency (<1 month). They are therefore one of the few platforms that can provide sample coverage on scales that can match the data collected by satellites. An increasing number of satellite missions are being planned for the measurement of biogeochemical properties in both the surface ocean (ocean colour sensors) and atmosphere (trace atmospheric species, including CO_2). VOS could be used to provide ground truthing data for the bio-geochemical properties of both the atmosphere and the ocean.

Major advantages of VOS for biogeochemical research include sampling at a wide-range of spatial scales (1-1000km) with moderately high frequency (<1 month). There are few restrictions on deployable measurement technologies. It is important that endurance requirements of equipment must be only of the order of weeks due to frequent port-calls. The risks of equipment loss are much less than with moored-systems making it possible to deploy expensive measurement systems. It is also possible to achieve human interaction with the measurement systems. Communication of data to shore is straightforward, and water samples can be collected and returned to shore-laboratories for variables that cannot be measured at sea.

There are some disadvantages in the use of VOS. These including difficult sampling of the subsurface ocean. Shipping routes are variable and do not always coincide with areas of highest scientific interest. Commercial vessels are not designed for scientific installations and some vessel modifications may be required. Commercial vessels are frequently sold, replaced, or reassigned to different routes. Scientific installations may disrupt ship operations for which the companies obtain no commercial compensation.

Recommendations for implementation: Recommendations for VOS would include the encouragement of shipping companies and naval architects to have science capability "built-in" at the design stage of new commercial vessels. Co-operative shipping companies could be given an incentive for supporting scientific research (e.g. tax benefits). Pilot studies for the adaptation of biogeochemical measurements to commercial ship operations should be encouraged, including immediate utilization of the existing European research vessel fleet for such observations.

New biochemical technology

For understanding plankton ecosystems and their links with global changes the implementation of existing and development of novel molecular techniques and assays is likely to yield crucial new insights into marine ecology and biogeochemical function. Moreover for experimentation at all scales, for time series observations on either fixed stations or Voluntary Observing Ships there is a great potential for use of both molecular probes as well as reliable, accurate sensors/analysers for a suite of biological and chemical variables. For example when considering the central importance of organic matter in marine biogeochemistry, biologists and chemists should be stimulated to develop the new analytical techniques to unravel the sources, molecular complexity and reactivity of dissolved and particulate organic matter in the sea.

Recommendations for implementation: The approaches described above require new technology to be applied for experiments, time series stations and VOS. One new technological approach is to develop and apply functional probes of genes and enzymes. However, there is also a requirement to develop and apply fast reliable sensors/analysers for a range of biological and chemical variables to be measured routinely at time series stations as well as ships of opportunity. Such variables should include probes for functional genes and enzymes, whenever possible.

The desired functional gene/enzyme probes may be summarized as follows.
i) Probes specific to biogeochemical function are required to characterise key organisms and processes. This will require a strong interface with biochemists and molecular biologists.
ii) The development of new techniques for rapid, quantitative assay of key rates such as with fluorescent hybridisation and immunological probes coupled with flow cytometry

Global and regional focus

With respect to global change in coming decades to centuries there is good scientific and strategic reason to pay attention to the North Atlantic, as well as the Polar Oceans. The North Atlantic in conjunction with the Arctic is pivotal for the conveyor belt of deep-ocean circulation. The North Atlantic is easily accessible also for joint research with North American counterparts. The Arctic Ocean is already experiencing change in sea-ice cover and circulation and is a long-term research area of the Nordic European countries. The Antarctic Ocean is pivotal in global climate change (Lancelot et al. 2000; Moore et al. 2000; Watson et al. 2000). Strong national Antarctic research programmes exist and will continue in major European nations, allowing partnerships in ocean biogeochemical and climate research within the European Union.

Among the regional seas, Europe has excellent opportunities to focus on the Baltic, North Sea, Black Sea, English Channel, Alboran Sea, Biscay, Norwegian Trench, Adriatic and Aegian Seas. These seas can be assigned Marine European Research Areas (ERA's) or 'chantiers' for integration of scientific research and stakeholder interests to achieve informed policy-making and sustainable development.

Concluding remarks

We anticipate considerable development in novel sensor technology to understand global changes in the ocean. Future global change of the surface ocean and lower atmosphere needs to be understood in context of the natural time scales of oscillations (seasonal, interannual to decadal), and over spatial ranges as large as ocean basins. This sampling of the surface ocean is also the ideal counterpart to satellite observations with a typical spatial resolution of one by one nautical mile. Long term stability and accuracy is at a premium in order to be able to distinguish long term trends. For example, the annual increase of pCO_2 in the atmosphere corresponds to an annual increase of about 1 micromole dissolved inorganic carbon (DIC) in the surface ocean, with a natural background of about 2000 micromoles. The parallel decrease of pH will also only be detectable with high accuracy detection at or beyond the third decimal pH unit. Much progress has been made on oceanographic pCO_2 and pH sensor/analyser packages where Europe has a leading position. Sensors for oxygen do exist but need regular servicing; existing analyser packages for major nutrients also need attendance.

The annual cycle of plankton blooms does have a greater amplitude than that of DIC and pH, but still requires accuracy in measurement. The monsoon drives seasonal upwelling of major (N, P, Si) and trace (Fe, Zn, Cu) nutrients but also the aeolian input of iron-rich dust from regions such as the Sahara. Fast repetition rate fluorometry (FRRF) and flow cytometry have been successful in providing virtual real-time information on cellular status as well as species composition of phytoplankton. For several other key variables like dissolved iron, zinc and other bio-essential metals, the dissolved organic carbon (DOC) pool or its molecular components, plankton molecular indicators (photopigments, DNA, enzymes) the development of sensors/analysers is required. Modern solid-state technology, miniaturized flow injection on microchips, true innovation, novel applications of physical principles (optics, laser optics, photoacoustics) to biochemical detection, joint ventures with bio/chemical entrepreneurs and analytical companies, and linkages with the fields of biotechnology, life sciences and biomedical research are envisioned.

Underway sampling from ships of opportunity (VOS) at high hull speeds of 10-25 knots has proven possible for surface waters, and at somewhat lower hull speeds also for the upper 200-500 metres water column by an oscillating platform (Seasoar, Scanfish) from which water is pumped to the ship. Recent examples are the successful use of a pumping torpedo towed at 15 knots with in-line filtration for continuous detection of dissolved iron, and high- rate data collection for DIC, pCO_2 and pH using an oscillating platform towed across the Ross Sea. However what is missing are the various sensors/analysers for collecting accurate data at a sufficiently high rate. Comparing such data with the ships hull speed one should be able to resolve at the same nautical mile resolution of a satellite, or even the 10-100 metres resolution of Langmuir circulation. Submersible sensor/analysers (see below) are conceivable but not

necessary. Shipboard sensors/analysers have the advantage of avoiding the harsh conditions within the sea, as well as allowing regular routine calibration, and checks on continuous accuracy and malfunctioning by a shipboard engineer or technician. For gradients in the lower atmosphere one envisions sensors and sampling from the ship superstructure. Pumping of air via filters for aerosol collection or into laboratories for trace gas analyses is already routine practice, but fast analysers are still the exception. For greater height above the air/sea interface a dedicated mast or a towed air balloon/kite platform can be used. Design of a latter type platform is a valid objective where its payload can be used to carry camera systems for looking down at or partly into (LIDAR) the sea, or analyser packages, as well as a suite of inlets along the vertical towards analysers on the ship itself.

For the existing ocean time series observation stations (e.g. ESTOC, BATS, HOTS) and various drifter buoy approaches, proven technology exists for moorings, assimilation and transmission of data via electronic conduits, satellites and computer networks. This has proven very useful recently for the few field parameters (conductivity, temperature, pressure, light spectrum) for which reliable sensors exist. However, here is also a need for development of rugged, reliable *in situ* sensors or analysers for a suite of biological and chemical parameters. To summarise, we can transfer data, but have little data to transfer! For the key variables mentioned above and for other parameters of ocean biology and chemistry, it is important to seek innovative development of techniques. Design and testing requirements must take into account the need for low power requirements, avoidance of biofouling, and ruggedness for long term deployment in a harsh, high pressure, corrosive environment. Development and testing could initially be done at existing time series observation programs like ESTOC. In coastal regions such as the North Sea, good opportunities exist and have occasionally been used on natural oil and gas production platforms. Newly developed sensors/analysers will be directly beneficial for providing GOOS observational approaches with key climate variables next to existing reliable physical parameters (temperature, conductivity, pressure).

Acknowledgements

The authors are most grateful and honoured at having been invited to the workshop Marine Scientific Frontiers for Europe, held at Bremen, 18-22 February 2001. Special thanks are for the hosts at Bremen for the excellent organization, hospitality and support in a most elegant environment. For drafting the above synthesis report, the various advance position papers published in this volume were most helpful. Proof-reading, discussions and contributions from other workshop participants W. Berger, A. Brandt, J. Boissonnas, L. D'Ozouville, G. Graf, J.-P. Gattuso, C. Heip, G. Herndl, R. Laane, J.W. de Leeuw, J. Patching, D. Thomas, P. Wassmann and V. Smetacek are gratefully acknowledged.

References

Andreae MO, Elbert W and DeMora SJ (1995) Biogenic sulfur emissions and aerosols over the Tropical South Atlantic. 3. Atmospheric dimethylsulfide, aerosols and cloud condensation nuclei. J Geophys Res 100(D6):11,335-11,356

Antia AN, Koeve W, Fischer G, Blanz T, Schulz-Bull D, Scholten J, Peinert R, Neuer S, Kremling K, Kuss J, Hebbeln D, Bathmann U, Conte M, Fehner U and Zeitzschel B (2001) Basin-wide particulate carbon flux in the Atlantic Ocean: Regional export patterns and potential for atmospheric CO_2 sequestration. Glob Biogeochem Cycl 15(4):845-862

Ayers G-P, Cainey JM, Gillet RW, Saltzman ES and Hooper M (1987) Sulfur dioxide and dimethlysulfide in marine air at Cape Grim, Tasmania. Tellus 49B:292-299

Boetius A, Ravenschlag K, Schubert CJ, Rickert D, Widdel F, Gieseke A, Amann R, Jørgensen BB, Witte U and Pfannkuche O (2000) A marine microbial consortium apparently mediating anaerobic oxidation of methane. Nature 407:623 - 626

Buitenhuis E, de Baar HJW and Veldhuis MJW (1999) Regulation of photosynthesis and calcification of *Emiliania huxleyi* by the different species of dissolved inorganic carbon in seawater. J Phycol 36:64-73

Charlson RJ, Lovelock JE, Andreae MO and Warren SG (1987) Oceanic phytoplankton, atmospheric sulphur, cloud albedo and climate. Nature 326:655 - 661

Clarke AD, Davis D, Kapustin VN, Eisele F, Chen G, Paluch I, Lenschow D, Bandy AR, Thorton D, Moore K, Maudlin L, Tanner D, Litchy M and Albercook G (1998) Particle nucleation in the tropical boundary layer and its coupling to marine sulfur sources. Science 282:89-92

Cooper DJ, Watson AJ and Nightingale PD (1996) Large decrease in ocean-surface CO_2 fugacity in response to *in situ* iron fertilisation. Nature 383:511-13

Cornell S, Rendell A and Jickells T (1995) Atmospheric inputs of dissolved organic nitrogen to the oceans. Nature 376:243-246

de Baar HJW and Boyd PM (2000) The Role of Iron in Plankton Ecology and Carbon Dioxide Transfer of the Global Oceans. In: Hanson RB, Ducklow HW and Field JG (eds) The Dynamic Ocean Carbon Cycle: A Midterm Synthesis of the Joint Global Ocean Flux Study. Chapter 4, International Geosphere Biosphere Programme Book Series, Cambridge University Press pp 61-140

de Baar HJW and Croot P (2001) Iron resources and oceanic nutrients; advancement of global environment simulation (IRONAGES). Abstract at IGBP Open Science Conference, Amsterdam, July 2001

de Baar HJW, La Roche J (2003) Trace Metals in the Ocean: Evolution, Biology and Global Change. In: Wefer G, Lamy F, Mantoura F (eds) Marine Science Frontiers for Europe. Springer, Berlin pp 79-105

Deuser WG, Jickells TD, King P and Commeau D (1995) Decadal and annual changes in biogenic opal and carbonate fluxes to the deep Sargasso Sea. Deep-Sea Res 42:1923-1932

Frankignoulle M, Abril G, Borges A, Bourge I, Canon C, Delille B, Libert E and Theate J-M (1998) Carbon dioxide emission from European estuaries. Science 282:434-436

IPCC (2001) Climate Change: The Scientific Basis. Report of the Intergovernmental Panel on Climate Change

Karner MB, DeLong EF and Karl DM (2001) Archaeal dominance in the mesopelagic zone of the Pacific Ocean. Nature 409:507-510

Kolber ZS, Van Dover CL, Niederman RA and Falkowski PG (2000) Bacterial photosynthesis in surface waters of the open ocean. Nature 407: 177-179

Lancelot C, Hannon E, Becquevort S, Veth C and de Baar HJW (2000) Modeling phytoplankton blooms and related carbon export production in the Southern Ocean: control by light and iron of the Atlantic sector in Austral spring. Deep-Sea Res 47:1621-1662

LaRoche J, Boyd PW, McKay RM and Geider R J (1996) Flavodoxin as an *in situ* marker for iron stress in phytoplankton. Nature 382:802-805

Moore JK, Abbott MR, Rihman JG and Nelson DM (2000) The Southern Ocean at the Last Glacial

Maximum: A strong sink for atmospheric carbon dioxide. Glob Biogeochem Cycl 14:455-475

Nixon SW, Ammerman JW, Atkinson LP, Berounsky VM, Billen G, Boicourt WC, Boynton WR, Church TM, Ditoro DM, Elmgren R, Garber JH, Giblin AE, Jahnke RA, Owens NJP and Pilson M (1996) The fate of nitrogen and phosphorus at the land-sea margin of the North Atlantic Ocean. In: Howarth RW (ed) Nitrogen Cycling in the North Atlantic Ocean and Its Watersheds. Kluwer Academic Publishers, Dordrecht (The Netherlands) pp 41-180

Paerl HW and Fogel ML (1994) Isotopic characterization of atmospheric nitrogen inputs as sources of enhanced primary production in coastal Atlantic Ocean waters. Mar Biol 119:635-645

Petit JR, Jouzel DJ, Raynaud D, Barkov NI, Barnola J-M, Basile I, Bender M, Chapellaz J, Davis M, Delaygue G, Delmotte M, Kotlyakov VM, Legrand M, Lipenkov VY, Lorius C, Pépin L, Ritz C, Saltzman E and Stievenard M (1999) Climate and atmospheric history of the past 420,000 years from the Vostok Ice Core, Antarctica. Nature 399:429-436

Ploug H, Stolte W, Epping EHG and Jørgensen BB (1999) Diffusive boundary layers, photosynthesis, and respiration of the colony-forming alga *Phaeocystis* sp. Limnol Oceanogr 44:1949-1958

Riebesell U, Zondervan I, Rost B, Tortell PD, Zeebe RE and Morel FFM (2000) Reduced calcification of marine plankton in response to increased atmospheric CO_2. Nature 407:364-367

Smetacek V (1999) Diatoms and the ocean carbon cycle. Protist 150:189-206

Timmermans KR, Gerringa LJA, de Baar HJW, van der Wagt B, Veldhuis MJW, de Jong JTM, Croot PL and Boye M (2001) Growth rates of large and small Southern Ocean diatoms in relation to availability of iron in natural seawater. Limnol Oceanogr 46:260-266

Wallace D (2003) Ocean-Atmosphere Exchange and Earth-System Biogeochemistry. In: Wefer G, Lamy F, Mantoura F (eds) Mraine Science Frontiers fpr Europe. Springer, Berlin pp 107-129

Wassmann P, Olli K, Wexels-Riser C, Svensen C (2003) Ecosystem Function, Biodiversity and Vertical Flux Regulation in the Twilight Zone. In: Wefer G, Lamy F, Mantoura F (eds) Marine Science Frontiers fpr Europe. Springer, Berlin pp 279-287

Watson AJ, Bakker DCE, Ridgwell AJ, Boyd PW and Law CS (2000) Effect of iron supply on Southern Ocean CO_2 uptake and implications for glacial atmospheric CO_2. Nature 407:730 - 733

Watson AJ, Liss P and Duce R (1991) Design of a small-scale *in situ* iron fertilization experiment. Limnol Oceanogr 36:1960-1965

Westbroek P, Brown CW, Van Bleiswijk J, Brownlee K, Brummer GJ, Conte M, Egge J, Fernandez E, Jordan R, Knapperbusch M, Stefels J, Veldhuis M, Van der Wal P and Young J (1994) A model system approach to biological climate forcing - the example of *Emiliania huxleyi*. Glob Planet Change 8:27-46

Wollast R (2003) Biogeochemical Processes in Estuaries. In: Wefer G, Lamy F, Mantoura F (eds) Marine Science Frontiers for Europe. Springer, Berlin pp 61-77

Operational Oceanography – the Stimulant for Marine Research in Europe

D. Prandle[1*], J. She[2] and J. Legrand[3]

[1]*Proudman Oceanographic Laboratory, Bidston Observatory, Birkenhead, UK*
[2]*Danish Meteorological Institute, Copenhagen, Denmark*
[3]*Institut Français de Recherche pour l'Exploitation de la Mer (Ifremer), Brest, France*
* corresponding author (e-mail): Dp@pol.ac.uk

Abstract: Goals of the European Marine Research Plan include promoting the quality of, and access to, the science needed for effective Coastal Zone Management – underpinning governance to ensure sustainable exploitation of this invaluable resource. The diversity in nature, usage and hence challenges in European coasts requires fostering of localised scientific expertise with an attendant range of approaches. Operational oceanography links science and decision making, by providing nowcasts and forecasts of the physical (tides, surges, waves, currents, temperature, salinity, ice and sediment transport), chemical (pollutants, tracers etc.) and biological (algae bloom, productivity, fish stocks and ecological indices) marine environment in different space-time scales (Fig. 1). Operational oceanography aims to maximise the value of predictions by optimal assimilation of information provided by observing networks with the systematic temporal and spatial resolution of holistic knowledge synthesized within numerical models. However, the success of both operational oceanography and applied marine science depends increasingly on access to major investments in technology. Linkages across ocean-atmosphere-seas-coasts and between physics-chemistry-biology-geology in marine sciences extends this dependence to national and international activities, e.g. data from meteorological agencies, satellites, international survey programmes, etc. A peculiar challenge in designing the GOOS is to perceive future global benefits, in addition to present local concerns. Here we examine related 'Scientific Frontiers' and associated technical issues.

EuroGOOS challenges

Euro-GOOS was established to facilitate collaboration between European agencies involved in operational oceanography, both to extend and improve existing services and to address related global issues such as the establishment of GOOS (Global Ocean Observing System). Expanding services is dependent on exploitation of existing science and stimulation of scientific research and development to address future 'end user' requirements. Since the wider scientific community interested in issues such as Global Climate Change constitute, perhaps, the most significant 'end-user', there is a common-cause between Operational Oceanography and Marine Science R & D.

Moreover, the organisational efficiency developed for Meteorological Science can be harnessed to attract the investment for observational networks, data services, computational facilities, training, etc. needed to stimulate parallel developments of marine science. EuroGOOS aims to promote and support Operational Oceanography, by: (i) developing operational modelling capabilities, (ii) facilitating international co-operation in Europe in establishing monitoring networks, exchanging data and model outputs, (iii) encouraging research and training.

The Science Advisory Working Group of Euro GOOS has identified a series of initiatives to foster effective development of diverse aspects of marine

From WEFER G, LAMY F, MANTOURA F (eds), 2003, *Marine Science Frontiers for Europe.* Springer-Verlag Berlin Heidelberg New York Tokyo, pp 163-173

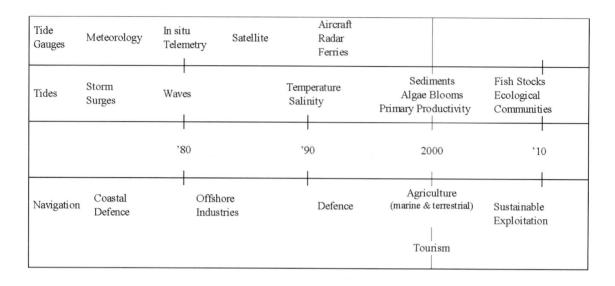

Fig. 1. Historical progress in operational services alongside related development of observation technologies.

science alongside effective interaction with the Operational Community. This has stimulated programmes aiming to provide: accurate fine-resolution bathymetry, routine standardised sampling along ferry routes, effective exchange of marine and meteorological data, specifications for future satellite missions, interaction between ocean-sea-coastal scientists, etc. Following a series of Euro GOOS International Workshops the following Initiatives have been recognised as high priority:

1. Development of Community Models (e.g. ERSEM, CARTUM, COHERENS)
2. Access to set-up data (bathymetry, prescribed climatic data)
3. Access to forcing data (Meteorological time series. Ocean boundary conditions)
4. Access to assessment/assimilation data (*in situ* and remote sensing)
5. Collaboration with Operational Agencies for participation in GODAE (2003-2005)
6. Collaboration with operational agencies on evaluation of new modules (extending scope of existing service).
7. Assessment of opportunities/constraints in developing coupled ocean-atmosphere-sea simulation systems.
8. Establishment of infrastructure requirements based on strategic review, etc.

9. Evaluation of requirements for a modus operandi of a High Performance Computing Network.
10. Observer System Sensitivity Experiments to optimise design of observation networks.
11. Assessment of provisions for training and Capacity Building.

Initiatives 2., 3., 4., 5. and 6. are being addressed by the Data and Science Panels of EuroGOOS, by international GODAE panels and by Framework V programmes. Further details on possible ways to progress the remaining initiatives are suggested below.

1. Development of Community Models - 'Total System Science' simulations such as encapsulated in ecological models involve many years of software construction supported by a much larger observational and theoretical backgrounds. Even the largest of marine research centres have to choose between which of the many 'modules' it leads development of, collaborates on or uses. For European centres not to be relegated to 'users', support for existing and new areas of leadership and collaboration is necessary.

7. Ocean-atmosphere-sea-coast coupling - Global Climate Change concerns focus attention directly on this issue. How might changes in Atlantic circulation impact on coastal climate (and related

dynamics to ecology)? How will this impact be conveyed (shelf-edge exchanges, internal circulations, atmospheric coupling) and modulated? Noting the much shorter 'half lives' of the shelf seas – are there first signs of Global Climate Change in shelf seas? Such scientific questions are foremost world wide. The integrated communications provided within the Operational Oceanographic approach allows individual, specialist scientists to interact with and impact on these issues.

8, 9 and 11 Establish European Infrastructure Requirements, Training and Capacity Building.

Initiatives 1. and 7. above could be achieved by various bi-lateral and multi-lateral agreements between major European Research Centres. However, for maximum benefits the full range of scientific expertise spanning the diverse range of habitats in Europe must be involved for both socio-political benefits and to ensure studies of global fluxes incorporate all such environments.

The models by which such integration can be facilitated need to be explored along with the needs for infrastructure investment in Very High Performance Computers, High Performance Data Networks, etc. The success of the ECMWF (European Centre of Medium-range Weather Forecast) in stimulating European research into meteorological, climate and oceanographic is noted – some (virtual) analogues in the Marine Community might be conceived.

10. Optimal Design of Observing Networks (ODON) - Optimal design of observing system means the design of the observing system is optimal in a sense of being cost-efficient. The issue is two folded: for a given amount of investment, a cost-efficient observing system means the observing system can provide a database with the best quality; for a given scientific quality requirements for an observing system, a cost-efficient design means the observing system needs minimum investment. Optimum observing system design has been a major concern in operational oceanography and climate change studies, such as GOOS, WOCE (World Ocean Circulation Experiment) and OOPC (Ocean Observation Panel of Climate). However, only a few preliminary studies have been conducted, due to lack of both quantitative methods and mature research strategy. Such studies need experts on system evaluation and design, data assimilation, error analysis, information theory, marine monitoring technology, field experiments, together with a funding mechanism. It is obvious that this kind of study can only be conducted at the community level.

It is the authors' view that the question of Optimal Design of Observational Networks is the outstanding medium-term challenge for Marine Science (in Europe and beyond) because (i) the success and development of forecasting systems is directly linked to observational capability and (ii) the cost of observational networks are orders of magnitude greater than modelling operations. This design must exploit synergy between remote sensing, *in situ*, shore-based and ship-borne instruments. The optimally designed networks will rely on internationally coordinated funding (GOOS) with rates of investment linked to indications of improved forecasting capabilities.

The above discussions are illustrated by Fig.1 (historical progress in operational services alongside related development of observation technologies), Fig. 2 (components involved in a marine forecasting system) and Fig. 3 (mis-match between resolution of monitoring systems and models). Moreover, since ODON embraces developments in both science and technology in all disciplines, it can act as a timely focus for integration of Marine Science R&D in FRAMEWORK 6. Major challenges in the optimal observing system design include:

C1 - Quantitative quality assessment of the European ocean observing system

C2 - Determination of optimal sampling distances and locations in coastal/shelf seas

C3 - New technologies to fill monitoring gaps in existing shallow water observing system (*in situ* and satellite)

C4 - Test-bed experiments to integrate design, technology and modelling

Perversely, despite their importance, these issues have received little attention in marine science. Recognising the reliance on satellite data, the promise of ARGO floats and AUV and the related needs for sea-truth vertical profiles etc., the remaining part of this paper introduces some of the underlying concepts (developed extensively in meteorology).

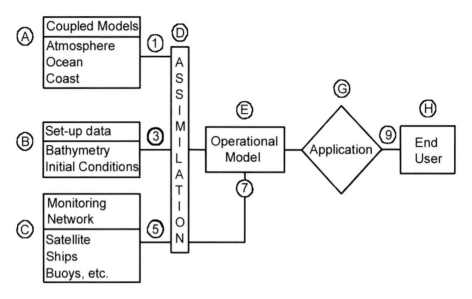

Fig. 2. Components involved in a marine forecasting system.

State-of-the-art of optimal design of observational networks

In the following sections we first summarise existing European coastal/shelf sea observing networks, and then describe concepts/strategies dealing with the challenges (C1 to C4) in the ODON.

Existing European coastal/shelf sea observing networks: In situ networks and remote sensing

Current European coastal/shelf sea observing systems include both *in situ* and remote sensing networks. *In situ* monitoring networks have been established in European coastal/shelf sea areas measuring water level, surface winds, waves, temperature, salinity and currents profiles such as SEANET, MAREL, Seawatch, BOOS (Baltic Operational Oceanography System) observing network etc.. These networks utilise tidal gauges, mooring and drifting buoys and platforms, and are operated on a routine basis. This is important data sources for operational modelling. However, most of the observations are limited to either the surface or near the coastline. Vertical profiles and offshore observations are rather sparse. There are fewer salinity and currents measurements than water level, winds, waves and surface water temperature measurements. In addition to these synthetic networks, research-related monitoring such as by research vessels also contributes to the ocean observing system. In general the existing European coastal/shelf sea observing networks cannot resolve either meso-scale or offshore ocean phenomena. Use of existing *in situ* networks has been limited in validating ocean models except for water level and sea surface temperature (SST), which have been used in data assimilation in combination with satellite measurements (Annan and Hargreaves 1999; Philippart and Gebraad 2000)

Remote sensing includes satellite, airborne and ground-based remote sensing (e.g. HF radar). Satellites constitute a major source of remote sensing

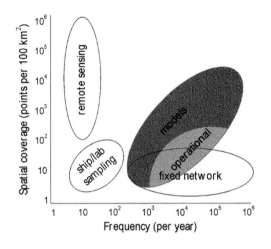

Fig. 3. Mis-match between resolution of monitoring systems and models.

measurements in operational ocean prediction. Currently SST is mapped by using AVHRR and VISSR in polar orbit (NOAA TIROS) and/or geostationary satellites such as Meteosat. After 2002, Meteosat Second Generation Scientific Application Facility (SAF) will provide near real time SST products four times a day with 2km resolution over the North Atlantic and regional seas by combining NOAA AVHRR and Meteosat SST. Clouds are a major constraint for the SST remote sensing which makes the SST data largely patchy. Active microwave sensors such as altimeter, scatterometer and Synthetic Aperture Radar (SAR) provide datasets on ocean topography, near surface winds, waves and meso-scale features. Passive microwave sensors (SSM/I) are used to monitor sea ice and wind speeds. These microwave sensors are installed in polar orbit satellites such as TOPEX/POSEIDON, ERS1/2, QuikScat and some planned satellite missions such as JASON. Existing near real time satellite products includes ERS winds/waves and QuikScat winds. Over the past two decades, it has been demonstrated that SSS can be measured from space because of the dependence of sea water emissivity on salt at low microwave frequencies (0.5 to 2 GHz). NOAA/NESDIS is studying instrument concepts for their next generation polar orbiting satellites. Coastal SSS has recently been measured and mapped using L-band

radiometers mounted on small aircraft. Real time data delivery is still a major concern in using satellite data in operational oceanography. Since satellite measurements are only available for the sea surface, it is sensible to have an *in situ* monitoring systems providing sub-surface measurements so that 3-dimensional ocean models can take full advantage of the data via data assimilation.

Quality assessment of existing ocean observing systems

A coastal/shelf sea observing system is an integrated system with many components (e.g. technologies, multi-variables monitoring, precisions and accuracy, and systematic performance etc.). Its function is diverse such as that data from the system can be used for data assimilation, model validation, forcing atmospheric models, forecast, nowcast, hindcast or for process studies in different interested scales. Definition of the "quality" of an observing system depends on the objectives of that system. Here we mainly focus on an observing system for improving operational marine service. Evaluation of an existing observing system means to provide information such as spatial-temporal distribution of data redundancy/insufficiency, data usefulness in data assimilation and coastal/shelf physical phenomena that the system can resolve. This is a premise for integrating and optimising an observing system. Currently a quantitative assessment of the European coastal ocean observing system is not available. Here we briefly describe principles and methods used in observing system evaluation. Generally three kinds of approach can be used: statistical evaluation, dynamic evaluation and semi-empirical evaluation.

Statistical evaluation means assessing the objective-dependent quality of an observing system by using sampling or reconstruction errors or information content determined from the observing system. The smaller the error or the larger the effective information of the system, the better the quality of it. The statistical evaluation can be conducted by using sampling error analysis, optimal interpolation (OI), scale analysis and information theory. She (1996) evaluated the ENSO observing system consisting of VOS XBT network, TAO and

TRITON buoy arrays. The study gave the spatial-temporal distribution of the sampling error and reconstruction error which clearly demonstrate the distribution of data redundancy and insufficiency in the ENSO observing system for the purpose of global climate change modelling.

Dynamic evaluation means assessing the observing system quality by using initial or forecasting errors based on the database of the observing system. The procedure is to reconstruct the initial field by assimilating the observations for given sampling strategies and then compare initial or forecasting errors resulting from different sampling schemes. The smaller the error the better the quality of the sampling scheme. This approach is also called Observing System Simulation Experiment (OSSE) and has been extensively used in atmospheric observing system design but only a few examples have been reported in oceanography by using simplified 3D ocean models (e.g. Le Traon et al. 1999 in OCEANOBS99). No usage of operational models for OSSE has been published for European coastal/shelf seas.

Semi-empirical evaluation views the quality of an observing system as a function of many factors, such as reconstruction error, information in resolving ocean phenomenon with certain scales, technology and cost considerations, etc. Weights can be given to each factor for a given sampling design (some of them are empirical) in calculating the quality function. This method has been used by Bailey et al. (1999, OCEANOBS99). The method itself is more comprehensive and flexible but needs to be further quantified and combined with statistical and dynamic evaluation.

The above methods should be developed and applied in the European coastal/shelf sea observing system. There are several issues that previous studies have not touched on. One is the assessment of mixed *in situ* and remote sensing networks. Another is that important physical processes have not been included quantitatively in the observing system evaluation such as ocean-atmosphere coupling, transport crossing the open boundaries of semi-enclosed seas etc. Existing *in situ* ocean observing systems only measure single ocean elements rather than measure systematic ocean phenomenon at appropriate scales. These aspects should be included in the assessment of European ocean observing systems. Here we suggest the following research topics in the quality assessment of the existing European Ocean Observing System (OOS) (workflow summarised in Fig. 4) for operational marine service in Europe:

• Statistical analysis of spatial and temporal representation effectively provided by the OOS, effective information content for interested scales resolved by the OOS, data redundancy and insufficiency distribution.

• Observing system simulation experiments to assess contributions from different observation networks to ocean nowcast and forecast.

• Cost-benefit analysis of the existing OOS

Network integration and new technology

Following assessment of an existing observing system, the spatial distribution of data redundancy and insufficiency can be quantified. This information can be used in integrating and rationalising existing observing systems; i.e. increasing observations in the insufficient areas and reducing those in the redundant areas. However, there is still a question of how and where these should be deployed. New technology development and optimal design theory are two major approaches to answer the question. This subsection deals with new technology needed and the next subsection discusses the optimal design theory.

Existing monitoring technology effectively measures sea surface values for most of physical variables in operational oceanography, such as surface waves, SST and water level. By optimally merging the existing *in situ* observations with satellite observations to rationalising existing monitoring networks, these variables can be assimilated into operational ocean models (Janssen 1999; Annan and Hargreaves 1999; Philippart and Gebraad 2000). Surface currents can be monitored by HF (High Frequency) radar and *in situ* sensors. Technology for assimilating surface currents in 3D ocean models has been developed in the EuroROSE (MAST III) project. However, assimilating surface currents in operational predictions has not yet been achieved since surface currents over a large area have not been available

Evaluation of Existing Ocean Observing System

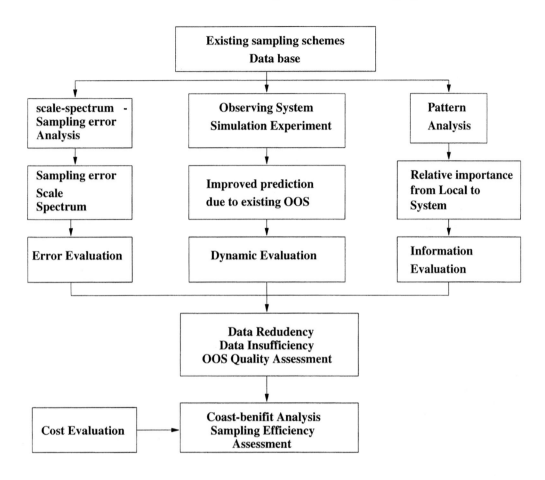

Fig. 4. Workflow of quantitative quality assessment of observing system.

in near real-time. Satellite remote sensing of surface salinity will be practical in the near future with a spatial resolution of 1degree or so and an accuracy lower than *in situ* measurement.

For assimilation purposes in coastal/shelf seas, a certain amount of *in situ* data are needed to supplement these satellite measurements. There are few sub-surface monitoring tools to provide sufficient data with 3D coverage for assimilation in the operational 3D coastal/shelf sea models. The ferry-box technology is currently only feasible for measurements near the surface. Mooring buoys have high quality for long-term profile measurement but are expensive, say 300K-600K € each, and of risky to

be lost in non-protected areas. VOS measurements are too sparse and not systematic. Current floating profilers (ARGO) are only suitable for deep water (2000-3000m). There is an urgent need for the community to develop cost-effective technologies or optimising existing technology for shallow water temperature and salinity (T/S) profile measurements so that sufficient T/S spatial sampling can be achieved over European coastal/shelf seas.

Some new approaches such as modified ARGO floats for shallow waters, bottom moored profilers and ferry-towed profilers could be attractive candidates for cost-effective shallow water T/S profile measurements. These technologies have already been in the test-phase but need more development. The shallow water ARGO floating profiler will provide T/S profiles at regular times (e.g. once or twice a day) at the price of the order of 10-15k Euro per profiler. The floating profiler can be surface drifting or bottom drifting type. Its autonomy is of the order of 100 cycles (to be optimised). As the profiler is located by ARGOS (also used for the transmission of data) and closer to the shore, the possibility of re-using it after recovery and refurbishment is greater than in the deep ocean. This kind of profiler, however, is not suitable for the shallow water area with muddy bottom where the profiler can be stuck in the mud and not be able to be popped up to the surface. Bottom moored profiler may be a better choice in the areas with muddy bottom or storm tidal currents and open boundaries of semi-enclosed seas. This profiler has the similar cost as ARGO floats (personal communication with Siegfried Kruger, IOW). Ferry-towed profiler is a natural and cost-effective way for T/S profile monitoring. The quality (mainly the precision), however, has not been satisfactory. The low cost of these technologies makes it possible to deploy them in a higher resolution than any other existing *in situ* monitoring technologies, for a given amount of investment. These developments will make 3D T/S assimilation realisable in coastal/shelf sea operational prediction.

Optimal deployment of observing networks

If technologies and funding become available to deploy new instruments, the subsequent questions are: where to deploy them and what are optimal sampling distances and frequency? Since the coastal/shelf seas are not homogeneous, it is important that we deploy the instruments in informative locations. The information content varies with locations. The characteristic scales, spectrum and spatial pattern are the statistical basis for optimum observing system design. Sampling density closely depends on these statistics. A comprehensive diagnosis of these indices is needed before the optimal design can be attempted.

Optimal sampling instruments, distances and locations should make the observing system cost-efficient. Here "cost" means the financial support

for the observing system including *in situ* and remote sensing components, which may be approximately estimated by summing the total costs of different kinds of instruments, data transmission and their maintenance budget. "Efficient" means that the observing system should be efficient in making benefits. The benefits mainly depend on the quality of the datasets derived from the observing system and their relevant products, i.e. how useful the observing system is for managing the coastal zone and protecting people and facilities. Note that the benefits may be different for applying the same datasets for different purposes and with different data dissemination systems. This indicates that the benefits also depend on how the information derived from the observing system is used. However, for simplicity we assume that the data will be used in as efficient a way as possible in the optimal design study.

Generally, we have two ways to measure the quality of the datasets and their relevant products. One is to use the integrated error of the observation system from the real ocean, which can be chosen as observation error (including sampling error and instrumental error), noise-signal ratio, analysed or assimilated initial field error, forecasting error or some weighted summation of these errors. The other is to use "effective information" resolving interesting scales in the observing system. The above two ways correspond to the following two categories of optimal design problems: one is for operational prediction and modelling usage and the other is for understanding the important physical processes in the atmospheric and oceanic systems.

In the first case, the observing system should have sufficient resolution and accuracy because errors in the input datasets are one of the most essential factors to decide the error in the modelling and prediction results. Its quality can be measured by an error function, such as measurement error and sampling error for raw datasets, analysis error in objective analysis and/or data assimilation and forecasting error. This gives rise to two issues for optimal design of observing systems in operational oceanography. One is to design an observation system with the highest quality (i.e. the system has minimum errors) for a given cost of the system and given objectives of the system. The other is to design an observing system with the minimum cost for an acceptable error criterion for the given objectives of the system.

In the second case, many observation systems in the ocean do not have high resolution as required in the operational prediction. The main purpose for designing such a sparse sampling ocean observing system is to extract the maximum information for appropriate scales within the constraint of limiting costs.

The optimum design methods have been partly developed by She (1996), She and Nakamoto (1996) and applied in the ENSO observing system. Fig. 5 shows optimal sampling distances (in degree) in the tropical Pacific for a SST sampling error of 4%, i.e.

this sampling scheme gives minimum sampling points required to reach a sampling error of 4%. The sampling distance distribution shown in the figure is the most economic sampling solution we can get for the given sampling error of 4%. Sampling error analysis based on the gridded 10 year SST data with 1degree by 1degree resolution and optimal design theory were used in this study. For European OOS optimisation we need to know:

• Relationships between sampling distances and quality of the OOS and its subsystems.
• Optimal sampling distances and instruments for required OOS performance.
• Spatial distributions of effective information content in European ocean and their relations with optimal OOS design.

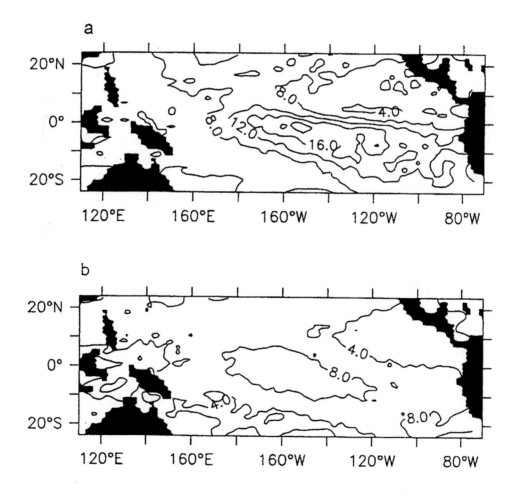

Fig. 5. Spatial distribution of optimal sampling distances (in degree) in tropical Pacific for a given SST sampling error criterion of 4% (after She and Nakamoto 1996). **a)** Longitudinal optimal sampling distance (in degree); **b)** Latitudinal optimal sampling distance (in degree).

Test-bed Experiment

Test-bed experiment means "to conduct optimal design study in a limited area in order to test the feasibility of the full design package". The experiment should include four modules: design, technology development, modelling and field experiment. The design module defines optimal sampling instruments, distances and locations. The technology module develops cost-effective technologies; field experiment deploys instrumentation (including new technologies), retrieves and transmits data (quality controlled). Modelling illustrates the benefits from using the data and how well this compares with existing OOS. Such a test-bed experiment is regarded as a necessary step to establish a cost-efficient and permanent European OOS and is hence strongly recommended.

Abbreviation list

ARGO - Global array of profiling floats
ARGOS - Advanced Research and Global Observation Satellite
AUVs - Autonomous Underwater Vehicles
AVHRR - Advanced Very High Resolution Radiometer
BOOS - Baltic Operational Oceanography System
CARTUM - Comparative Analysis and Rationalization of Second-Moment Turbulence Models
COHERENS - Dissemination and exploitation of a COupled Hydrodynamical Ecological model for REgioNal Shelf seas
CTD - Conductivity and Temperature Depth profiler
ECMWF - European Center of Medium-Range Weather Forecast
ENSO - El Nino and South Oscillation
ERS - Earth Resource Satellite
ERSEM - European Regional Seas Ecosystem Model
EuroGOOS - European Global Ocean Observing System
GODAE - Global Ocean Data Assimilation Experiment
EuroROSE - The EUROpean Radar Ocean SEnsing project
GOOS - Global Ocean Observing System

HF - High Frequency radar
IOW - Baltic Sea Research Institute
MAREL - Network automated measurements for coastal environment monitoring
Meteosat - European Geostationary Operational Meteorological Satellite
NOAA - National Oceanic and Atmospheric Administration
NESDIS - National Environmental Satellite, Data, and Information Service
OOPC - Ocean Observation Panel of Climate
OOS - Ocean Observing System
OSSE - Observing System Simulation Experiment
QuikScat - Quick Scatterometer satellite
SAF - Satellite Application Facility
SAR - Synthetic Aperture Radar
SEANET - Monitoring NETwork in the North SEA region
SEAwatch - Seawatch buoy network
SSM/I - Special Sensor Microwave/Imager
SST - Sea Surface Temperature
T/S - Temperature/Salinity
TAO - Tropical Atmosphere Ocean
TIROS - Television Infrared Observation Satellites
TRITON - TRIangle Trans-Ocean buoy Network
VISSR - Visible and Infrared Spin Scan Radiometer
VOS - Volunteer Opportunity Ships
WOCE - World Ocean Circulation Experiment
XBT - Expendable BaThythermograph

References

Annan JD and Hargreaves JC (1999) Sea surface temperature assimilation for a three-dimensional baroclinic model of shelf sea. Cont Shelf Res 19:1507-1520

Bailey R, Thomas S and Smith N (1999) Scientific evaluation of the global upper ocean thermal network. Vol. 2, Proceedings of International Conference of the Ocean Observing System for Climate, publ by CNES, France

Janssen PAEM (1999) Wave modelling and altimeter wave height data. ECMWF Technical Memorandum No. 269, European Center for Medium Range Weather Forecasts

Le Traon P, Dibarboure G and Ducet N (1999) Mapping capability of multiple altimeter mission. Vol. 2, Proceedings of International Conference of the Ocean

Observing System for Climate, publ by CNES, France

Philippart and Gebraad (2000) Assimilating satellite altimeter data in operational sea level and storm surge forecast. Proceedings of EuroGOOS Conference, Rome, 1999

She J (1996) Optimal Evaluation and design study for upper ocean observing system. Japan Marine Science Technology Centre Tech Rep 70 p

She J and Nakamoto S (1996) Optimal network design based on spatial sampling error study. International Workshop on Ocean Climate Variations from Season to Decades with Special Emphasis on Pacific Ocean Buoy Network, Published by Japan Marine Science and Technology Centre pp 79-106

"Sea of Substances"
Pollution: Future Research Needs

R.W.P.M. Laane

National Institute for Coastal and Marine Management/RIKZ, P.O.Box 20907,
2500 EX Den Haag, NL
corresponding author (e-mail): r.w.p.m.laane@rikz.rws.minvenw.nl

Abstract: Due to a lack of knowledge by policymakers and the general public, all kinds of doom and gloom stories exist when it concerns substances in the sea. This resulted in a strong negative image of substances in the marine environment, which is fed by feelings and emotions. Facts and figures do not seem to count. Proper definitions for contamination (increase above background level by man) and pollution (introduction by man of substances or energy leading to deterious effects) are set to facilitate the scientific discussion on the pollution of the sea. During the last 100 years more than 150,000 substances are produced by man and will ultimately end up in the marine environment. Although it was not possible to analyse and detect all these substances a few decades ago, nowadays it is possible to technical means are available to establish and quantify the presence of all these substances. However, in national and international monitoring studies a relatively small amount of 200 substances are under study. It is not possible for reason of time and costs to analyse and assess all substances. More and more signals appear also that the effects on organisms are not caused by this relatively small amount of substances, but a pool of up to now unknown substances.To protect the marine environment from deterious, harmful or hazardous effects several national and international targets were set for the input to and the concentration in the marine environment. However, it is increasingly recognised that these targets based on toxicological data are indicating a potential effect and that there is a shortage of toxicological data to set targets for all substances. The wide range in targets for substances set in different countries indicates that no international consensus exists concerning the method to assess a potential risk of a substance in the aquatic environment. For the future it is recommended that exposure to substances must be the key in future toxicological effects and risks studies. To optimise policy measures for substances, first of all negative effects on organisms and ecosystems must be traced and than the substances responsible for these effects must be identified. It will be more realistic when the judgement of an impact of substances on different levels in the ecosystem is performed on a risk assessment base and not on a hazard assessment base on a catchment level.

Man's impact

It is well recognised by scientists, policymakers and the general public that the exploitation by man on coastal zones, seas and oceans has increased enormously over the last 100 years. Conflicts concerning the use of the coastal and marine environment became manifest already when the global population rose to circa one billion whereas currently there are about 6 billion inhabitants. Coastal zones all around the world are under pressure: 37% of the world population is living within 100 km of the coast and even now coastal populations are growing faster than others. Twelve of the twenty largest urban areas of the world are located within 160 km of the coast and of the cities with more than 2.5 million inhabitants, 65% are located along the coasts (Anonymous 1998).

The causes of the impact of man's activities on land and in the sea on the marine environment can be summarised as follows: 1) Physical interventions: constructions, land reclamation, dikes, disap-

From WEFER G, LAMY F, MANTOURA F (eds), 2003, *Marine Science Frontiers for Europe.* Springer-Verlag Berlin Heidelberg New York Tokyo, pp 175-188

pearance of mangroves etc., 2) Removal of matter: biomass (e.g. fish), sand and gravel, mining etc. and 3) Input of substances above natural background levels: nutrients, metals and organic micropollutants.

The aim of this paper is to summarise the progress in science made during the last decades related to the input of substances and to define future needs and research to establish a proper protection and management of healthy marine ecosystems.

Doom and gloom stories

Four hundred years ago men were afraid of the sea. Their activities were mainly restricted to small scale fishing in the coastal zone and it for most people it was completely unknown what lied behind the horizon and beneath the surface. Also, men could not cope with the power of the waves and the impact of storms, destroying their boats and coastline. At that time men created many legends and bizarre creatures to hide their fear and ignorance of the sea. One of the first unresolved stories is that of Jonah, who was swallowed by the whale whereas later on all kind of serpents and mermaids appeared (Ellis 1995).

Nowadays, scientists have unravelled many of these stories and have gained knowledge concerning the sea and the ocean. However, all kind of doom and gloom stories still exist when it concerns substances in the sea. Slogans as: Men are on their way out, the oceans are dead and on the critical list and terms like toxic chemicals, persistent pollutants are quoted regularly by the media and even in scientific journals. This resulted in a strongly negative image of substances in the marine environment by the general public, which is fed by feelings and emotions. Facts and figures do not seem to count.

Definitions

To discuss marine pollution, properly the definitions must be clear.

Contamination has been defined as the (significant) increase in the concentration of a substance above natural background level by men in particular area. Pollution is the introduction by men, directly or indirectly, of substances and energy to the environment leading to deterious effects (Clark 1992).

Nowadays, when we refer to marine contamination and pollution the main focus is on chemicals introduced by men. However, it must be recognised that by definition, contamination concerns not only metals and organic substances such as polychlorinated biphenyls and dioxins. It includes also the increase in concentration of organic matter, debris, fresh water, particles, organisms (e.g. bacteria, viruses and alien species) and even nutrients in the coastal zones and the marine environment by men.

The main focus of this paper will be on substances as such and not on the contamination caused by organisms, as mentioned above.

Pollution – management cycle

There has always been an input of substances to the marine environment: a natural input. In early days when there were no living organisms on earth, this input was, compared with today relatively low, but not constant. In that time the concentrations and the inputs of substances were restricted to natural physical (e.g. changes in heat fluxes, erosion and weathering) and chemical processes (e.g. lighting and ignition).

Due to the increased activity of life (bacteria, plants, animals and man) the inputs and concentrations of certain substances changed in the aquatic environment. The changes of input of substances in relation to their impact have been presented in the pollution-management cycle (based on the ideas of Stigliani and presented by Salomons et al. 1995) (Fig.1).

In the pre-industrial phase (I) when the human activities were modest, the processes on earth can be described as natural. The human-induced excess of substances and their physical impacts were relatively low. Natural environments are characterised by a high biodiversity and high resilience and enough buffer and self clearing capacity to cope in time with changing inputs of relatively small amounts degradable (in)organic waste.

With the start of the industrial phase (II), the input of substances increased considerably as well as the physical disturbance, for instance in a way as happening now in the Asian countries. As a re-

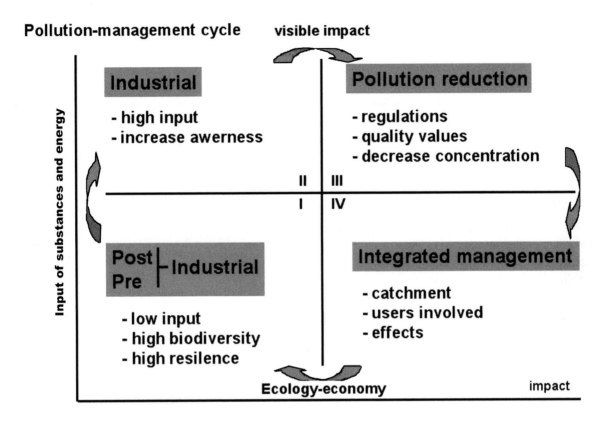

Fig. 1. Pollution – management cycle (modified with permission after Salomons et al. 1995). For explanation see text.

sult, the assimilative or buffer capacity of the systems could not cope (e.g. self clearing capacity) anymore with the increased input of substances. For instance, the input of degradable waste and bacteria in sewage increased in the fifties in the Netherlands, leading to nearly anoxic conditions in the main rivers. In this phase, impact is not restricted anymore to the vicinity of the activities (phase II) but is spread out over wider areas. The coastal system can not assimilate and adept itself anymore in time with respect to the increased input of substances. The impact on the riverine and marine environment became visible (e.g. litter, oil victims), tangible and smellable (e.g. anaerobic fresh water). Jeremiahs of science publicly predicted in concert with the press the losses of natural resources and the mass mortality of organisms, the extinction of

the human race and other scientists showed little concern for environmental alterations and insisted and trusted that technology and adaptation of the earth system itself could overcome these inconveniences. A positive effect of this perceptible impact and doom stories is that public awareness of the health of the ocean increased.

To reduce the fluxes of substances, political actions are taken on a national and international scale (phase III). On a national scale objectives are set, emissions are regulated, industry is forced to use best available technology and increasingly more substances are recycled. The main source of anthropogenic substances are, as now in the western world, diffuse sources as agriculture, atmospheric deposition and the heritage of contaminated sediments.

Due to these actions and public awareness, the flow of substances decreased continuously and phase IV is entered. Thus, men's activities and the input of substances and their related chronic and sub-lethal effects have to be reduced further to reach a post-industrial phase.

In this phase again the input of substances is low, and is characterised by high biodiversity, resilience, buffer and self clearing and carrying capacity. This is only possible through the involvement of producers and users of substances in the decision making process. An increased collaboration between natural science and socio-economics is necessary for a more effective transfer of knowledge to policymakers and the general public. The results of such interactions may help to answer societal needs within an international context.

Marine contamination

The Greeks already knew that the sea is the end of everything, because the rivers run in to the sea,

leaving there everything, except water (Ball 1999). Much later, in the seventeenth century, Edmund Halley - a comet was mentioned after him - explained by evaporation that the sea level was not rising due to the input of riverine water. Today it is well recognised that seas and oceans act as a sink for natural and man-made compounds (Goldberg 1972; Laane et al. 1999). During the last hundred years the concentration of numerous compounds has increased both at the sources and in marine waters.

Historical development

It is interesting to trace through time how the quantification of substances in the marine environment evolved (Fig.2). Although Aristoteles was aware of the differences in density between fresh and salt water, Robert Boyle (1672-1691), the godfather of chemical oceanography, was the first to publish about the salt content of sea water. This became possible because the content of salts in sea water

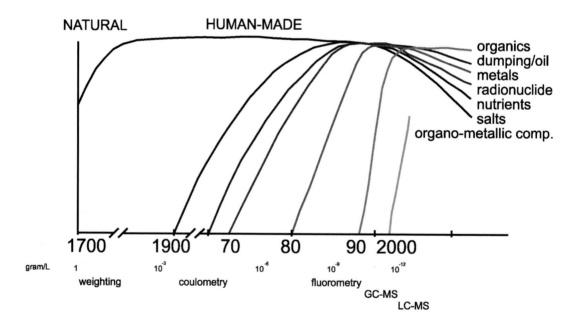

Fig. 2. Development of procedures to study certain substances in the aquatic environment. On the vertical axis the intensity of attention is given relatively.

is relatively high (grams) and they could by weighted by a simple balance. In 1772 Antoine Lavoisier (1743-1794) published the first results on the analyses of different salts in sea water, obtained by fractionated crystallisation and the dissolution of salt in different solvents. It was Gay-Lussac (1735-1784) who measured the salt content of seawater at different locations and he concluded that the salt content of the open ocean was rather constant. He also established the total amount of suspended matter by filtering and weighing.

It must be recognised that changes in the salt contents in the coastal zone can be defined as contamination, since these changes are induced by man. Human activities in the catchment area, for instance the large scale cutting of to create better skiing tracks, increased urbanisation and more asphalted roads leads to less water retention by soil. During a certain period the rivers receive more water and the salinity gradient in the estuaries is pushed more into the coastal zone. This also happens through canalisation of rivers whereas occurs the opposite by damming. These physical alterations in the catchment area, together with the construction of harbours in estuaries and the related dredging activities, also changes the concentration of suspended matter in the river and the output to the coastal zone in many places around the world (Goudie 1990). Nowadays, in the industrial countries (phases II and III), men and nature are already adapted to these physical changes in the river part of the catchment area and the associated changes in fresh water and suspended matter run off to the coastal zone. Restoration works are planned to further restore the rivers or to cope with the flooding of land due to the increased run off of fresh water to the rivers. It is recommended that in these restoration studies, the impact on the estuarine and coastal system must be an integral part of the study.

In the third world countries, erosion of land and damming are presently taking place on a large scale, respectively increasing or decreasing the input of suspended matter to the coastal zone.

Although the results of the discharge of easy degradable human waste was already noticed in the Middle ages by the foul smell and dead fishes, it was only in the beginning of the twentieth century that it became possible to measure reliably the oxygen and nutrient concentrations in water by titration and by applying an UV spectrophotometer. With these instruments concentrations of dissolved inorganic nitrate and phosphate could be measured from milligrams to micrograms per litre.

As a result of the growth of maritime shipping trade, from $800*10^6$ tonnes in 1955 to circa of $5700*10^6$ tonnes in 2000 (Anonymous 1998), the amount of oil on beaches, in the coastal zone and at open sea increased dramatically in the fifties and sixties. Later on, due to regulations of the International Maritime Organisation (IMO) and the responsibility of ship owners, the amount discharged decreased. Presently, 20-30 % of the total amount is still discharged legally or illegally directly into the sea.

The analysis of concentrations of different compounds in oil mixtures became possible by Infra Red spectroscopy and later on by Gas Chromatography–Mass Spectrometry (GC-MS).

Even today, many vessels are not properly designed with adequate capacity to store garbage and sailors do not like to bring their trash ashore. In this way more than 5 million tonnes of plastic debris enters the ocean yearly. Organisms are entangled or snared in this plastic debris, giving rise to starvation, exhaustion, inflammation or drowning..

The increased input of radio nuclides, due to atmospheric testing of atom bombs, in the sixties was measurable with the Geiger teller.

It is only recently that the increase in metal concentrations in the coastal zone and in the open sea can be measured reliably. Even more recently it became possible to measure reliably hydrophobic organic substances such as PCB's and PAH's with capillary GC-MS. Concentrations up to 10^{-9} gram per litre are reached. These relatively low concentrations could also be measured with a fluorometer. Over the last decade it became possible through Liquid Chromatography - Mass spec-trometry (LC-MS) to measure also hydrophilic substances, such as surfactants, in water. Presently, it is even possible to measure individual enantiomers of pesticides and drugs (Jørgensen and Jørgensen 2000).

In essence, now it is possible to detect the presence of each arbitrarily substance in the marine environment.

PCB story

It should be realised that the ability to establish and quantify the presence of a certain substance is dependent on the technical means availability. This does not imply that before the instrumental possibility the substance was not present. This is clearly demonstrated by the PCB story (Bernes 1999). Industrial production of PCB's started in 1927 and by accident Sören Jensen (see Bernes 1999) noticed the presence fourteen unknown peaks in a GC-trace of a fat sample of a white-tailed eagle in the mid sixties. It was only in the beginning of the seventies that the substances related to these unknown peaks were identified as PCB's by GC-MS. Hence, for PCB's, a gap of nearly 50 years exist between production, occurrence in the aquatic system and unambiguous identification (Fig. 3a). The drop in production of PCB's become evident over the last two decades by a decreasing concentration of PCB's in, for instance, the surface sediments of the Great Lakes in the US and the Dutch coastal zone (Laane et al. 1999).

For certain categories of compounds under study, e.g. the flame retardant PBDE's, the production and concentration in surface sediments and mother milk still increase (Fig. 3b) (Bernes 1999). However, due to the political attention for Persistent Organic Pollutants (POP's) and the increased technical skills to detect these substances, the time gap between the start of production, environmental impact and political action, has decreased considerably.

Ranking substances

The amount of substances under study (ca. 200) is relatively small compared to the total amount of compounds produced by man ($\geq$ 150,000). It is not possible, for reasons of time and costs, to analyse and assess all these substances. Several protocols have been developed to set priorities for substances to be studied (e.g. Wezel and Kalf 2000). To screen and rank substances, different methods and criteria are applied and different (inter)national lists of priority substances are available (Rodan et al. 1999). These classification systems vary in complexity. Some only consider toxicity, others include only exposure factors such as persistence and fate. In general, protocold are believed to be better predictors of risks when both toxicity and exposure indicators (e.g. bioaccumulation) are included.

More and more signals appear in the literature that the effects, measured in the laboratories or field, are not caused by the relatively small amount of compounds monitored and under study, but by a pool of up to now unknown substances (e.g. Hendriks et al. 1994).

a)   b)

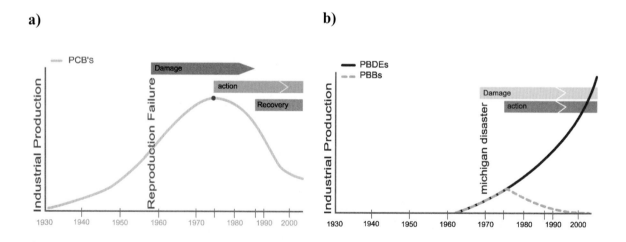

Fig. 3. Simplified scheme of the delay in time between the annual production of PCB's **a)** and PBB's and PBDE's **b)** and the sequence in environmental and political events (modified with permission after Bernes 1999).

In future, it can be expected that the concentrations of well studied substances (e.g. PCB's, PBB's, dioxins and metals) will reduce further. Political goals, such as nearly zero emissions and back to natural concentrations in 2020, can not be achieved due to the contamination heritage of the past sequestered in the river and coastal zone sediments. From a thermodynamically point of view these political goals are not possible either; the ultimate result of every physical and biological activity is that substances and energy will end up in the aquatic environment.

More attention will be given to substances which are used by man to initiate a wanted biological effect (such as pharmaceutical and pesticides).

In general, it must be recognised also that the concentration of a substance does not imply that there will be a deterious effect. This was already stated in the sixteenth century by the Swiss alchemist Theophratus, alias Paracelcus:" Only the dose makes things not a poison".

Marine pollution

To protect marine organisms and ecosystems against deterious, harmful or hazardous effects several national and international political actions were taken in the seventies and eighties under the auspices of the world concept of sustainable development and sustainability. To reduce harmful effects of substances it was realised (inter)nationally that measures had to be taken to reduce the input of substances into the coastal zones and the marine environment. Moreover, in the eighties objectives were set in several countries: often expressed as quality values (QVs) for a certain set of substances on the so called black and grey lists.

The terms toxic, hazardous and harmful are often used synonymously. Hazard is defined as the potential for harm (e.g. toxic effect). Its magnitude is related to the severity of the expected consequence. Risk is a function of the magnitude of a hazard and the probability that it occurs. It is a human concept: taking into account an understanding and consideration of causes, identification of victims, degree of control and personal values.

Most of the assessments of substances are made by (model)studies comparing the (predicted) field concentration with a quality value (QV), target or an objective. These quality values are derived from no effect concentrations (NOEC) derived mainly from acute toxicity lab tests and waterborne concentrations. Due to the lack of no effect concentrations for sediments, the sediment quality values are deduced from water concentrations using the method of equilibrium participation.

Potential effect

It is increasingly recognised that for the protection of organisms using quality objectives, the NOEC concentration for acute effects is not sufficient and that a chronic criterion must be taken into account.

However, for the 200 substances studied there is already a severe shortage of NOEC data. It has been calculated that at least 3 billions EURO is required to obtain for all substances reliable NOEC values for all substances. Reliability is very important, because many published studies reporting NOEC data, are not validated from an environmental chemistry point of view. Almost no budget studies were made, concentrations were not monitored during the experiment, etc.

It can be questioned if it is necessary to assemble all these data. It should be kept in mind, however, that the comparison of a concentration with a QV indicates a chance on an effect only and does not indicate any risk.

To transfer NOEC data for different organisms to a Quality Value (QV), different methods are applied by different countries. Chapman et al. (1999) summarised the sediment quality values for some metals and metalloids which have been calculated using different methods in more than 30 countries. They found a large range in the quality objective for metals (more than several orders of magnitude, see Figure 4).

This wide range in QV's clearly indicates that no international consensus exists concerning the method to assess a potential risk of a compound in the aquatic environment. Apart from the different methods applied the major reason for these differences is speciation and complexation of the substance, determining the bioavailable fraction, differences in analyses to establish the effect concentration and genetic differences of organisms.

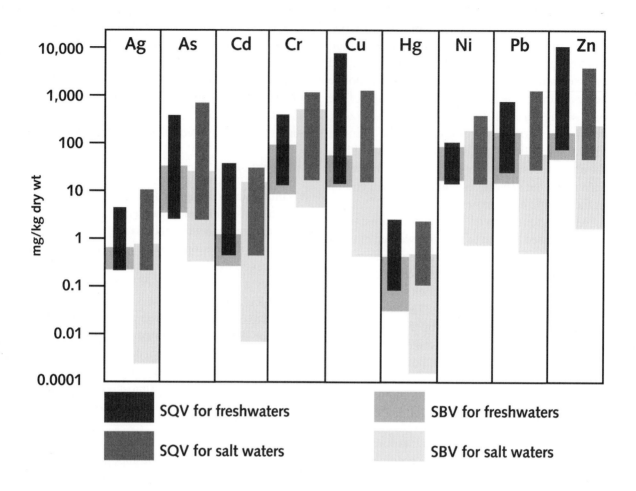

Fig. 4. Range in sediment quality values (SQVs) and sediment background values (SBV) for metals and metalloids in fresh and salt waters (with permission from Chapman et al. 1999).

In lab experiments aming to determine a no effect concentration, it is assumed that during the entire procedure the substance is 100% bioavailable. Presently, it is well recognised that substances in the aquatic environment are not 100% bioavailable. Metals and organic substances have the tendency to bind to organic matter, i.e. dissolved and particulate organic matter present on the milligram per litre level. In this way, the bioavailability is less (Cornelissen et al. 1998). Distribution studies of substances to organic matter describe the organic matter in a totally undefined pool of organic matter. To improve our understanding of distribution process, it is necessary to know more about the

chemical characteristics of the organic matter involved, for instance through identification with pyrolysis mass spectrometry.

The separation of organic matter in dissolved and particulate matter is fully operational. The fraction smaller than 0.45 μm is often referred to as dissolved and what is retained on the filter is called particulate. However, in the dissolved fraction there are, apart from amorphous matter, living and dead micro bacteria, micro algae and viruses. These biota have other adsorption capacities than the detrital dissolved organic matter. The same holds for particulate organic matter: depending on the season it exists of a pool of detritus and living organisms such as algae and bacteria. In sediments, on top of algae and bacteria, nematodes are present in the particulate organic matter pool.

In studies of substances adsorption on and desorption of organic matter, the biological and chemical characteristics of organic matter must be taken into account.

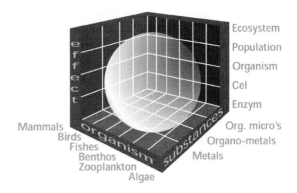

Fig. 5. Relation between different substances and their effects and organisms.

Acute and chronic effects

Harmful effects of substances on organisms can be divided in acute and chronic effects. Acute toxicity effects appear over a relative short time of period, usually a few days and are very often lethal. Chronic effects become then apparent after a relative long time interval, and can be lethal or sublethal. Sub- lethal effects modify or interfere with reproduction, growth and behaviour processes of organisms.

Nowadays, the chance of acute effects of substances on organisms is relatively small, because the concentrations are decreasing slowly and men are more aware of the negative effects of substances. However, during an accidental spill of large amounts of oil or other transported chemicals into the sea, the chance on acute effects is still real.

Most marine toxicological studies are presently focused on sublethal effects. They can be classified in studies describing the effects on the cellular (e.g. interactions of substances with DNA or enzymes), the organisms, the population and the ecosystem level (Fig. 5). The expression of the effect can be for instance narcotic, teratogenic, mutagenic and estrogenic.

At the moment there is a strong growth in the development in the application of bio-effect methods for lab studies. Bio-effect methods include biomarkers (e.g. enzyme activities as cytochrome P450), bioassays (whole organisms) and rapid screening methods (e.g. cell lines) by which environmental samples or extracts are tested. The advantage of these tests aiming to assess deterious effects, is that the effect of all substances is measured. A disadvantage of such is that the end point is death; there are not many tests available to establish chronic effects. The tests can be performed by lab-controlled conditions. However, there is an urgent need for the environmental chemical validation of these experiments. Up to now too much tests are performed without giving attention to the fate of the substance during the test. Furthermore lab experiments are difficult to relate to the field situation at the moment. A concentrated extract of a field sample is often used in lab tests to identify the effect. However, during extraction more than the bioavail-able fraction can be set free, and by concentrating the extract, no realistic situation is created.

It will be necessary to perform lab tests under field conditions and to study in the field deterious effects directly.

Field effects

Harmful effects of organic substances in the field, like DDT, were first addresses by Rachel Carson in the beginning of the sixties in her book Silent Spring (Carson 1963). Decreasing shell thickness, suppression of immune systems leading to a higher risk of illness and inflammations, disruption of the endocrine system were reported in many areas all around the world (Colborn et al. 1999). Birth effects and less reproductive potential of fish eating birds and reproductive impairment of seals are closely linked to POP's, especially PCB's (Walker and Livingstone 1992; Jones and de Voogt 1999). However, it is still very difficult to assign a substance to an effect; this relation is very vague in most cases (Fig. 5). Only in the case of imposex of snails it is concluded, based on lab experiments, that tributyltin compounds(TBT) cause these effects. Even for this case, this relation is not completely clear; trifphenyltin (TFT) compounds could also be responsible for the effect.

Due to strong regulations and responsible care, it seems that the sub lethal effects by substances on birds and mammals are diminishing in the western world. However, due to fast industrialisation of the third world countries (phase II, Fig. 2), the input of potential harmful substances will increase again.

Exposure is the key to toxicological effects and risks. Exposure is the external contact of a substance with an organism, however from a toxicological view point, the dose (the total amount of substances that is in contact with the outside world of an organisms) is more important (Crosby 1998).

To protect aquatic organisms on the longterm against chronic effects induced by substances it is important to consider bioaccumulation and dietary pathways, rather than defining a strict water borne concentration and rely on that. In particular this is necessary for warm-blooded predators, e.g. mammals. They are known to be specifically at risk when persistent and bioaccumulating compounds are involved. Hence, bioaccumulation and chronic effects must be taken into account to protect the marine environment against hazardous effects.

Risk assessment

Up to now, most effects are detected on a cellular and organismal level. Impact of substances on the population and the ecosystem level are much more difficult to assess. First of all it is difficult to discriminate the changes caused by substances from changes that are naturally occurring (e.g. strong winters; Zijlstra and de Wolf 1988) and from changes induced by man (e.g. beam-trawl fishing with and exploitation of fish, sand and gravel). In future, it will be more realistic when the judgement of an impact of substances on different levels in the ecosystem is performed on a risk assessment base and not on a hazard assessment base.

A major step forward in the assessment of substances is made thanks to the Toxicity Identification Evaluation (TIE) method. First an effect is

measured with bioassays or rapid screening methods and than the toxic sample is fractionated, for instance into a water and a fat soluble part. In the separated fractions the effect is measured again and by continuous fractionation the substances are identified that cause the effect. In this way a more realistic view on the impact of substances will be obtained and specific measures can be taken to reduce the emission of these specific substances.

In the production - decomposition cycles of carbon and other compounds (Fig. 6), the impact studies of substances focuses on the production part at different trophic levels

Little is known about the impact of substances on the decomposition part of the cycle: for instance on bacteria and externally active enzymes. Some studies do take into account a faster decomposition of substances by co-metabolism during the degradation of organic matter.

In soil studies it is already noticed that the organic matter concentration in the top layer is increasing slowly over time (Kirschbaum 2000). This might indicate that the decomposition process is slowed down. It is well known that this may be due to changes in pH and other parameters. However, it can also be caused by substances. Especially exo-

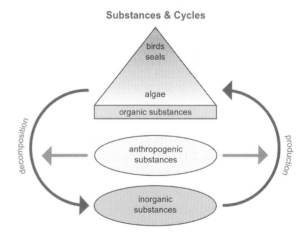

Fig. 6. Production and decomposition cycle of carbon compounds indicating the interaction with anthropogenic substances.

and ecto-enzymes, degrading large organic molecules, are sensitive for substances (Lee and Tay 1998). It is to be expected that a small change in the decomposition process world wide, as already noticed in soils, could be enough to explain the missing pool of carbon in the global carbon cycle (Hansen et al. 2000; Rustad et al. 2000).

Water quality models

Environmental concentrations of different compounds in the marine environment have been measured and modelled for already more than a decades (Kullenberg 1982). The existing environmental fate models describe mostly the fate of different compounds in the abiotic environment and relate this concentration to a quality value expressed as the so called PEC-NEC (predicted environmental concentration – no effect concentration). The state of the art of the environmental fate models is that processes and parameters are generically described and Quality Structure Activity Relations (QSARs) can be implemented to make the application of the model accessible for more compounds.

Different models for eutrophication, metals and organic substances in different countries around the North Sea are benchmarked against each other (OSPAR 1998 a and b). The major conclusions was, that the performance of the models is comparable but that there is a lack of data to calibrate and validate the models. Differences in performance between the models are not due to differences in the parameter setting and description of processes; they are mainly due to different approaches in the hydrographical model and the way suspended matter transport is described.

A large step forward in water quality modelling can be made when the underlying hydrographical and suspended matter sub models are improved. Smith et al. (1996) already noticed, for instance, that the amount of water entering the North Sea in different models differs considerably. In the NO-MADS (North Sea Model Advection-Dispersion Study) study it was shown that the trajectory of a particle in different models in terms of distance and direction also showed big discrepancies (Salden et al. 1996).

Management

In most marine studies, concerning the impact of substances, the sources of these substances are defined as direct (e.g. rivers) and indirect sources (e.g. atmosphere). An estimation of the relative contributions of all potential sources from various human activities to the marine environment is summarised in table 1 (Anonymous 1998).

Catchment approach

However, from a management point of view, concentrations in and impact on the marine environment must be connected to activities on land and not, rivers etc. Measures, to reduce emission of

Offshore production	1
Maritime transportation	12
Dumping	10
Subtotal ocean-based sources	23
Run-off and land-based discharges	44
Atmosphere	33
Sub-total land based sources	77
Total sources	100

Table 1. Relative contribution (%) of different sources of substances to the marine environment (Anonymous 1998).

substances must be taken through human efforts on land. Hence, for a proper management of substances, a connection must be made between the impact on the marine environment and the responsible activity on land. Catchment approaches, in which the socio-economical developments on land determine the input of substances into rivers and estuaries and ultimately to the coastal seas and oceans, are thus required to optimise effectively the measures necessary to protect the marine environment and to ensure healthy ecosystems. A good example of such a study is the nitrogen impact study for the Baltic Sea; here environmental improvements are connected to social-economical developments, ensuring cost-effective measures (Gren et al. 2000). To come up with solutions for eutrophication effects around the coasts of the US, the description of the status, local action and long term development were integrated to achieve the optimal environmental result (Anonymous 2000).

Healthy marine ecosystems

Up to now, nearly all studies concerning marine pollution address (potential) problems. Concentrations of substances are measured in different compartments and laboratory impact studies indicate potential hazard for the system. Lethal and sublethal effects on organisms in the field, induced by substances, are studied as well. However, all these studies are reinforcing the (potential) problem. "Solutions", presented by policymakers strive for an almost total reduction of the input of all substances and close to background values for natural substances. It is already becoming clear that society and stake holders do not agree anymore with such strict measures. The ultimate international aim: a healthy marine ecosystem, with no effects of substances is clear for everybody. However, there will always be effects of substances and how to define a healthy ecosystem.

During the Brend Spar discussion, it became clear that societies have an image of what a healthy marine ecosystem is. Dumping into the deep ocean was not at all an option. However, these viewpoints were mainly based on abdominal and emotional feelings, partly fed by our problem stories. Scientific arguments do not seem to play a role in these discussions. But, do we have enough scientific knowledge to describe a healthy marine ecosystem, on the short and long term?

Man will progressively to use the marine environment for all kind of activities. As a consequence the marine environment will change further, in one way or the other. It will be a challenge to identify indicators for healthy marine ecosystems and to monitor them world wide.

In a report of the FAO in 1959 (cited by Ruivo 1972) the concern about negative effects of pollution on aquatic organisms was already emphasized.

In 1970 FAO organised a meeting of 40 well-known environmental and natural scientists to identify substances posing threats on the marine environment, to evaluate and discuss existing analytical methods, to design an international monitoring system and to define further research needs. One of their conclusions was that marine science was ready to measure the health of the ocean. Can we really do that thirty years later? An urgent requirement was delineated for the gathering of data on the world production of substances as a guide to evaluate their potential threat to the marine environment (Butler at al. 1972).

One ocean – one world

In the last thirty years, large amounts of money have been spent world wide to set up local, national and international monitoring programs in coastal and marine waters. On a global scale, these activities can be compared with that of a multinational. However, it is still difficult to tell if the marine system is healthy or not. We are data rich and information poor is often concluded for some coastal system. However, we are data poor because of a lack of permanent ocean observation.

Before we can better define the impact of substances on the marine ecosystem, it is necessary to know the functioning, the future development of the natural ecosystem and the natural variability of ecosystems.

The benefits derived through the use of the oceans and via the exploitation of marine resources are received by nations with up-to-date scientific, technological and financial capacity. Their recognition of the importance of the ocean as such with respect to global climate and to men has led to a reduced input of most substances to the marine environment of the western world. However, it is not enough to manage our "western" waters. The oceans are connected to each other, all around the globe. It is to be expected that the industrial development and the increase in population in third world and developing countries, will increase the input of substances to their coastal waters. Our present knowledge of healthy marine ecosystems must be transferred to developing countries and capacity building over there will be necessary. Societal, economical and cultural differences between the western world and the developing countries, determining for a large extent their viewpoints on a healthy coastal zone, sea and ocean, will make the discussion more complicated though fruitful for the future.

Acknowledgements

Prof. Dr. J. de Leeuw, Dr. D. Vethaak and Drs R. Bosman and gratefully acknowledged for their comments which helped to improve the final manuscript.

References

Anonymous (1998) The Ocean our Future. Cambridge, London, 247p

Anonymous (2000) Clean Coastal Waters: Understanding and reducing the effects of nutrient pollution. National Research Council, Washington DC

Ball P (1999) H_2O; a Biography of Water. Weidenfeld and Nicolson, London

Bernes C (1999) Persistent Organic Pollutants: A Swedish View of an International Problem. Monitor 16, Swedish Environmental Protection Agency, Stockholm, Sweden pp 9-12

Butler P, Andrén LE, Bonde GJ, Jernelöv AB, Reish DJ (1972) Test, monitoring and indicator organisms. In: Goldberg ED (ed) A Guide to Marine Pollution. Gordon and Breach Science Publishers, New York pp 147-159

Carson R (1963) Dode Lente. H. Brecht's Uitgeversmaatschappij NV, Amsterdam, The Netherlands

Chapman PM, Wang F, Adams WJ, Green A (1999) Appropriate Application of Sediment Quality Values for Metals and Metalloids. Environ Sci Techn 33(22):3937-3941

Clark RB (1992) Marine Pollution, third edition. Clarendon Press, Oxford, 5-6

Colborn T, Dumanoski D, Myers JP (1999) Our Stolen Future. Penguin Books, USA

Cornelissen, G, van Noort PCM, Govers HAJ (1998) The mechanism of slow desorption of organic compounds from sediments: A study using model sorbents. Environ Sci Techn 32:3124-3131

Crosby DG (1998) Environmental Toxicology and Chemistry. Oxford University Press, New York

Degens ET, Kempe S, Richy JE (1991) Biogeochemistry of Major World Rivers. SCOPE 42, J Wiley & Sons, New York

Ellis R (1995) Monsters of the Sea. Alfred A. Knopf, New York

Goldberg ED (1972) A Guide to Marine Pollution. Gordon and Breach Science Publishers, New York 168 p

Goudie A (1990) The Human Impact on the Natural Environment. Blackwell, Oxford

Gren I-M, Turner K, Wulff F (2000) Managing a Sea: The Ecological Economics of the Baltic. Earthscan Publications Ltd, London 138p

Hansen BR, Ducklow HW, Field JG (2000) The Changing Ocean Carbon Cycle. Cambridge University Press, England

Hendriks AJ, Maas-Diepeveen JL, Noordsij A, Van der Gaag MA (1994) Monitoring response of XAD-concentrated water in the Rhine delta: A major part of the toxic compounds remains unidentified. Water Res 28:581-598

Hood DW (1971) Impingement of Man on the Oceans. Wiley Interscience, New York

Jones KC, de Voogt P (1999) Persistent organic Pollutants (POPs): State of the science. Environ Pollution 100:209-221

Jørgensen SE, Jørgensen BH (2000) Drugs in the environment. Chemosphere 40:691-699

Kirschbaum MUF (2000) Will change in soil organic carbon act as positive or negative feedback on global warming? Biogeochem 48:21-51

Laane RWPM, Sonneveldt HLA, van der Weyden AJ, Loch JPG, Groeneveld G (1999) Trends in the spatial and temporal distribution of metals (Cd, Cu, Zn and Pb) and organic compounds (PCBs and PAHs) in Dutch coastal sediments from 1981 to 1996: A model case study fro Cd and PCBs. J Sea Res 41:1-17

Lee K, Tay KL (1998) Measurement of Microbial Exo-enzyme Activity in Sediments for Environemntla Impact Assessment. Microscale Testing in Aquatic Toxicity 15, Chapter, CRC Press, Boca Raton, Florida, pp 219-236

OSPAR (1998a) Report of the ASMO Modelling Workshop on Eutrophication Issues, 3-8 November 1996, the Hague, The Netherlands

OSPAR (1998b) Reprot of the ASMO Workshop on Modelling Transport and Fate of Contaminants, 4-7 November, the Hague, The Netherlands

Rodan BD, Pennington DW, Eckly N, Boethling RS (1999) Screening for Persistent Organic Pollutants: Techniques To Provide a Scientific Basis for POPs Criteria in International Negotiations. Environ Sci Techn 33(20):3482-3488

Ruivo M (1972) Foreword. In: Goldberg ED (ed) A Guide to Marine Pollution. Gordon and Breach Science Publishers, New York pp 9-10

Rustads LE, Huntington TG, Boone RD (2000) Controls on soil respiration: Implications for climate change. Biogeochem 48:1-6

Salomons W, Bayne B, Heip CH, Turner K (1995) Changing Estuarine and Coastal Environment; sustain-ability and biodiversity in relation to eco-nomical aspects. EERO Workshop Report, Geesthacht pp 1-3 and pp 7-9

Smith JA, Damm, P, Skogen D, Flather RA, Patch J (1996) An Investigation into the variability of Circulation and Transport on the North-West European Shelf using Three Hydrographical Models. Deutsche Hydrographische Zeitschrift 48(3/4):325-348

Walker CH, Livingstone DR (1992) Persistent Pollutants in Marine Ecosystems. SETAC special publication series. Pergamon Press, Oxford

Wezel AP van, Kalf D (2000) Selection of Substances. National Institute of Public Health and the Environment, RIVM report 601503017, Bilthoven, The Netherlands, 40 p

Zijlstra JJ, de Wolf P (1988) Natural Events. In: Salomons W, Bayne BL, Duursma EK, Förtsner U (eds) Pollution of the North Sea; an Assessment. Springer Verlag, Berlin pp 164-182

Long-Term Habitat Changes and their Implications for Future Fisheries Management

M.J. Kaiser[1*], J.S. Collie[2], S.J. Hall[3] and I.R. Poiner[4]

[1]School of Ocean Sciences, University of Wales-Bangor, Menai Bridge,
Gwynedd, LL59 5EY, UK
[2]Graduate School of Oceanography, University of Rhode Island,
Narragansett, Rhode Island 02882, USA
[3]Australian Institute for Marine Science, PMB 3, Townsville MC,
Queensland 4810, Australia
[4]CSIRO Diversion of Marine Research, PO Box 120, Cleveland,
Queensland 4163, Australia
* corresponding author (e-mail): m.j.kaiser@bangor.ac.uk

Abstract: Recovery rate estimates for a variety of habitats would suggest that some areas of the seabed will continue to be held in a permanently altered state by the physical disturbance associated with fishing activities. What is clear from the studies undertaken to date is that there exist communities and habitats that are so sensitive to physical disturbance that all forms of bottom-fishing with towed gear should be excluded from these areas forthwith. While it would appear that gear restriction management regimes have the added benefit of conserving habitats, target species and benthic fauna within the management area, at present it is not possible to determine whether there are any wider benefits for the fishery that exploits target species outside the management area.

Introduction

Traditional fisheries management has tended to focus on the conservation of stocks of single species or more recently in its most sophisticated form has addressed the problems of multi-species fisheries (Jennings and Kaiser 1998). However, there is a growing appreciation of the need to consider the wider ecosystem effects of fishing activities on the marine environment and that these effects should be taken into consideration in any future management plans. This approach was highlighted most recently at a meeting held in Montpellier, France in 1998 that focused exclusively on the ecosystem effects of fishing. This meeting convened by the International Council for the Exploration of the Sea (ICES) and the Scientific Committee for Oceanographic Research (SCOR), attracted 331 delegates and produced a symposium volume with 35 peer reviewed papers and provides an excellent reflection of the growing importance of this area of research (ICES Journal of Marine Science, Volume 57, No. 3). The ecosystem effects of fishing change predator-prey relationships, lead to shifts in community structure and alternative stable states, alter body-size composition of species, lead to genetic selection for different body and reproductive traits, affect populations of vulnerable non-target species (seabirds, marine mammals and reptiles), reduce habitat complexity and perturb benthic communities (Jennings and Kaiser 1998). Increasingly, the latter two issues have assumed scientific importance as they have become conservation issues. The effects of towed bottom fishing gear on benthic communities and habitats has received considerable media attention in both the trade and popular media press and this has mirrored a rapid increase in the research effort that has ad-

From WEFER G, LAMY F, MANTOURA F (eds), 2003, *Marine Science Frontiers for Europe.* Springer-Verlag Berlin Heidelberg New York Tokyo, pp 189-201

dressed these issues (for reviews see Dayton et al. 1995; Jennings and Kaiser 1998; Kaiser 1998; Auster and Langton 1999; Kaiser and De Groot 2000).

Disturbance from fishing activities

Towed bottom fishing gears are used to catch those species that live in, on or in association with the seabed. Typically, these gears are fished in such a way that they remain in close contact with or dig into the seabed. The passage of towed fishing gear over the seabed has a number of immediate effects on the benthic habitat and associated fauna, these include:
• Disturbance of the upper 1 cm – 20 cm of the substratum causing short-term resuspension of sediments, remineralisation of nutrients and contaminants, and re-sorting of sediment particles.
• Direct removal, damage, displacement or death of a proportion of the endo- and epibenthic biota.
• A short-term (0-72 hrs) aggregation of scavenging species.
• The alteration of habitat structure (e.g. flattening of wave forms, removal of rock) and biogenic reefs.

These short-term effects are well documented in recent studies (see Jennings and Kaiser 1998; Kaiser and De Groot 2000 for reviews). The results from short-term studies are informative and often have confirmed our expectations of the type of changes that might occur as a result of fishing activity. Nevertheless, the usefulness of each study on its own is limited by factors such as the specific location, type of gear used and season during which the study in question was undertaken. Collie et al. (2000) overcame this problem by extracting summary data from a population of fishing impact studies and undertook a meta-analysis of the combined data set to ask the following questions:
• Are there consistent patterns in the responses of benthic organisms to fishing disturbance?
• How does the magnitude of this response vary with habitat, depth, disturbance type and among taxa?
• How does the recovery rate of organisms vary with these same factors?

Meta-analysis is the summary of multiple, independent studies to detect general relationships (e.g. Gurevitch and Hedges 1999). Meta-analysis has considerable statistical power because the results from each study can be regarded as independent replicates, permitting ecological questions to be examined on a much larger, and perhaps more relevant, scale than would otherwise be possible.

Collie et al. (2000) found that the magnitude of the immediate response (i.e. change in abundance or biomass) to fishing disturbance varied significantly according to the type of fishing gear used in the study, the habitat in which the study was undertaken, and among different taxa.

Immediate effects of bottom-fishing

Effects of different gears

The initial impacts of different fishing gears were mainly consistent with expectations. Intertidal dredging had more marked effects than scallop dredging, which in turn had greater effects than otter trawling (Fig. 1a). Although at first sight, the apparent lack of effect from beam trawling is somewhat surprising, we suspect that the relative paucity of data for this gear is almost certainly part of the explanation. It should also be borne in mind, however, that beam-trawling studies were generally conducted in relatively dynamic sandy areas, where initial effects may be less apparent or are less easily detected. Fishing disturbance effects of intertidal dredging are likely to have the greatest initial effects on the biota because fishers are able to use the harvesting machinery accurately, working parallel lines along the shore. In contrast, fishers using towed nets in subtidal areas are unable to actually see precisely where their gear is fishing although technological advances in positioning systems are making it increasingly easier to achieve very accurate positioning of fishing gear on the seabed. It is also significant that it is easier for a scientist to accurately collect samples from intertidal c.f. subtidal areas where sampling error is undoubtedly introduced with potential detrimental effects for the statistical power to detect change. Otter trawling appears to have the least significant impact on fauna compared with other gears, however it is necessary flag a few warnings about this observation. While it is the otter doors that hold the wings of the otter trawl open that have the greatest impact on the sediment habitat. They constitute a small proportion of the total width of the gear (ca. 2 m c.f. 40 – 60 m).

Effects in different habitats

Several authors have suggested that the relative ecological importance of fishing disturbance will be related to the magnitude and frequency of background of natural disturbances that occur in a particular marine habitat (Fig. 2; Kaiser 1998; Auster and Langton 1999). Certainly, it makes intuitive sense that organisms that inhabit unconsolidated sediments should be adapted to periodic sediment resuspension and smothering. Similarly, it seems plausible that organisms living in seagrass beds rarely experience repeated intense physical disturbances or elevated water turbidity as created by bottom fishing gears (Fig. 2). Indeed, such intuition has been the cornerstone of hypotheses about impacts and recovery dynamics for benthos (e.g. Hall 1994; Jennings and Kaiser, 1998). However, Collie et al. (2000) found that their initial impact results with respect to habitat were somewhat inconsistent among analyses. While the initial responses to fishing disturbance of taxa in sand habitats were usually less negative than in other habitats, a clear ranking for expected impacts did not emerge (Fig. 1b). Such inconsistencies may reflect interactions between the factors arising from the unbalanced nature of the data, with many combinations of gear and habitat absent. For example, the relatively low initial impact on mud habitats may be explained by the fact that most studies were done with otter trawls. If data were also available for the effect of dredgers on mud substrata a more negative response for this habitat may have been observed. Nevertheless, it should be borne in mind that initial effects of disturbance may be hard to detect in mud communities that often have low abundances of biota which tend to be burrowed deep (10 – 200 cm) within the sediment. Presumably, the deeply burrowed fauna would be relatively well protected from the physical effects of disturbance although the passage of the gear will cause their burrows to collapse. Whether these inconsistencies can be explained in this way can only await further study.

Immediate effects on biota

Collie et al. (2000) found that the most consistently interpretable result within their meta-analysis was the vulnerability of fauna, with a ranking of initial impacts that concurred broadly with expectations based on morphology and behaviour. The lugworm *Arenicola* spp. had the greatest initial response to disturbance which is not surprising given that this was the target of a commercial fishery (Fig. 1c).

Collie et al. (2000) also undertook a regression tree analysis that perhaps provides the first quantitative basis for predicting the relative impacts of fishing under different situations. Following the tree from its root to the branches we can make predictions, for example, about how a particular taxon would be affected initially by disturbance from a particular fishing gear in a particular habitat. Thus, trawling would reduce anthozoa (anemones, soft corals, sea ferns) by 68%, whereas asteroid starfishes would only be reduced by 21%. Similarly, repeated (chronic) dredging is predicted to lead to 93% reductions for Anthozoa, Malacostraca (shrimps and prawns), Ophiuroidea (brittlestars) and Polychaeta (bristle worms), whereas a single (acute) dredge event is predicted to lead to a 76% reduction. This approach might ultimately provide a useful quantitative framework for predicting fishing impacts that can be updated and refined as new data emerge.

Recovery rates after trawl disturbance

Soft sediments

In an environment that is open to disturbance by fishing gear, the short-term effects of bottom-fishing disturbance on habitats and their biota are of interest but of far less ecological importance than the issue of the potential for recovery or restoration. Unfortunately, relatively few studies of trawl disturbance have included a temporal component of sufficient duration to address longer-term changes that occur as a result of bottom fishing disturbance.

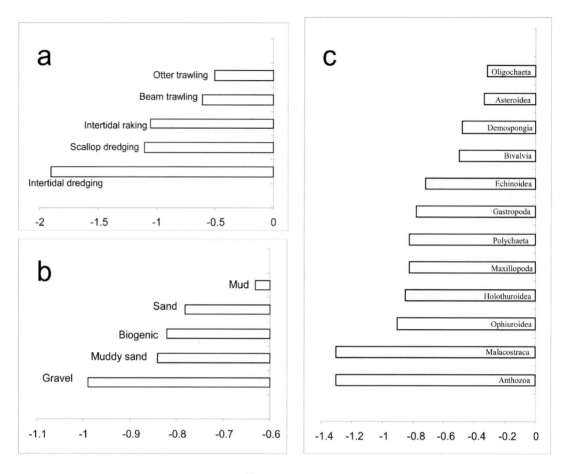

Fig. 1. The predicted mean short-term response of taxa to physical disturbance. **a)** the response of invertebrate abundance to different gear types; **b)** the response of invertebrate abundance or biomass in different habitats; **c)** the response of different taxa to physical disturbance that occurred immediately after that disturbance had occurred. Data are on a transformed scale where values correspond to % declines from controls as follows: -0.1= 10%; -0.22=20%; -0.35=30%; -0.5=40%; -0.68=50%, -1.35=75%;-4.61=100%. In all cases the initial response of the fauna was negative. Adapted from Collie et al. (2000).

This is almost certainly a result of the conflict between financial resources, project duration, statistical and analytical considerations. Nevertheless, Collie et al. (2000) were able to incorporate studies that included a recovery component into their analysis. This permitted them to speculate about the level at which physical disturbance becomes unsustainable in a particular habitat. For example, their study suggested that sandy sediment communities are able to recover within 100 days which implies

that they could perhaps withstand 2-3 three incidents of physical disturbance per year without changing markedly in character (Fig. 3). This level of fishing disturbance is the average predicted rate of disturbance for the whole of the southern North Sea. However, when fishing effort data is collected at a fine spatial (9 km²) resolution (Rijnsdorp et al. 1998) it becomes clear that effort is patchily distributed and that some relatively small areas of the seabed are visited by >400 trawlers per year. This level of

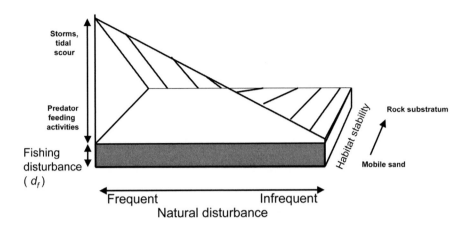

Fig. 2. A simplistic model to illustrate the relative importance of a constant level of fishing disturbance in different habitats (habitat stability) that are subjected to different frequencies or levels of natural disturbance. As levels of natural disturbance decline, fishing disturbance accounts for a greater proportion of the total disturbance experienced and becomes increasingly important. Adapted from Jennings and Kaiser 1998.

fishing equates to a total disturbance of approximately 8 times per year (Rijnsdorp et al. 1998). If Collie et al's (2000) recovery rate estimates for sandy habitats are realistic, this would suggest that these areas of the seabed are held in a permanently altered state by the physical disturbance associated with fishing activities.

At this point, there are some important limitations within the data that should be considered. First, the small spatial scale (the maximum width of most of the disturbed areas examined was < 50 m) of most of the trawl impact studies make it likely that much of the recolonisation was via active immigration into disturbed patches rather than reproduction within patches. We found recovery to be slower if the spatial scale of impact was larger, as it would be on heavily fished grounds due to the additive effects of an entire fleet of trawlers. Second, it should be noted, that while we might accurately predict the recovery rate for small-bodied taxa such as polychaetes, which dominate the data set, sandy sediment communities often contain one or two long-lived and therefore vulnerable species. Note, for example, the occurrence of the large bivalve *Mya arenaria* in the intertidal zone of the Wadden Sea. While the

majority of the benthos in this environment recovered within 6 months of lugworm dredging, the biomass of *M. arenaria* remained depleted for at least 2 years (Beukema 1995). This delayed recovery of larger-bodied organisms is no doubt even more important in habitats that are formed by living organisms (e.g. soft corals, sea fans, mussels) as the habitat recovery rate is directly linked to the recolonisation and growth rate of these organisms (see below).By now there is sufficient evidence in the literature to suggest that under conditions of repeated and intense bottom-fishing disturbance a shift from communities dominated by relatively high biomass species towards dominance by high abundances of small-sized organisms will occur.

Biogenic habitats

It is clear that intensively fished areas are likely to be maintained in a permanently altered state, inhabited by fauna adapted to frequent physical disturbance. These effects will be most apparent for stable types of habitats that contain structural biogenic components. Presumably, such habitats will have the longest recovery time compared with less

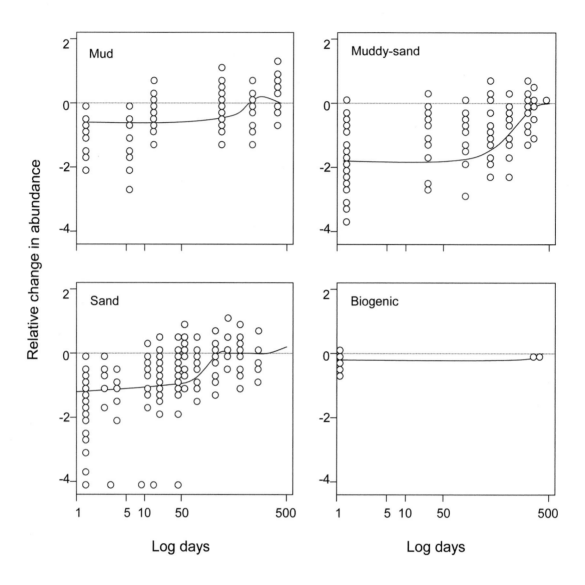

Fig. 3. Results from a meta-analysis of the effects of fishing disturbance on benthic communities. The scatter plots of the relative change of all species (each datapoint represents the relative abundance of a different species on each different sampling date) in different habitats at time intervals after the occurrence of a fishing disturbance. The fitted curves show the predicted time trajectory for recovery to occur. On the y-axis, 0 shows no change in abundance, negative values show a relative decrease in abundance. Adapted from Collie et al. (2000).

stable substrata. Yet it is for these habitats that the paucity of data is most apparent. While it would appear that none of the habitats included in Collie et al.'s (2000) study fall into this category, some new data are beginning to emerge. Hall-Spencer and Moore (2000) examined the effects of fishing disturbance on maerl beds. Maerl beds are composed of highly dichotomous calcareous algae. This

forms a complex substratum with a high degree of 3-dimensional complexity. Not surprisingly, the associated assemblages have high diversity and many of the associated species are large-bodied and slow-growing. Hall-Spencer and Moore (2000) showed that four years after the occurrence of an initial scallop-dredging disturbance had occurred, certain fauna, such as the nest building bivalve *Limaria hians*, had still not recolonised trawl tracks (Fig. 4). Similarly, work by Sainsbury (1987) and Sainsbury et al. (1997) suggests that recovery rates may exceed fifteen years for sponge and coral habitats off the western coast of Australia. The presence of such habitats was important for fish species of commercial importance.

Deducing the effects of chronic disturbance

The perceived problems that might be associated with intense and prolonged bottom-fishing disturbance have only been examined in the last 20 years.

However, the bottom-fishing fleets have been in operation much longer. For example the beam trawl fleet in the southern North Sea expanded dramatically through the 1960s and 1970s. Consequently, many present-day studies have been undertaken in what is already a considerably altered environment. Despite our efforts to predict the outcome of fishing activities for existing benthic communities, we are often unable to deduce the original composition of the fauna because data gathered prior to the era of intensive bottom fishing are sparse. This is an important caveat because recent analyses of the few existing historical datasets suggest that larger bodied organisms (both fish and benthos) were

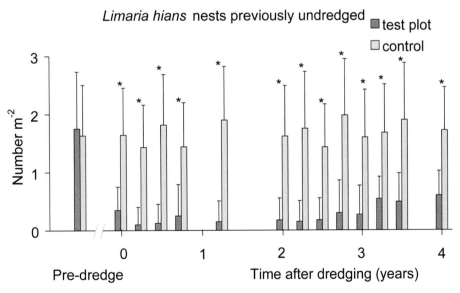

Fig. 4. Numbers of byssus nests of the bivalve *Limaria hians* recorded within test and control plots prior to and over 4 years after experimental scallop dredging on the previously undredged ground. Error bars are ± SE (n = 20), significant differences (ANOVA on log transformed data) are indicated by *(P < 0.05) (Photograph copyright Jason Hall-Spencer, from Hall-Spencer and Moore (2000)).

more prevalent prior to intensive bottom trawling (Greenstreet and Hall 1996; Frid et al. 2000; Rumohr and Kujawski 2000). Moreover, in general, epifaunal organisms are less prevalent in areas subjected to intensive bottom fishing (Collie et al. 1997; Sainsbury et al. 1997; Kaiser et al. 2000a, b; McConnaughy et al. 2000; Rumohr and Kujawski 2000). An important consequence of this effect is the reduction in habitat complexity (architecture) that accompanies the removal of sessile epifauna. Nevertheless, it has been hard to convincingly demonstrate that towed bottom fishing activity has been responsible for changes in bottom fauna and habitats. Often, effort data are lacking at a scale or over a time period that is relevant to ascertain the disturbance history of a particular area of seabed. In the few instances when such data have been available, observations have indicated consistently a shift from dominance by high biomass organisms towards communities dominated by small-bodied opportunistic species (Collie et al. 1997; Engel and Kvitek 1998; Bradshaw et al. 2000; Kaiser et al. 2000a,b; McConnaughey et al. 2000). It is becoming increasingly apparent that habitat modification appears to have important consequences for fish communities (Sainsbury et al. 1987; Auster and Langton 1999; Kaiser et al. 1999).

Essential Fish Habitat

Recent amendments to the *Magnuson-Stevens Act* require fisheries managers to define 'essential fish habitat' (EFH) and address the impact of fishing gear in their management plans (Benaka 1999). This is probably one of the first legislative steps taken in fisheries management that will require the assimilation and application of the scientific knowledge outlined in the paragraphs above. In many ways, this legislation is one of the first measures to embrace an ecosystem perspective in fisheries management. In some instances, it is fairly simple to identify those habitats that might be considered essential to the life-history of some species. Such habitats include spawning and nursery areas, many of which are protected from fishing activity in European waters. However, of equal relevance are the habitat quality issues that affect the acquisition of food and the avoidance of predators. Hence there is an urgent need to identify those habitats that have an important or 'essential' functional role for particular species or types of fish (e.g. piscivores/herbivores/omnivores or flatfish/roundfish) at other stages of their life-history.

Habitat complexity is a product of the surface topography of the substratum and the sessile epifauna that grow upon it. Reef forming organisms can result in habitats of very high complexity providing a multitude of refuges for a diverse range of species. More subtle features such as sand ridges and pits created by the feeding or burrowing action of benthic fauna may provide shelter for bottom-dwelling fish species (e.g. Walters and Juanes 1993; Auster et al. 1997). Bottom fishing activities are capable of greatly reducing habitat complexity by either direct modification of the substratum or removal of the fauna that contribute to surface topography (Auster and Langton 1998; Jennings and Kaiser 1998). Hence, degradation of habitat complexity by fishing activities may lead to changes in the associated fish assemblages (e.g. Sainsbury et al. 1997). Alteration of habitat feature has been shown to have important consequences for freshwater fishes and this is the caveat that underpins much of the ecological restoration projects centred on salmonid habitats. An initial study of habitat/fish assemblage relationships indicated that even subtle alterations in habitat characteristics can cause a shift in the dominance of certain fish species within the assemblage (Kaiser et al. 1999). Presumably, a good understanding of the link between fish and their habitat would enable us to predict the consequences of habitat alteration. For example, for certain species such as sole *Solea solea* that preferentially live in relatively uniform sandy areas (e.g. Gibson and Robb 1992; Rogers 1992), the exclusion of towed bottom fishing gear from an area of the seabed could permit the growth of emergent sessile fauna that make the environment better suited to flatfishes such as plaice *Pleuronectes platessa* and dab *Limanda limanda* (Kaiser et al. 1999). Thus in the case of the sole fishery, the fishing activity may maintain the seabed habitat in a condition that favours the target species. Quite clearly the opposite is would be true for any species of fish that favour more complex habitats.

Integrating habitat conservation objectives into fisheries management

It would appear that with sufficient scientific information it should be possible to formulate a regime of fishing effort (= physical disturbance for towed bottom-fishing gear) that would be environmentally sustainable. Here we define environmentally sustainable as the process by which the habitat and its associated biological assemblage can recover before a subsequent disturbance event. For example, in shallow sandy areas of the seabed, two to three physical disturbances of the seabed every year may have little or no net effect on the habitat or resident assemblage. However, at present, the definition of sandy areas is too imprecise a habitat criterion on which to base such a management plan. We know, for example, that sand flats that are dominated by tube-building spionid worms take much longer to recover if these worms are removed through physical disturbance, as the worms normally have a stabilizing effect on the habitat (Thrush et al. 1996). Nevertheless, the complete exclusion of bottom-fishing disturbance from sandy habitats that are fished at present may actually have a negative effect on the fishery. Physical disturbance will, to some extent, promote dominance by opportunistic species such as small polychaetes that form a major component of the ecosystems of many commercially important flatfish species (Rijnsdorp and van Leeuwen 1996). More importantly, the exclusion of fishers from areas that they have fished for many years may displace fishing activity into areas previously free of fishing activity.

What is clear from the studies undertaken to date is that there exist communities and habitats that are so sensitive to physical disturbance that all forms of bottom-fishing with towed gear should be considered for exclusion from these areas forthwith. As a matter of urgency, there is need to identify other habitats that have long recovery times and that are exposed (or might in the future be exposed) to towed bottom-fishing gear – the most likely candidates are those that contain a high proportion of structural fauna. In European waters examples of such habitats would include:

• Deep-sea coral reefs of *Lophelia pertusa*.

• Maerl beds.
• Reefs of mussels (*Modiolus modiolus*), oysters, *Limaria hians* and *Sabellaria* spp.
• Areas of the seabed with aggregations of sea fans.
• Beds of fan mussels (e.g. *Atrina fragilis*).
• Sea grass meadows.

It is important at this point to define what we mean by sensitive fauna or habitats. Sensitive fauna may be defined by their physical attributes (e.g. fragility of body structure), their reproductive strategies (e.g. infrequent recruitment or low reproductive output) or remaining population size (e.g. the lower the population size, the more vulnerable to extirpation that species will be). Sensitive, non-sensitive, structured and non-structured fauna and habitats – all will be affected to some degree by towed bottom-fishing gear. However, for effective management, we need to define and identify sensitive species or habitats, and the extent (in area) of these resources, and the management approach(es) that should be used. This usually requires clearly defined management objectives (that can be measured and monitored) and data on fishing effort (level and spatial distribution), impact and recovery times. From this information, management strategies can be developed and tested against the objective(s). We can then start defining what we mean by a sustainable fishery.

Rotational closures and gear exclusion zones

The total exclusion of certain forms of fishing activities from areas of the seabed will inevitably lead to opposition from the fishing industry and by its nature such measure are extreme. Nevertheless a recent large-scale study on the NW Atlantic coast of North American has demonstrated elegantly the effectiveness of such large-scale closures. Alternatively, inshore fisheries lend themselves to the partitioning of seabed resources such that certain areas of the seabed can continue to be exploited using gears that cause minimal environmental damage. The two following examples illustrate the potential of such approaches.

In New England, USA, seasonal closed areas have been an important component of fisheries management since the early 1970s but had little impact on the groundfish stocks that they were designed to protect. In 1994, three large areas that in total covered 17 000 km² of the seabed were closed all year to all fishing gears that might retain groundfish (trawls, scallop dredges, hooks etc). These closed areas were maintained for five years and were found to incidentally protect the more sedentary components of the assemblage such as flatfishes, skates and scallops. Although less protection was afforded to cod *Gadus morhua* and haddock *Melanogrammus aeglefinus* additional legislation to protect specific important juvenile habitat lowered stock-wide mortality rates. Scallop dredgers were excluded because they took a by-catch of groundfish species. The relaxation of fishing effort on scallops had dramatic effects and led to a 14-fold increase in scallop biomass within the closed areas during 1994-1998 (Fig. 5). A portion of the closed areas was re-opened to scallop dredging in 1999 (Murawski et al. 2000). The returns of scallops during this period were so encouraging that managers are now contemplating a formal 'area rotation' scheme for this fishery presumably on a time-scale of 4-5 years.

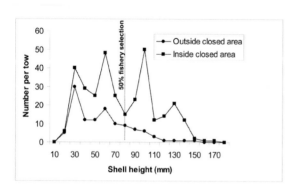

Fig. 5. Standardised abundance of sea scallops (numbers per dredge tow) by shell height, taken in the July National Marine Fisheries Service dredge survey on Georges Bank. Data are presented separately for the areas closed and those open to scallop dredging. Harvestable animals are indicated by the 50% selection line. Adapted from Murawski et al. (2000).

The second example comes from an inshore fishery off the south coast of England. When two commercially important species co-exist in the same habitat conflict may arise between different sectors of the fishing industry. A good example of this situation is when fishers using towed bottom fishing gears (scallop dredges, beam trawls and otter trawls) operate in the same areas in which fixed bottom gear (crab pots) are deployed. Kaiser et al. (2000b) examined an area subject to a voluntary agreement between these two sectors of the fishing industry such that some areas are used exclusively by fixed gear fishers, some are shared seasonally by both sectors, and others are open to all methods of fishing year-round. This agreement was enacted to resolve conflict between the two sectors of the industry. An additional perceived benefit of this agreement was the possible protection of the seabed from towed bottom fishing gear. Kaiser et al. (2000b) undertook comparative surveys of the benthic habitat and communities within the area covered by the agreement and compared different areas subjected to a range of fishing disturbance regimes. Communities found within the areas closed to towed fishing gears were significantly different from those open to fishing either permanently or seasonally. Abundance/biomass curves plotted for the benthic fauna demonstrated that the communities within the closed areas were dominated by higher biomass and emergent fauna that increase habitat complexity (Fig. 6). Areas fished by towed gear were dominated by smaller-bodied fauna and scavenging taxa. While it would appear that gear restriction management regimes have the added benefit of conserving habitats, target species and benthic fauna within the management area, it is at present not possible to determine whether there are any wider benefits for the fishery that exploits the target species outside the management area.

Future research priorities

With respect to the design of future studies, we feel that experimentalists wishing to address the fishing impacts issue will be best served by abandoning short-term, small-scale pulse experiments. In-

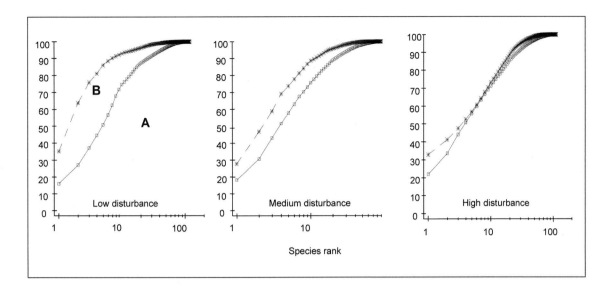

Fig. 6. Abundance/biomass curves of samples collected from areas protected from towed bottom fishing gear (low disturbance), areas open seasonally to towed bottom fishing gear and those areas that are fished all year with towed bottom fishing gear (high disturbance). As the level of bottom fishing disturbance increases the biomass curve (B) converges with the abundance curve (A) which is a typical response in stressed communities. Adapted from Kaiser et al. 2000b.

stead, the scientific community should be arguing for support to undertake much larger scale press and relaxation experiments. One half of the experiment has already been done - since fishing activity has been providing the press for many years, what we now require are more carefully designed closed area contrasts. There are two principal advantages to this approach. First, the results obtained are clearly interpretable in terms of real world intensities of fishing disturbance. Second, the spatial scale of the protected areas could be relatively small (and hence replicated, to fulfil the requirements for sound experimental design) without compromising unduly the interpretation of recovery dynamics: estimates of recovery in small protected areas in a sea of disturbance are likely to be conservative, while recovery in small deliberately disturbed patches are not. Third, the experiments would be conducted in the very habitats (i.e. real fishing grounds) about which the question of recovery is actually being posed. An examination of data collected prior to the start of intensive bottom-

fishing could form the background against which we judge the condition of present-day benthic assemblages and habitats (e.g. Rumohr and Kujawski 2000).

Our current understanding of the functional role of many of the larger-bodied long-lived species is limited and should be addressed to predict the outcome of permitting chronic fishing disturbance in areas where these animals occur. In addition, our understanding of the ecosystem services that many of these species provide is limited by a paucity of scientific understanding (e.g. Hall-Spencer and Moore 2000).

To date the majority of studies that have addressed fishing impacts on the seabed have concentrated on the biota with little reference to gear/sediment interactions (Pilskaln et al. 1998). As fishing gears disturb soft sediment they produce sediment plumes and remobilise previously buried organic and inorganic matter. Presumably this increases the release of nutrients into the water column and has important consequences for rates of biogeochemical cycling. To date, this issue has received little attention with the exception of one or two as yet unpublished studies.

References

Auster P, Malatesta R, Donaldson C (1997) Distributional responses to small-scale habitat variability by early juvenile silver hake, *Merluccius bilinearis*. Env Biol Fish 50:195-200

Auster PJ, Langton RW (1999) The effects of fishing on fish habitat. Fish habitat: Essential fish habitat and restoration. Benaka L Bethesda, Maryland, American Fisheries Society. Symposium 22:150-187

Ball BJ, Fox G, Munday BW (2000) Long- and short-term consequences of a *Nephrops* trawl fishery on the benthos and environment of the Irish Sea. ICES J Mar Sci 57:1315-1320

Benaka L (1999) Fish habitat: Essential fish habitat and rehabilitation. Bethesda, Maryland, American Fisheries Society

Beukema JJ (1995) Long-term effects of mechanical harvesting of lugworms *Arenicola marina* on the zoo-benthic community of a tidal flat in the Wadden Sea. Neth J Sea Res 33:219-227

Bradshaw C, Veale LO, Hill AS, Brand AR (2000) The effects of scallop dredging on gravelly seabed communities. In: Kaiser MJ and SJ De Groot (eds) Effects of Fishing on Non-Target Species and Habitats: Biological, Conservation and Socio-Economic Issues. Oxford, Blackwell Science pp 83-104

Collie JS, Escanero GA, Valentine PC (1997) Effects of bottom fishing on the benthic megafauna of Georges Bank. Mar Ecol Prog Ser 155:159-172

Collie JS, Hall SJ, Kaiser MJ, Poiner IR (2000) A quantitative analysis of fishing impacts on shelf-sea benthos. J Anim Ecol 69:785-799

Dayton PK, Thrush SF, Agardy MT, Hofman RJ (1995) Environmental effects of marine fishing. Aquatic Conserv 5:205-232

Engel J, Kvitek R (1998) Effects of otter trawling on a benthic community in Monterey Bay National Marine Sanctuary. Conserv Biol 12:1204-1214

Frid CLJ, Harwood KG, Hall SJ, Hall JA (2000) Long-term changes in the benthic communities on North Sea fishing grounds. ICES J Mar Sci 57:1303-1309

Gibson RN, Robb L (1992) The relationship between body size, sediment grain size and the burying ability of juvenile plaice *Pleuronectes platessa* L. J Fish Biol 40:771-778

Greenstreet SPR, Hall SJ (1996) Fishing and groundfish assemblage structure in the northwestern North Sea:

An analysis of long-term and spatial trends. J Anim Ecol 65: 577-598

Gurevitch J, Hedges LV (1999) Statistical issues in ecological meta-analysis. Ecol 80:1142-1149

Hall SJ (1994) Physical disturbance and marine benthic communities: Life in unconsolidated sediments. Oceanogr Mar Biol Annu Rev 32:179-239

Hall-Spencer JM, Moore PG (2000). Impact of scallop dredging on maerl grounds. In: Kaiser MJ and SJ De Groot (eds) Effects of Fishing on Non-Target Species and Habitats: Biological, Conservation and Socio-Economic Issues. Oxford, Blackwell Science pp 105-118

Jennings S, Kaiser M (1998) The effects of fishing on marine ecosystems. Adv Mar Biol 34:201-352

Kaiser MJ (1998) Significance of bottom-fishing disturbance. Conserv Biol 12:1230-1235

Kaiser MJ, De Groot SJ (2000) Effects of Fishing on Non-Target Species and Habitats: Biological, Conservation and Socio-Economic Issues. Oxford, Blackwell Science

Kaiser MJ, Ramsay K, Richardson CA, Spence FE, Brand AR (2000a) Chronic fishing disturbance has changed shelf sea benthic community structure. J Anim Ecol 69:494-503

Kaiser MJ, Rogers SI, Ellis JR (1999) Importance of benthic habitat complexity for demersal fish assemblages. In: Benaka L (ed) Fish Habitat: Essential Fish Habitat and Restoration. Bethesda, Maryland, American Fisheries Society 22:212-223

Kaiser MJ, Spence FE, Hart PJB (2000b) Fishing gear restrictions and conservation of benthic habitat complexity. Conserv Biol 14: 1512-1525

McConnaughey RA, Mier KL, Dew CB (2000) An examination of chronic trawling effects on soft-bottom benthos of the eastern Bering Sea. ICES J Mar Sci 57:1377-1388

Murawski SA, Brown R, Lai H-L, Rago PJ, Hendrickson L (2000) Large-scale closed areas as a fishery-management tool in temperate marine systems: The Georges Bank experiment. Bull Mar Sci 66:775-798

Overholtz W, Tyler A (1985) Long-term responses of demersal fish assemblages of Georges Bank. US Fiseries Bulletin 83:507-520

Pilskaln CH, Churchill JH, Mayer LM (1998) Resuspen-sion of sediment by bottom trawling in the Gulf of Maine and potential geochemical consequences. Conserv Biol 12:1223-1229

Rijnsdorp AD, Buijs AM, Storbeck F, Visser E (1998) Micro-scale distribution of beam trawl effort in the southern North Sea between 1993 and 1996 in relation to the trawling frequency of the sea bed and the impact on benthic organisms. ICES J Mar Sci 55: 403-419

Rijnsdorp AD, Leeuwen PI van (1996) Changes in growth of North Sea plaice since 1950 in reslation to density, eutrophication, beam-trawl effort, and temperature. ICES J Mar Sci 53: 1199-1213.

Rogers SI (1992) Environmental factors affecting the distribution of Dover sole (*Solea solea* L.) within a nursery area. Neth J Sea Res 29:151-159

Rumohr H, Kujawski T (2000) The impact of trawl fishery on the epifauna of the southern North Sea. ICES J Mar Sci 57:1389-1394

Sainsbury KJ, Campbell RA, Lindholm R, Whitelaw AW (1997). Experimental management of an Australian multispecies fishery: Examining the possibility of trawl-induced habitat modification. In: Pikitch K, DD Huppert and MP Sissenwine (eds) Global Trends: Fisheries Management. Bethesda, Maryland, American Fisheries Society 20:107-112

Sainsbury KJ (1987) Assessment and management of the demersal fishery on the continental shelf of northwestern Australia. In: Polovina JJ and S Ralston (eds) Tropical Snappers and Groupers - Biology and Fisheries Management. Boulder, Colorado, Westview Press pp 465-503

Thrush SF, Whitlatch RB, Pridmore RD, Hewitt JE, Cummings VJ, Wilkinson MR (1996) Scale-dependent recolonization: the role of sediment stability in a dynamic sandflat habitat. Ecol 77:2472-2487

Walters CJ, Juanes F (1993) Recruitment limitation as a consequence of natural selection for use of restricted feeeding habitats and predation risk taking by juvenile fishes. Canadian J Fish Aquat Sci 50: 2058-2070

More Sand to the Shorelines of the Wadden Sea
Harmonizing Coastal Defense with Habitat Dynamics

K. Reise

Alfred-Wegener-Institut für Polar- und Meeresforschung, Wattenmeerstation Sylt,
D 25992 List, Germany
corresponding author (e-mail): kreise@awi-bremerhaven.de

Abstract: In the Wadden Sea, dynamic shoreline habitats have suffered substantial losses, squeezed between a rigid coastal architecture and a rising sea level. Shore replenishments with sand taken from the bottom of the North Sea would have a wide potential to restore lost habitats and species diversity seaward of dykes, bulkheads and eroding shorelines. Such an artificial sand supply has multiple advantages. It dissipates wave energy and may appease the hunger for sand caused by sea level rise. At sheltered shores, the longevity of nourished sand deposits will be sufficient to allow for a development of natural geomorphics and biotic successions. The aesthetic value and the touristic reputation of the coast will improve. It is assumed that about half of the shore length may be suitable for replenishments, preventing erosion, covering shoreline petrifications and replacing land reclamation fields. However, before such a large-scale management action is initiated, assumptions on the coastal sediment budget and implications on coastal ecology need to be scientifically explored and experimental replenishments need to be analyzed. The project may become a key issue of an integrated coastal management at developed sandy coastlines.

Introduction

Sedimentary shores are dynamic and their shape and position is influenced by wind, water and their resident biota (Davis 1985; Ginsburg 1975; Short 1999). Humans have reacted to the undesirable effects of the dynamics of the marine environment by modifying the coast according to their needs (Carter 1988; Viles and Spencer 1995; Nordstrom 2000). They have created firm coastal architectures along sedimentary shores that differ greatly from natural structures, facilitated by the concurrence of two global trends: the desire to settle as close as possible at the sea and by a rise in sea level. Attempts to maximize the safety of people and their property against shore dynamics entailed losses in natural habitats and ecotone functions. This conflict may be resolved, based on analyses of past developments, sediments budgets, and the potential for restoring shoreline dynamics with shore replenishment using marine derived sediments.

The shore of the shallow inner North Sea is liable to extensive flooding by storm tides. Inunda-tions have caused the death of many people and permanent flooding of land. This has prompted the evolution of ever improving coastal defense struc-tures as well as the reclamation of land from the sea. The Wadden Sea, located at the southeastern coast of the North Sea, is ca. 500 km long and up to 30 km wide. It is adjoined by intensive agriculture and is an important recreation area for the populations of The Netherlands, Germany and Denmark. It comprises one of the largest coherent intertidal areas in the world with reed and salt marshes, mudflats and sandflats, deep tidal inlets, 58 islands and high sands with dunes and beaches (Wolff 1983).

The three bordering countries cooperate in the protection of nature in the Wadden Sea, and adopted a common management plan (Anonymous 1998; de Jong et al. 1999). It has been agreed to increase natural morphology and dynamics of beaches, salt marshes and tidal flats, to allow for natural vegeta-tion succession, and to provide favorable habitats

From WEFER G, LAMY F, MANTOURA F (eds), 2003, *Marine Science Frontiers for Europe.* Springer-Verlag Berlin Heidelberg New York Tokyo, pp 203-216

for birds. These so-called 'targets' often are in conflict with coastal defense. It is the aim of this paper to initiate research for a solution to this conflict.

Here, I explore the potential of shore replenishments with sand dredged from offshore. First, I give a brief history of geomorphological, ecological and human developments in the Wadden Sea region, with an emphasis on coastal protection and its ecological implications. I then proceed with a suggestion to employ shore replenishments not only at the exposed sandy beaches of seaside resorts but also along sheltered shorelines that are retreating or are defended by hard structures. Sand nourishment on a sheltered shore has the potential to restore coastal dynamics and the associated habitat diversity. It may serve to combine the needs and demands of coastal and nature protection, shoreline aesthetics and tourism. However, science still needs to explore the drawbacks and benefits as well as immediate and long-term implications in the framework of an integrated coastal management.

History of the Wadden Sea

The primordial coast

The Wadden Sea is of recent origin. In the course of glacial retreat, the North Sea rapidly flooded the shallow region south of the Dogger Bank, and the Wadden Sea with its chain of barrier islands and extensive tidal flats originated at its southeastern coast about 6,000 years ago (Bartholdy and Pejrup 1994; Flemming and Davies 1994; Oost and Boer 1994; Streif 1989). Sea level continued to rise, albeit slower and with some intermittent declines. Accordingly, the Wadden Sea changed in size and shape, mostly in response to sea level, storm events and sediment supply (Fig.1). The areal extent and the relative proportions of freshwater, brackish and marine marshes, mud and sand flats were quite variable through the course of time.

The Wadden Sea is subject to severe storm tide surges. Low pressure areas approach the coast from the North Atlantic with strong onshore winds.

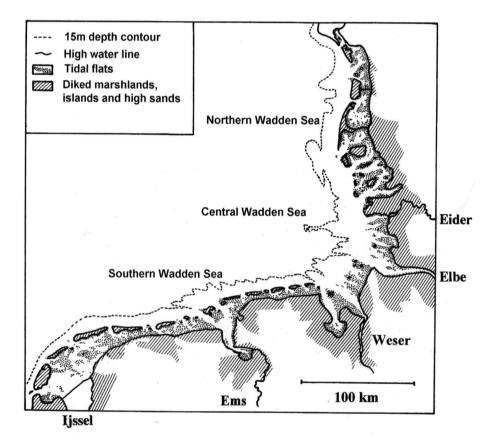

Fig. 1. The Wadden Sea

These meet a shallow sea and often generate in the funnel-shaped eastern North Sea a steep increase in sea level towards the coast with 2 m or more above mean high tide. Extensive floodings of the flat coastal area occurred as a consequence of these conditions in the past. Since inundations also gave rise to fertile soil, there was a strong incentive for the agricultural use of low-lying marshlands in spite of the associated dangers.

Dyking at the mainland

This situation improved with the invention of dykes that protected against floods from the sea about a thousand years ago which evolved into a very effective coastal defense system (Kramer and Rohde 1992; Reise 1996; Wolff 1992). High levels of precipitation required an elaborate irrigation of the dyked lands (polders) to avoid flooding with freshwater. Irrigation of the often peaty soils caused these to shrink. Peat was also excavated. This not only lowered the land but reduced its capacity to store rain-water. Thus, irrigation had to be improved further, and the dyked marshland subsided gradually below mean sea level. When severe

storms struck the weakly dyked coast in medieval times, severe loss of life and arable land occurred.

Subsequently, efforts in dyking increased and the techniques used improved considerably. Most of what had been lost to the sea in the middle ages was regained by progressing embankments. Sediments were trapped in land reclamation fields in front of new dykes, preparing the terrain for the next embankment which was then was formed with a still better dyke design to provide more safety. To save the costs of maintaining extensive dykelines along estuaries, sluices and storm surge barriers were built that close during storm tides. To avoid the inundation of arable land during heavy rains, coinciding with onshore winds, tidal flats were embanked for the purpose of storing freshwater runoff until ebbing tides were again low enough to allow for discharge into the sea.

On average, these stepwise developments shifted the dykeline of the mainland 5 km seaward, decreasing the distance towards the barrier islands by one third (Fig.2). The coastline became rather straight because the inlets were embanked most easily. High water levels in the Wadden Sea coast increased by 0.2 to 0.3 m per century since the

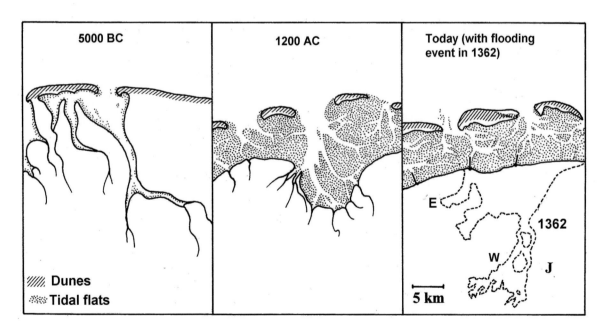

Fig. 2. Reconstruction of coastal change in the Wadden Sea between the islands of Langeoog and Wangerooge with the former Harlebucht. E,W and J: Esens, Wittmund and Jever. Modified from Flemming and Davis (1994) and Kramer and Rhode (1992).

embankments progressed. The level to which salt marsh deposits accumulated in front of the dykes increased accordingly. Therefore, salt marshes which became embanked early remained lower in level than those embanked later. The marsh between the mainland glacial deposits and the sea attained a concave profile which is unfavourable for irrigation. At present no further embankments are intended because the demand for arable land has ceased and the value assigned to natural tidal flats was increased (Wolff 1992).

Before marshlands became dyked, people settled on artificial mounds to avaoid flooding during storm tides. When confidence in the protection provided by dykes increased, houses were built on level terrain. Requirements on the safety standard of the most seaward dyke increased accordingly. Modern dykes attain a width of 60 to 100 m at the base and a height of 5 to 7 m above mean high tide level. They have a sandy core, covered with clay which in turn is overgrown by a dense turf of grass maintained by grazing sheep. Seaward of the dykes, forelands are developed by ditched land reclamation fields with brushwood groins (Fig.3).

These forelands serve to dissipate much of the wave energy during storm tides, relieving pressure from the dykes. The accumulating deposits are mostly fine grained. They originate from adjacent mudflats or constitute imported sediments from the North Sea. Where such a foreland is narrow or not developed, the lower seaward slope of the dykes is armoured with boulders, concrete and asphalt to prevent damage by waves.

Dredging in the estuaries

In the larger estuaries and tidal inlets leading to seaports the situation is complicated by the maintenance of shipping routes. These are routinely dredged to adjust their depth to the increasing size of the vessels that use the channels. The deepening of the shipping canal along with an embankment of estuarine wetlands entailed an increase in tidal range and of sediment influx into the estuary. Dredged sands have been discharged from the estuaries further offshore but have partly been dumped ashore to provide new areas for industrial developments. Also sand bars or islands have been

Fig. 3. Embanked marsh with channel and sluice for irrigation and a ditched land reclamation field with brushwood groins in front of the dyke. Südwesthörn in the northern Wadden Sea, July 1996.

created (e.g. Pagensand inside the Elbe estuary) or outside upon tidal flats to get rid of the dredged sand (e.g. Minsener Oldeoog at the outer Jade). In the estuaries an accelerating rise of sea level may ease shipping but also requires immense financial expenditure on dykes, sluices and irrigation of the adjacent low-lying marshes.

Protecting the islands

On the islands in the Wadden Sea, agriculture was less important than in the marshes of the mainland. Here fisheries and seafaring were an alternative way of making a living, and during the last century an increasing number of tourists was attracted, now dominating the economy. Sea level rise and storm tides changed the shorelines of the barrier islands. Most shifted landwards. When this endangered villages and tourist resorts, various measures of coastline stabilization were employed. Mobile dunes were stabilized with planted marram grass and sand traps. Some dunes have been remolded into sand dykes to prevent overwash. At the beach, groins and bulkheads have been built. More recently, beaches have been replenished with sand taken from between the islands or from offshore. An increase in storm occurrence during the last three decades has particularly affected the islands.

Implications of coastal architecture

Embankments and forelands

There are several effects of the modern coastal architecture in the Wadden Sea. First of all, the safety of people and of their property has increased behind dykes. On the other hand, costs of coastal defense have become high. Hence governmental institutions have become responsible for coastal protection.

Irrigation and peat excavations have resulted in the conversion of former wetlands, brackish reed marshes, salt marshes and tidal flats into arable land. Exceptions occurred when in the 1970s wetland areas were created behind new dykes to compensate for the loss of marine habitats. These partly brackish wetlands are free from inundations with seawater and lack the imports and disturbances by

storm tides of the North Sea. Therefore they do not resemble former coastal habitats. Nevertheless these are rich in wildlife.

Seaward of the dykeline, foreshore is gained by trapping deposits with a labyrinthine pattern of ditches and by brushwood groins which break some wave energy (Fig.3). Here the vegetation and fauna is composed of salt marsh species, intermittently disturbed during the regular renewal of ditches and groins. Drainage is much better than in natural salt marshes. These forelands lack the stagnant water bodies and the irregular topography with meandering and branching channels typical of salt marsh creeks. Thus, forelands distinctly differ from naturally developed salt marshes.

Loss of mud and hunger for sand

As a consequence of embanking sheltered embayments and forelands, tidal flats became gradually depleted of fine-grained sediment (Flemming and Nyandwi 1994; Dellwig et al. 2000). The shorter the distance between mainland dikes and barrier islands became, the less of the fine mud remained. Thus the remainder of the Wadden Sea tidal flats became more sandy. In comparison to holocene sediments deposited before modern coastal architecture came into place, the propotion of fine-grained particles decreased considerably. This is not only because fine sediments were trapped in forelands and were embanked and thus removed from the system. The tidal zone also became more narrow, resulting in a higher input per unit area by the hydrodynamic energy of the North Sea. Furthermore, the slope of the coast became steeper because the dykeline had shifted seaward and the high water level had increased.

There may be a sand sharing system between the barrier islands, tidal inlets, tidal flats and salt marshes (Louters and Gerritsen 1994). An equilibrium configuration of these components is assumed, maintained by relocations of sand after disturbances. For example, when the size of a tidal basin has been reduced by embankments, the cross section of tidal inlets adapts to smaller volumes of tidal waters by attracting sand from landward or seaward areas. In reaction to a rising sea level, natural shorelines would become displaced landwards. In the case of the Wadden Sea, this would involve mainland shores as well as island shores. Sediment that is thus eroded and mobilized would allow the coastal system to grow upwards with the level of the sea. It is assumed that there is a tendency for shallow areas to maintain their depth. Sea level rise therefore creates a hunger for sand that is moved landwards. However, the mainland is protected by a solid dykeline preventing shoreline retreat and the delivery of sand. The only source to meet the resulting hunger for sand are the tidal inlets with their seaward sand bars and the barrier islands. It is not known to what extent and with what time lag sand becomes transported towards the coast from offshore to meet deficiencies within the Wadden Sea. In the case of an acceleration of sea level rise (IPCC 1996, World Climate Conference at The Hague, Nov. 2000) the hunger for sand will further increase. Where sediment supply from offshore is sufficient, tidal flats may grow with the sea while tidal channels become deeper or wider. Where there is a lack of sediment supply, tidal flats will become drowned. Future developments may differ between individual tidal basins that vary in size and shape (Reise and de Jong 1999).

Stronger westerly winds during the last three decades associated with a climatic anomaly in the North Atlantic (Rodwell et al. 1999) increased sediment mobility, particularly on the outer shore of the Wadden Sea, at sand bars in the tidal inlets, at high sands and the barrier islands (i.e. Hofstede 1999a,b). At the inner shore of the Wadden Sea, the frequent inundations of salt marshes and managed forelands allowed these to accumulate sufficient sediment to keep pace with rising high tide levels. However, erosion at cliffed marsh edges often caused a concomitant loss in salt marsh area (Dijkema et al. 1990).

Ecological cascades and coastal squeeze

The concurrent developments of embanking, sea level rise, island retreat, steepening of the coastal slope with losses of mud and hunger for sand caused cascading ecological effects. Coastal habitats decreased in area along the mainland coast. The rate of embankment was higher than the accretion of new marshes (Dijkema 1987). No sea-

ward growth of tidal flats compensated for the losses at the landward side. The gradual transition between terrestrial, brackish and marine habitats gave way to an abrupt boundary between the land and the sea. Natural shoreline dynamics are superseded by shores fixed with groins and petrified seawalls. Coastal habitats became squeezed and lost much of their diversity.

Ecosystem functions were also affected. The gradual transition zone between the supratidal and intertidal presumably acted as an efficient filter for runoffs from the land, served to trap organic material imported with the storm tides, and was a site of nutrient regeneration. The decline of this zone together with the steeper slope might have increased the turbidity in the tidal waters. The function of shore habitats for breeding birds also lost their importance.

The loss of nearshore muddy tidal flats must have affected benthic primary production by epipelic diatoms. These are food to invertebrates such as species of corophiid ampipods, hydrobiid snails or juveniles of bivalves and arenicolid lugworms, which in turn are the food for juvenile flatfish and for birds like redshank and various plovers (Reise 1998). Higher mobility of sediments at seaward sandy flats may shift the benthic fauna towards a dominance of smaller forms and thus to lower biomass (Lackschewitz and Reise 1998). Establishment of mussel beds and seagrass beds may be hampered. In conclusion, the modern coastal architecture has affected geomorphology and ecology of the entire region.

Maintaining dynamic shorelines

Managed shoreline retreat

When sea level rises, sedimentary shorelines will move and this is in conflict with most human uses of the coast. The dyked or otherwise defended shore of the Wadden Sea is a prerequisite for the people to stay in this coastal region. However, a discrepancy is emerging between a natural state corresponding to sea level change and a fixed coastal architecture corresponding to safety and property rights of the citizens (Fig.4). This necessitates an increasing effort in coastal defense and

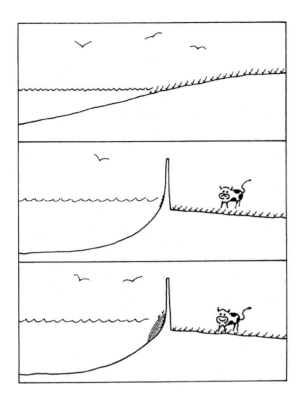

Fig. 4. Cartoon depicting coastal change in the Wadden Sea with profiles from a primordial shore (above) to an embanked, subsided marsh with an increased high water level (middle), and on to a sand replenishment (dotted) restoring dynamic shoreline habitats (below).

causes a reduction in supra- and intertidal habitats with their associated biodiversity and ecological functions. As a countermeasure, a managed coastal retreat might be envisioned. However, the potential for shifting the dykeline landwards (depoldering) is limited given the intensive use of most embanked areas.

A managed coastal retreat might relieve the conflict between defensive structures and the rising sea, but older dykelines further inland would have to be enforced. Farm houses and other buildings must be set upon artificial mounds as is still practiced on undyked marshy islands (Halligen). The seaward dyke may be maintained but should include sills of a lower height to let seawater through during storm tides and sluices to let it out again when the storm has passed. Similar to land reclamation fields, deposits may be trapped and the marsh has a chance to grow upwards with

the rising sea level. Such semi-poldered marshlands would be of limited agricultural use but tourism might be an alternative in such a wetland landscape.

Given the required safety standards that have to be met, a managed landward retreat would be rather costly. If acceptance by the people that inhabit the coastal area can be achieved at all, substantial subsidiary or compensatory payments would be necessary. Therefore, coastal retreat may be possible only for the few uninhabited polders.

Shore replenishments at sheltered coasts

An alternative to managed retreat may be realized in seaward direction: the creation of a gentle shoreline slope with sand dredged in the North Sea (Fig.4). This is already practiced along the outer coast of the Wadden Sea since about 50 years, partially in combination with structural defenses (groins and bulkheads). The aim is to prevent shoreline retreat and to maintain beaches for recreation at the seaward side of the barrier islands. At the outer coast, currents are strong and wave energy is high. Consequently, these shore replenishments have to be renewed frequently. At the island of Sylt, about 1×10^6 m³ of sand are added to the beach every year (Kelletat 1992).

Erosional sites on the sheltered shores of the Wadden Sea have rarely received artificial additions of sand. Here groins and stonewalls have been employed. These protective structures merely displace the locus of erosion, either longshore to adjacent unprotected areas, offshore towards the tidal flats or both (Nordstrom 2000). Longshore displacement of erosion entails more groins and seawalls, and offshore displacement of erosion entails the need to strengthen these structures. As a consequence, the shoreline will deviate more and more from its natural state, becomes increasingly armoured with stones and concrete, habitat diversity declines and organisms restricted to the high tide zone will eventually be lost.

As an example, of the 68 km of the sheltered, eastern shoreline at the island of Sylt in the northern Wadden Sea, an inventory revealed that 35% is armoured with stones or other hard materials and an additional 18% is protected by groins and artificial sedimentation fields (Reise and Lackschewitz (in press); Fig.5). As Sylt is a very slender and densely populated island with settlements and roads along most of the shore, one may expect all retreating shorelines to become eventually 'petrifed' if the current practice is continued. Thus, to the already armoured 24 km another 28 km will have to be added, together 76% of the Wadden Sea shoreline of the island. At the downdrift ends of groin fields and seawalls, erosion tends to become more pronounced. Therefore, these protective structures will have to be extended even beyond the present length of the eroding shoreline, and may ultimately encompass the entire length of the side of the island facing the Wadden Sea. At the other side of the island, the exposed eroding beach has been supplied with groins which have no or even negative effects on the net sediment budget. For the last 25 years recurrent sand replenishments have been carried out that slowed down a further retreat of coastal cliffs.

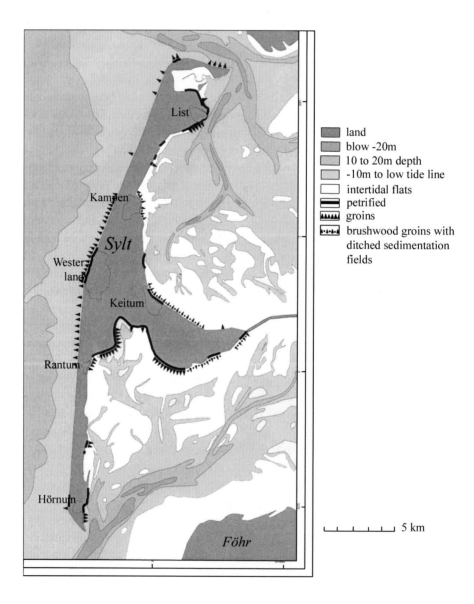

Fig. 5. Island of Sylt with structural protections along the shore indicated schematically. Most groins at the exposed sandy beach (left) have fallen into decay and shore protection is accomplished by regular additions of sand, while hard structural defences dominate the sheltered shore (right).

Similar trends as on the island of Sylt can be observed on other barrier islands in the Wadden Sea. Of the marsh islands most have an entirely armoured shoreline. Along the mainland, most of the shoreline is protected by brushwood groins or by boulders. A general inventory of the extent of unmodified shorelines and natural shoreline habitats, and of those supplied with protective structures is not yet available for the Wadden Sea. The addition of sand to the shore is a direct countermeasure to meet the hunger for sand in response to sea level rise. The articially supplied deposits of sand along the shore may serve to dissipate wave energy during storm tides similar to the managed forelands. It is suggested to extend shore nourishments to those sheltered coasts which are subject to erosion or are defended by groins and bulkheads including armoured dykes.

Added sand

The hunger for sand will gradually erode the added deposits away. Their longevity will widely differ between sites, depending on the local hydrodynamics. At the Wadden Sea shore of Sylt, there are two unintended examples. An islet of 15 ha was created artificially 60 years ago, and since then has only lost 20% of its original area. Three other islets were created three decades ago. Their shape changed but they hardly reduced in size (Reise and Lackschewitz submitted). Vegetation developed on all of these islets and breeding colonies of birds became established. No protective structures were employed, even though the adjacent shorelines of Sylt are under erosion or are armoured. These examples may indicate that the longevity of replenished sand deposits at erosive sites in the inner Wadden Sea shores may be a century or even more. However, this needs site specific investigations as there may be localities where sand deposits are readily washed away.

Sand is an ideal choice of material to be added. It is not an alien material to the Wadden Sea shores. The offshore North Sea is a convenient nearby source of sand. Added sand should be rather coarse (>250 μm) to increase the longevity of deposits. Interstices between coarse sand grains may soon fill with ambient grain sizes. Supplying finer grain

sizes would increase turbidity and the deposits might vanish within a shorter period of time. Deposits of sand will be continuously reshaped by currents and waves, and during this process will resemble more and more naturally shaped beaches, sandy hooks, sand bars and islets.

Locations for sand additions should be in the first place eroding beaches or eroding seaward edges of salt marshes where this retreat is considered to be intolerable and structural defense would be employed otherwise. Sand deposits would serve to prevent further retreat and also to appease the hunger for sand. Secondly, sand should be supplied to armoured shores for appeasing the hunger for sand which otherwise would affect adjacent shores and seaward tidal flats. Thirdly, sand replenishments could substitute land reclamation fields with their brushwood groins and ditches. In addition to trapping fine grained material imported by the North Sea, these fields probably grow at the costs of other shore areas, causing a sediment deficit there. They are repeatedly disturbed by the repair of groins and by the ditching machines. They create a foreland of a rectangular shape with straight parallel ditches, differing in quality and outer appearance considerably from natural salt marshes. More natural shorelines may be achieved with artificially added sand deposits.

Research is needed to estimate the required amounts of sand and the proper size and shape of deposits, adapted to individual locations. To provide an estimate on the possible magnitude of sand to be added in the Wadden Sea, the following assumptions may be considered. In many cases an amount of 200 to 400 m³ m⁻¹ of shoreline may be sufficient to generate sizable shore habitats of various shapes and a longevity of several decades (Fig. 6). The total length of shoreline protected by dykes or seawalls is 1140 km. This figure includes dams but excludes the estuaries because here the policy is to remove sediment to ease shipping. Any artificial sand additions would be in conflict with this aim.

Outside the estuaries, coastal squeeze may be combated with supplying sand. However, there are many shores with extensive forelands with no signs of erosion, others may be too difficult to reach with sand supplying techniques, and still others may have mudflats which should not be covered with

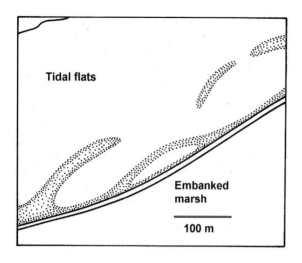

Fig. 6. Sketch for the design of artificially added sand deposits (dotted) in front of a dyke to restore dynamic and diverse shoreline habitats with about 200 to 400 m³ of sand per m of shore. Mud may accumulate in the shelter of sand bars.

sand. The length of shoreline actually suitable for adding sand needs to be investigated in detail. As an informed guess, I assume that half of the protected length may be suitable (but see the island of Sylt as an example for more than half of the shore length). Then a total volume of some 170 x 10⁶ m³ is needed. This is 3 to 4 times as much as has been supplied to the exposed beaches of the barrier islands in the past. If these shore nourishments are spread over three decades, about twice as much sand is required annually than is currently supplied to the exposed beaches. Assuming 6 € per m³ of sand, total costs might be in the order of 1 to 30 x 10⁶ € annually over a period of 30 years. From this, the running costs for the maintenance of land reclamation fields, groins and stonewalls might be subtracted.

Areas for sand extraction should be located outside or at the outer edge of the sand sharing system of the tidal basins. In most cases this would be the sandy bottom of the North Sea below 10 to 15 m depth. From offshore delivery sites, vessels may transport the dredged sand-water mixture inshore, and from there a mooring barge with a pipeline may distribute the sand to the desired locations. This nourishment technology considerably improved over the last decades and is now used routinely throughout the world, and nourishment has become very cost-effective relative to hard structures used for shore protection (Nordstrom 2000).

Implications of shore replenishments

For coastal protection, replenished sand deposits have the extra advantage of meeting the hunger for sand which results from sea level rise and of dissipating wave energy that would otherwise impact upon the dykes or seawalls. Given the uncertainties in the predictions of an acceleration in sea level rise or of long-term trends in storminess, the transient nature of added sand may be viewed as an advantage. If other measures of coastal defense are deemed to be necessary in the future, no permanent obstacles remain that may hinder further operations.

Disturbances

Shore replenishments entail three types of disturbance. Firstly, the seabed of the coastal North Sea is disturbed at the sites from which the sand is excavated. Such donor areas will be located on fine to medium sand seabeds between 10 to 30 m depth, generally occupied by the *Tellina-fabula* association that comprises more than half of the benthos in the eastern North Sea (Salzwedel et al. 1985). The abundance and biomass of this macrofauna are near the mean found in this area with a species density that is somewhat lower than for the entire region. It follows from the above figures that about 5 donor pits of 6 x 6 km need to be excavated to a depth of 1 m on average. Naturally this provides only a rough idea of the scale of disturbance caused in the offshore area. Donor pits will disturb less than 1% of this type of seabed.

Secondly, the added sand deposits will mostly smother upper intertidal flat benthos along the shore of the Wadden Sea. Benthic assemblages that live there will be displaced seaward in the tidal zone. If we assume again that about half of the presently defended shoreline inside the Wadden Sea might be suitable and feasible for shore replenishments with 300 m³ m⁻¹ of shoreline, then between 2 and 4% of the intertidal zone will become covered with sand deposits from the North

Sea. In front of dykes, sand deposits should be placed at some distance to avoid wind blown sand disturbing the dense turf of the upper dyke.

Finally, there will be an immediate disturbance caused by the shore replenishment operations. These involve the mooring of a large vessel in narrow tidal channels, installation of pipelines across the intertidal zone, and mechanical machinery to modulate the deposits according to an intended design. Even if donor sites are at a sandy seabed, the sand-water mixture supplied through pipelines will cause an increase in turbidity in the tidal waters during the operation.

Ecological benefits

These inherent disturbances must be compared to the subsequent ecological advantages created by the deposited sand. These depend on size, shape and longevity of the deposits. In the simplest case, the application may be designed as a flat beach in front of an armoured or erosional shoreface. However, other designs may be ecologically more rewarding (Fig.6). In the shelter of sandy spits, mudflats and salt marsh vegetation may develop. A semicircular arrangement of a sand bar in front of a shoreline may enclose stagnant and temporary brackish water with marine inputs during storm tides. Sand bars and sandy islets parallel to the shore may offer breeding sites for coastal birds safe from mammalian predators.

The replenished sands restore dynamic shoreline habitats of the supra- and upper intertidal zone. These were once extensive in the Wadden Sea but have been squeezed between embankments, by coastal protection measures and sea level rise. The supplied sand deposits are subject to hydrodynamic mobilization and biogenic stabilization. They will continuously change and adapt a more and more natural shoreline configuration. Colonization of pioneer vegetation will be a spontaneous process. Shoreline plants characteristically have a wide ranging dispersal of seeds and vegetative parts. Deposited sands offer potentials for plant successions characteristic for dunes and upper beaches as well as for salt marsh plants in the shelter of sandy ridges. The invertebrate fauna of this zone comprises a mixture of marine, brackish and

terrestrial species and is highly diverse. Mudflats with their benthic assemblages including microalgal mats and seagrass beds may develop in the shelter of sandy deposits. Such sites will provide rich feeding grounds for coastal birds, and sandy spits and islets will provide ideal breeding sites. In particular endangered populations such as Kentish plover and little tern (Rasmussen et al. 2000) will find undisturbed breeding sites which are presently a limiting factor. Other plovers and terns, oystercatcher, shelduck and various gulls are likely to establish breeding colonies on such sandy and at first scarcely vegetated ridges more or less separated from land.

As the deposited sands will be eroded by waves and intermittently stabilized by vegetation, a diverse array of dynamic habitats may develop, providing niches for shoreline species that otherwise might run out of habitat in the Wadden Sea. Spreading shore replenishments along the coast over a long period of years will provide different phases of habitat succession following the initial sand supply at the different sites. This pattern will be favourable for diversity in the sense of the mosaic-cycle concept (Remmert 1991). Species diversity will increase when added sand deposits are not too small and sites not too far apart. If shore replenishments are performed on a large scale, there will be also implications for the function of the entire coastal ecotone. Numbers of birds foraging in the tidal zone are likely to increase because of an extra provision of breeding and resting sites. Temporarily, organic matter will be deposited and stored in these shoreline habitats. Eroded material will be deposited on adjacent tidal flats before it is eventually transported through the tidal inlets and then back into the North Sea.

Shore aesthetics and recreation

Shoreline aesthetics will improve considerably where armoured shorelines, groins and ditched sedimentation fields become concealed by the replenishments. Sharp boundaries and steep supra- and intertidal slopes have a chance to develop into picturesque, dynamic and diverse shores. Although longevity of the supplied deposits may be a century or more at most of the sheltered shores in the

Wadden Sea, there will be an apparent change in geomorphology and vegetation on a time scale within a human life-span. These dynamics may be more attractive to people than the present protective structures. The abundance of birds, diverse vegetation, and the variety of geomorphic forms may increase recreational usage. Where shore replenishments are carried out near populated places, recreational beaches may be created in front of armoured seawalls. In general, the amenability of the Wadden Sea shore is likely to improve considerably.

Research on coastal dynamics

Shore replenishments within the Wadden Sea with sand extracted from the bottom of the North Sea will be a large-scale operation and a change in coastal protection measures which should not be started without a thorough scientific analysis of the assumptions and implications.

Sea level rise and sediment mobility

There should be a sound knowledge of sediment budgets and of long-term developments in the morphodynamics of the individual tidal basins in the Wadden Sea. The assumed sand sharing system between the tidal basins and the outer coast needs empirical testing and further modelling. Particularly, the hunger for sand in the face of an accelerating sea level rise needs site specific quantifications and predictions on the spatial pattern of erosion. Are seaward sources of sand sufficient and will these deliver the required amounts? To what extent does the hunger for sand in the tidal basins withdraw sand from the barrier islands? How is the ecology of the Wadden Sea affected by increased sediment mobility and a decline of mud? Presumably, the transport and supply of sand in the shallow coastal zone of the North Sea seaward of the Wadden Sea is important for sediment supply within the Wadden Sea. No research has been done on this hypothesis.

Artificial sand deposition within the Wadden Sea

Optimal sites for donor pits in the coastal North Sea need to be located, and their subsequent development with implications on the offshore sediment budget will have to be predicted. Sediment mobility and composition in front of existing protective structures should be analyzed before nourishments with sand from the North Sea are performed. The sources of the deposits accumulating in the land reclamation fields are not well understood. To what extent do they trap material from the North Sea or do they cause deficits of fine-grained material on adjacent tidal flats? This question has to be answered before these reclamation structures are replaced by added sand.

In the case of shore nourishments, coastal geomorphologists and sedimentologists are requested to predict the longevity of the added sand at particular sites and in the Wadden Sea in general. Shorelines should be categorized according to their suitability for replenishments. Criteria need to be developed to arrive at a priority list of sites based on costs of transport, potential protective effects, longevity, disturbances, ecological benefits, aesthetic and recreational aspects.

A crucial question is the size and the design of the supplied deposits at sheltered shores. Placements directly onto or very close to the shore may be preferable in most cases because of the immediate effects on shore protection and shore habitats. However, deposits may also be supplied into tidal channels which may then distribute the sand according to the prevailing hunger for sand. These divergent options have to be explored.

Ecological effects with and without nourishments

Ecologists have to provide models regarding the course of biotic successions, and the prospects of habitat and species diversity. Experimental shore replenishments would be ideal for this purpose. The few sand nourishments done in the past at sheltered shores and upon tidal flats may be evaluated retrospectively. Not only natural sciences should get involved. Management may take advice from the humanities to judge aesthetic and recreational values of the added diversity and dynamics to the shoreline by sand nourishments. The disturbances caused at donor and deposition sites need a careful evaluation as well. Ecologists have to predict the magnitude of the impact and to compare this with long-term prospects of the shoreline biota without shore replenishments. Much attention has been given to the ecology of the extensive tidal flats, while the narrow but geomorphically and biotically more diverse upper margin of the intertidal and lower margin of the supratidal has received less studies. In particular, no quantitative inventory is available for these biota.

Neither is there an inventory of erosional versus depositional sectors of the Wadden Sea shoreline. Locally, data are available on types and abundances of protective structures but no time line and maps have been prepared for the Wadden Sea. Predictions should be made on possible proliferations of these structures if sand replenishments are not done and sea level rise accelerates.

Costs and acceptance of shore replenishments

Effects for coastal protection and the costs of shore replenishments in the Wadden Sea will have to be compared to the costs of further development and maintenance of those structural protections which may be covered and substituted by sand deposits. This has to be augmented by the difficult valuation of the shoreline habitats to be created and being otherwise in short supply. Moneterizing might be based on protective functions, ecological functions and by the willingness of people to pay for such a change shoreline development.

Acceptance of shore replenishments by the coastal populations and by the touristic visitors has to be explored. There exist several proverbs which imply that one shall not trust a sandy ground and that valuables and hopes might get easily lost in treacherous sand. Sound predictions on the longevity of the sand deposits are important. Protective structures made out of boulders, iron, concrete, asphalt or wood are regarded as more trustworthy. Pejorative feelings associated with sand might be balanced by an improved scenary of the

shorescape and the evolving habitat diversity and abundance of birds at the sites of former structural protections. However, these aspects need careful evaluation by socio-economists and socio-psychologists.

Nature conservationists might argue against creating natural sites with an enormous technical, energetic and financial effort and might refrain from such an interference. This option must be discussed on the basis of scenarios on the future of natural shoreline habitats in the face of predicted accelera-tions in sea level rise.

References

Anonymous (1998) Trilateral Wadden Sea Plan. Ministerial declaration of the eigth Trilateral Governmental Conference on the Protection of the Wadden Sea. Stade 1997, Common Wadden Sea Secretariat, Wilhelmshaven, Germany

Bartholdy J, Pejrup M (1994) Holocene evolution of the Danish Wadden Sea. Senckenbergiana maritima 24:187-209

Carter RWG (1988) Coastal environments. Academic Press, London

Davis RA Jr (ed) (1985) Coastal Sedimentary Environments. Springer-Verlag, New York

De Jong F, Bakker JF, Berkel van CJM, Dankers NMJA, Dahl K, Gätje C, Marencic H, Potel P (1999) Wadden Sea Ouality Status Report. Wadden Sea Ecosystem 9. CWSS, Wilhelmshaven, Germany

Dellwig O, Hinrichs J, Hild A, Brumsack H-J (2000) Changing sedimentation in tidal flat sediments of the southern North Sea from the Holocene to the present: A geochemical approach. J Sea Res 44:195-208

Dijkema KS (1987) Changes in salt-marsh area in the Netherlands Wadden Sea after 1600. In: Huiskes AHL, Blom CWPM, Rozema J (eds) Vegetation Between Land and Sea. Junk Publishers, Dordrecht, Netherlands: 42-49

Dijkema KS, Bossinade JH, Bouwsema P, Glopper de RJ (1990) Salt marshes in the Netherlands Wadden Sea: Rising high-tide levels and the accretion enhancement. In: Beukema JJ et al. (eds) Expected Effects of Climatic Change on Marine Coastal Ecosystems. Kluwer Acad Publ 173-188

Flemming BW, Davis RA Jr (1994) Holocene evolution, morphodynamics and sedimentology of the Spiekeroog barrier island system (southern North Sea). Senckenbergiana maritima 24:117-156

Flemming BW, Nyandwi N (1994) Land reclamation

as a cause of fine-grained sediment depletion in back-barrier tidal flats (southern North Sea). Neth J Aquatic Ecol 28:299-307

Ginsburg RN (ed) (1975) Tidal Deposits. Springer-Verlag, Berlin

Hofstede JLA (1999a) Regional differences in the morphologc behavior of four German Wadden Sea barriers. Quarternary International 56:99-106

Hofstede JLA (1999b) Process-response analysis for Hörnum tidal inlet in the German sector of the Wadden Sea. Quarternary International 60:107-117

IPCC (1996) Climate change 1995 – Impacts, adaptations and mitigations of climate change: Scientific-technical analysis: Contribution of working group II to the second assessment report of the governmental panel on climate change. Watson RT, Zinyowera MC, Moss RH (eds). Cambridge University Press, Cambridge

Kelletat D (1992) Coastal erosion and protection measures at the German North Sea coast. J Coast Res 8:699-711

Kramer J, Rohde H (eds) (1992) Historischer Küstenschutz. Verlag Konrad Wittwer, Stuttgart

Lackschewitz D, Reise K (1998) Macrofauna on flood delta shoals in the Wadden Sea with an underground association between the lugworm Arenicola marina and the amphipod *Urothoe poseidonis*. Helgoländer Meeresunters 52:147-158

Louters T, Gerritsen F (1994) The riddle of the sands. A tidal system's answer to a rising sea level. Ministry of Transport, Public Works and Water Management. National Institute for Coastal and Marine Management (RIKZ), Den Haag

Nordstrom KF (2000) Beaches and Dunes of Developed Coasts. Cambridge University Press, Cambridge

Oost AP, Boer PL de (1994) Sedimentology and development of barrier islands, ebb-tidal deltas, inlets and backbarrier areas of the Dutch Wadden Sea. Senckenbergiana maritima 24:65-116

Rasmussen LM, Fleet DM, Hälterlein B, Koks BJ, Potel P, Südbeck P (2000) Breeding birds in the Wadden Sea in 1996 – Results of a total survey in 1096 and of numbers of colony breeding species between 1991 and 1996. Wadden Sea ecosystem No. 10. CWSS, Wilhelmshaven, Germany

Reise K (1996) Das Ökosystem Wattenmeer im Wandel. Geogr Rdschau 48:442-449

Reise K (1998) Coastal change in a tidal backbarrier basin of the Wadden Sea: Are tidal flats fading away? Senckenbergiana maritima 29:121-127

Reise K, de Jong F (1999) The tidal area. In: De Jong F, Bakker JF, Berkel van CJM, Dankers NMJA, Dahl K, Gätje C, Marencic H, Potel P (1999) Wadden Sea Quality Status Report. Wadden Sea Ecosystem 9. CWSS, Wilhelmshaven, Germany: 187-190

Reise K, Lackschewitz D (in press) Combating habitat loss eroding Wadden Sea shores by sand replenishment. Proceedings of the 10th Wadden Sea symposium, Groningen NL

Remmert H (ed) (1991) The Mosaic-Cycle Concept of Ecosystems. Ecol Stud 85. Springer-Verlag, Berlin

Rodwell MJ, Rowell DP, Folland CK (1999) Oceanic forcing of the wintertime North Atlantic Oscillation and European climate. Nature 398:320-323

Short AD (ed) (1999) Handbook of Beach and Shoreface Morphodynamics. J Wiley & Sons, Chichester

Streif H (1989) Barrier islands, tidal flats, and coastal marshes resulting from a relative rise in sea level in East Frisia on the German North Sea coast. Proc KNGMG Symp 'Coastal Lowlands, Geology and Geotechnology' 1987. Kluwer, Dordrecht: 213-223

Viles H, Spencer T (1995) Coastal problems. Erdward Arnold, London

Wolff WJ (ed)(1983) Ecology of the Wadden Sea. Balkema, Rotterdam

Wolff WJ (1992) The end of a tradition: 1000 years of embankment and reclamation of wetlands in the Netherlands. Ambio 21:287-291

Sustainability and Management: Coastal Systems

P.R. Burbridge[*] and J. Pethick

Centre for Coastal Management, University of Newcastle upon Tyne,
Newcastle upon Tyne NE1 7RU, UK
[*] corresponding author (e-mail): p.r.burbridge@ncl.ac.uk

Abstract: This paper examines the economic concepts of "Weak" and "Strong" sustainability and their application to the assessment of policy, investment and human uses of coastal areas. These interpretations of sustainability are applied to the reclamation of estuaries and coastal defense. These examples demonstrate that poor understanding of the dynamic processes and functions of coastal systems has resulted in continuing intervention in the coastal system that has forced it to re-adjust so frequently that it remains almost permanently in a state of dis-equilibrium. It is argued that this results in "weak" sustainability where human activities can only be sustained through major investments of man-made capital. The paper concludes that we must begin to manage human activities and man-induced and natural change at the coast in a more positive manner. For this to happen we need a European Marine Policy that: links terrestrial and marine management perspectives, and encourages cross-disciplinary research to both create new knowledge that will help to inform coastal policy and strengthen its utility for management.

Introduction

In 1998 the British House of Commons Select Committee on Agriculture report on Coastal Flooding and Erosion called for a fundamental change in the way we plan for and manage human activities because massive human intervention in coastal systems can not be sustained and is resulting in increased hazards to life, property and investment (House of Commons 1998). The prospects for achieving this basic change are limited by poor conceptual linkages between natural sciences such as Geomorphology and social sciences such as Economics. They are also weakened by sector based institutional systems which govern human development activities and which have neither the spatial or temporal perspectives to deal with the dynamics of coastal change.

The longer-term nature of coastal change is also difficult for politicians to deal with as sustainable and equitable human uses of coastal systems require a radical shift away from focusing on short-term sectoral development objectives and towards more integrated systems based planning and management. All of these limitations will have to be addressed by innovative scientific research by a number of disciplines and the integration of the knowledge gained to form more robust guidelines to help politicians, planners and mangers formulate more appropriate policies, investment strategies and natural resources management plans to guide coastal development. These issues are highlighted in the findings of the EU funded demonstration programme on Integrated Coastal Zone Management (ICZM) (EU 2000). The need for more integrated scientific information is also recognized in the draft EU Strategy to promote more integrated approaches to coastal management and the sustainable use of resources (EU 2000).

The basic thesis of this paper is that comprehensive knowledge and understanding of the dynamic process and functions of coastal systems are required to avoid inappropriate policies, investment and human use of coastal areas that can only be sustained through major investment of man-made capital. The implications of such subsidies of man-made capital are examined using the concepts of "Weak" and "Strong" sustainability set out by

From WEFER G, LAMY F, MANTOURA F (eds), 2003, *Marine Science Frontiers for Europe.* Springer-Verlag Berlin Heidelberg New York Tokyo, pp 217-228

Turner et al. (1998) in the LOICZ Focus Four programme and examples drawn from coastal geomorphology concerning the dynamics of natural and man induced change in coastal systems.

Based on these examples, we attempt to demonstrate that failure to understand powerful and dynamic coastal processes can results in a loss of natural capital, and inefficient allocation of man-made capital that could be used in better ways to improve human welfare. This forces us towards "weak" sustainability and effectively forecloses options for developing more sustainable human uses of natural and man-made capital.

The meaning of sustainability

The concept of "Sustainable Development" gained broad public attention following the publication of the World Commission on Environment and Development report "Our Common Future" (WCED 1987). Sustainable development was defined in this report as that which "meets the needs of the present without compromising the ability of future generations to meet their own needs" (WCED 1987, p8). This concept gained further public recognition as a result of the United Nations Conference on Environment and Development (UNCED) and the production of "Agenda 21" the global plan of action designed to promote more sustainable forms of development. Coastal management was given explicit recognition in Agenda 21 Section 17 dealing with the marine environment. In this section Integrated Coastal Management was given priority as the single most important means of reconciling a number of non-sustainable marine resources development issues. In a recent publication from the Land-Ocean Interchange in the Coastal Zone (LOICZ) programme sustainable coastal development was described as "the proper use and care of the coastal environment borrowed from future generations (Turner et al. 1998).

The term "Sustainability" is as much an expression of social choice as it is an expression of the ability of the environment to continue to support human needs and aspirations. It is possible to grow bananas in Iceland and production can be sustained—at a cost. To bear that cost is a question of social choice and not one of the technical feasibility of using geothermal energy to heat greenhouses in which the banana plants are grown. Apart from the novelty of growing bananas in Iceland, the low cost and abundance of the energy may preclude the need to consider whether it would be better to use the energy to generate electricity. Choosing between the relative benefits derived from growing bananas versus generating electricity is a matter of social choice as it is technically possible to use the geothermal energy to do both.

However, most coastal management challenges are more complex than growing bananas. What at first may appear rational social choice can lead to unsustainable coastal development where such choices are based on inadequate knowledge of coastal processes. A good example is the choice of a dynamic shingle beach at Dungeness for the location of an atomic power station. The benefits of a cheap site away from major human settlements with easy access to abundant cooling water at first seem to form good reasons to choose the site. However, as we have learned to our great cost, the site is on an eroding shore and to avoid damage to the plant we have to invest millions of pounds to truck shingle from the down current side of the shoreline to the upstream side of the plant to counteract the erosion which would otherwise undermine the plant's foundations.

In the above case, the economic benefits derived from a coastal location for the generation of electricity have been eroded because insufficient attention was paid to the long-term dynamics of the Dungeness beach system. This is by no means an isolated case.

The need to defend coastal infrastructure such as power stations from natural coastal processes could be termed "Weak Sustainability" because of the need to allocate natural resources (shingle and sand = natural capital) and man made capital (money and machines to truck the shingle and sand) to sustain the viability of the power plant. Weak sustainability is also associated with the assumption that there can be unlimited substitution possibilities between different forms of capital via technical progress. For example, the reclamation of inter-tidal mudflats and marshes has proceeded on the basis that there was little capital value represented by these coastal ecosystems and greater man-made capital could be achieved by their transformation into dry land. However, advances in scientific knowledge have identified highly valuable functions performed by wetlands as well as grave risks to the sustainable use of reclaimed areas resulting from both the destruction of the functions of wetlands and other coastal systems, and rising sea level. We will return to this point later in this paper.

"Strong sustainability" is associated in economic thinking with the conservation of different forms of capital (man-made, human, natural, and social/moral) in respect to meeting the needs of human populations over time. Strong sustainability applied to marine and coastal systems would mean that their natural capital expressed in terms of biological diversity, generation or renewable resources, and maintenance of natural processes and functions would remain constant or increase.

Management

Over the past thirty years there has been increasing recognition that coastal areas and the complex, productive yet fragile systems they contain require new and innovate forms of management. This has come about partly in response to increased understanding of the strategic importance of coastal resources in sustaining the expansion and diversification of economic development. Politicians, planners and managers have also become more aware that lives property and public and private investment can be put at great risk by ignoring the natural hazards associated with the powerful and dynamic processes that form and constantly modify coastal ecosystems. We have learned the hard way that poorly planned and managed development can make us more vulnerable to the risks of flooding, erosion and other natural processes.

These lessons have led to a progressive transformation in the way we attempt to manage development processes. From an ecological and environmental management perspective we have given increased emphasis to the inter-relationships between terrestrial and marine systems. We now recognise that there is a zone of transition between terrestrial and marine ecosystems where unique habitats and species have evolved in response to the high energy fluxes, salinity gradients, nutrient

fluxes and other features common to coastal systems. The term 'the Coastal Zone' was introduced in order to denote this special zone of transition and to make planners and managers aware that administrative arrangements, planning processes and management tools need to be adapted in order to achieve sustainable use of coastal areas and resources.

Our coastal management perspectives have evolved rapidly since the adoption of the Coastal Zone perspective in the 1960s. Initially, concerns in coastal management focused on improving the management of specific coastal resource based activities such as fisheries; recreation and tourism based on improvements within their individual economic sectors. These improvements led to new knowledge on the interactions among the many activities that seek exclusive rights of access to and use of coastal land and waters. Two specific management issues became more evident. The first is that the economic viability of many activities is dependent upon maintenance of the quality of the coastal environment and the continuous production of renewable resources from the highly productive yet fragile coastal ecosystems. Second, poor planning in the location and then on-going management of individual activities has degraded the environmental performance and economic viability of other activities. This led to increased emphasis upon integrated approached to development planning within the coastal zone and the adoption of the term Integrated Coastal Zone Management (ICZM) during the 1980s.

The term ICZM is still commonly used and is currently the theme for the development of a new European Union Strategy designed to encourage Members States to address coastal development issues in a more sustainable manner. However a different term has gained favour during the last 10 years and that is Integrated Coastal Management, the term Zone being given less emphasis. This has come about for two basic reasons. The first is that we now recognise that many problems and issues common to coastal areas have links to poor development planning and management in areas beyond individual national coastal zones. For example, the management of the Wadden Sea inter-tidal ecosystem is heavily influenced by pollution resulting from poor industrial management in the upper watersheds of rivers that cross a number of independent states. The second reason is that we have not achieved strong progress in developing integrated and cross-sectoral management procedures.

We would also make the point that there has been and continues to be a lack of emphasis in management on maintaining the functional integrity of the coastal ecosystem that sustains human development activities. This has forced us into a position where options for sustainability are weakened by the consequence of past management. Therefore the cost of re-establishing a more sustainable base for managing human activities has been increased.

The role of natural functioning coastal systems as the basis for sustainable management of human activities

We have argued that the utility of economic concepts of sustainable coastal management are constrained by limited recognition of the temporal scales over which dynamic coastal processes form and maintain the functions of coastal systems. We have also argued that current administrative arrangements and sectoral management systems are based on too narrow a spatial perspective to be able to promote more sustainable use of available coastal resource systems.

Too often the scale of coastal interventions is constrained by political and institutional considerations and devolves down to the lowest common denominator: local pressures placed upon short lengths of coast over short time periods. The proprietorial interests shown by local people in 'their' section of the coast is an extremely powerful force and one which democratic systems find difficult to pass over. Yet natural coastal systems, both physical and ecological, are driven by processes that transcend the local scale and the short time period. Educating coastal users to consider impacts of their actions in coastal areas remote in time and space from themselves is perhaps the most important principle in any coastal management programme (Pethick 2001).

Extreme, if infrequent, natural events such as hurricanes can pose a hazard to human life, property and investment. However, poor planning and management can ignore the possibility of such events and may cause human activities to become more prone to the risk damage from natural hazards. The limited time scales that characterise coastal management in countries such as Britain cause people to view erosion, flooding and transition in coastal landforms as threats to sustainable human occupation and use of coastal areas. We argue that this perception is misguided. Natural coastal systems generally operate to reduce such risks. For example, cliff erosion results in widening abrasion platforms upon which wave energy is increasingly dissipated so that erosion rates decrease; flooding results in deposition of sediment and elevation of low lying surfaces so that flood frequency decreases. Both these natural negative feedback mechanisms operate in our interest so that, instead of accepting risk as an integral part of a coastal location, we should ask ourselves why the coast has not adjusted more efficiently to the forcing factors of wave, tide and sediment supply.

We further argue that our massive and continuing intervention in the coastal system has forced it to re-adjust so frequently that it remains almost permanently in a state of dis-equilibrium, a state in which risk to ourselves is maximised. In essence we have created conditions which only allow weak sustainability.

Our arguments are based on the proposition that coastal form – comprised of estuaries, marshes, dunes and beaches and so on - is adjusted to its function. The function of the coastal morphology is to dissipate energy, principally from waves and tides. This dissipation is accomplished mainly by spreading the energy loading out over a wide area so that the stresses per unit area are reduced to below the critical threshold for erosion. This leads to medium- to long-term stability in the shore morphology (although short-term episodic storm events produce minor, oscillatory, adjustments) in which wave and tidal energy is reduced leading to low tidal range and wave height – in short to lower risks for coastal users. The main requirement for such adjustment is space - wide areas of intertidal beach, mudflat and salt marsh are a necessity for a stable and efficient shore.

Each individual component of this coastal morphology, for example a sand beach, occupies an energy niche on an energy gradient that reduces from high values in the nearshore to zero in the coastal hinterland. The reduction in energy produced by each coastal component means that its landward neighbour occupies a lower energy level and consequently exhibits a different morphology, for example sand beaches at the mouth of an estuary give way to mudflats further landward.

Taking a macro-scale perspective: coastal landform morphology responds to extrinsic forces such as those imposed by climate change, mean sea level rise, subsidence and tidal range evolution. (Bird 1993; Bijlsma 1997). The mosaic of coastal land-forms acts as an holistic system, that is the whole is greater than the sum of its parts, and each component depends upon the others. Changes in the energy regime, for example, as sea level rises, means that the entire mosaic shifts its absolute location but the relative positions of each component remain constant. Thus, as sea level rose after glaciation the entire coastal landform mosaic moved landwards, the so-called Holocene transgression, maintaining the coastal function and form but adjusting continuously to energy changes.

The predicted changes in the rate of sea level rise, as a result of global warming, will have important impacts on the coastal zone, displacing ecosystems, altering geomorphological configurations and their associated sediment dynamics, and increasing the vulnerability of social infrastructure (Chappell 1990; IPCC 1996; Crooks and Turner 1999). This should provide the stimulus for us to re-examine the effects of past and current coastal management actions and to assess the constraints they impose on moving toward a position of strong sustainability. To support this view we present two historical perspectives on the impact of past interventions that have imposed adverse impacts on coastal functions and which are weakening options for sustainable management of coastal systems. The first is the impact of reclamation and the second deals with coastal defence.

Reclamation- substitution of man-made capital for natural capital

Our management of the complex pattern of landform mosaics and energy niches has largely proved disastrous. We failed to see that the energy dissipation role of the coast was one that acted for our own good in reducing risks from flood and erosion, and that wide intertidal areas were a necessity. Instead we grossly undervalued the natural function of inter-tidal systems such as mudflats and saw them solely as low value coastal space whose value would be enhanced by transformation to dry land. For example, during the past 300 years some 2000 km^2 of intertidal land on the East Coast of Britain has been reclaimed to form new land for agriculture, urban development and industry. These intertidal areas of mudflat and marshes, including the major systems of the Thames and Humber, were an essential component of the landform mosaic, and removing them has reduced the energy dissipation capacity of the coast.

The results of this loss within the estuaries were catastrophic, involving increased tidal range, wave heights and therefore in erosion and flooding. Results from model simulations of the Humber estuary show that the loss of $500km^2$ of intertidal area due to reclamation produced a 1m rise in tidal amplitude. If similar impacts were felt in other estuaries of the British coast, including the Thames, it may explain the increased in flood events and in flood defence construction during the latter part of the 18th century when reclamation was at its peak (Pethick 2000).

In the same way reclamation of intertidal areas in the Bremerhaven area and straightening and channelization of the river leading to Bremen may well have caused the increased incidence and severity of flooding in the Bremen area.

Impact on tidal deltas

Perhaps the most important change produced by the reclamation of the estuaries was so large-scale that we have failed to connect cause and effect. The discharge from the tidal estuaries into the open sea results in interruptions of long-shore sediment transport on the open coast and the development of extensive tidal deltas (see for example Fitzgerald and Penland 1987). These deltas represent a type of crossroads between the movements of sediment along the shore and the movement of sediment into and out of the estuaries. They consist of sandy sediment, which form extensive sub-tidal sand waves and sand ramparts that offer considerable protection to the open coast shoreline. In fact the deltas act as a form of natural wave break, held in place by the force of the tidal waters in the estuaries. Thus the form of the open coast is a dynamic one representing a balance between the force of the estuaries and the force of the incoming wave energy.

The reclamation of intertidal areas over the last 300 years resulted in a massive reduction in tidal discharge from the estuaries into the open sea. This led to a collapse of the tidal deltas that formerly fronted the coastline and led to erosion of the cliffs and beaches that had sheltered behind the deltaic sandbars. The scale of the change was considerable; the loss of over half of the intertidal area of the east coast estuaries in Britain meant that approximately 100km of delta frontage was lost – and leads to the tentative conclusion that, prior to reclamation, coastal erosion on this coast was practically unknown. The changes produced by reclamation not only were widespread spatially, but were long-term changes whose full effects have not yet been experienced. Erosion of the coast today could therefore be the result of changes in the estuaries introduced centuries ago.

Coastal defence

The changes to the coastline resulting from erosion can be seen as a natural response of the coastal landform system to imposed changes, in many cases an attempt to recover the level of energy dissipation which was lost as the intertidal area decreased due to reclamation. Left to itself the coast would have recovered, and in some places it has done so. However, in most cases the coast was not left to adjust due to our further intervention.

Our reaction to the catastrophic changes that we had instigated merely compounded the problem. First, we failed to realise that the flooding and erosion were the results of our own ineptitude and we continue, to the present day, to believe that these risks are the outcome of natural phenomenon. Second, we attempted to stifle the natural response of the coast to the imposed changes. The floods and erosion which marked the adjustment of the coastal morphology to estuarine reclamation were reduced by a series of hard engineering structures, the flood embankments and wood, rubble or concrete revetments which today front much of the coastline.

The effect of these engineering structures has been if anything even more catastrophic than the initial reclamation. The coast has been immobilised, starved of sediment supply from formerly eroding cliffs and with no space left for energy dissipation or landward transgression. The coast has been literally petrified and can only be maintained through massive injections of man-made capital.

Inappropriate development

The immobilisation of the British coast by elaborate coastal defence systems during the last 150 years, resulting from the initial reclamation era, has itself resulted in yet another human response, which further reduces the sustainability of the coast. Although the initial reclamation was almost entirely for agricultural purposes, during the 19th and 20th centuries the reclaimed land has increasingly been used for industrial and urban development. The opportunity to develop on flat land close to the navigational pathways provided by the estuaries and which also provided water and waste disposal, outweighed any consideration of increased risks in these low lying areas. Sea level rise and land compaction has steadily increased those initial risks so that today much of this development is at least 1m below high tide levels. This inappropriate development includes extensive areas of London fronting the Thames, industrial areas along the Humber, and port infrastructures along many of the estuaries.

Whereas flood embankments protecting agricultural land may be regarded as temporary features of the coastal landscape, these urban and industrial sites are seen as permanent and the attitude has gradually developed that they must, indeed should, be protected against any risk from flood or erosion. This attitude was reinforced after the disastrous

1953 surge tide that flooded almost all of the reclaimed areas along the east coast of Britain (Fig. 1) and resulted in 186 deaths. The response to the flood event, however, was not to recognise the risk and begin to move people and property away from such a high risk area, but rather to improve the defences on-line. The result has been to reinforce the locational drift towards these areas, a prime example being the massive holiday and retirement home complex along the Lincolnshire coast where some 15,000 people, mainly elderly, now live in often single-storied buildings at elevations below high water level. A programme of sand recharge has been initiated on this coast, in conjunction with a

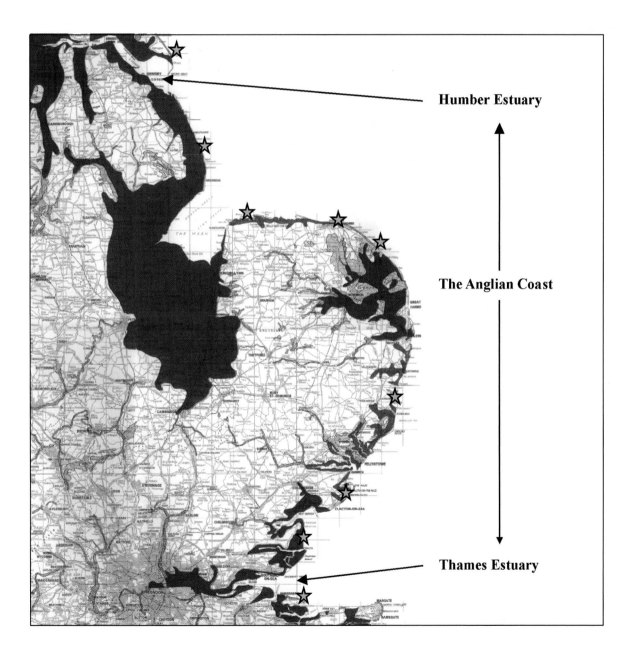

Fig. 1. Location of the Anglian coastline discussed in the text. Shaded areas indicate reclaimed intertidal areas now liable to tidal flooding. Stars indicate principle areas of coastal erosion.

concrete flood embankment, that offers protection against a 1:250 year flood, but the residents of this and other areas appear to believe that their protection is absolute. More important for long term coastal planning, the fact that they and the industrial complexes along this coast cannot easily be re-located means that the coastal system is immobilised permanently.

The response of coastal systems and constraints on sustainable human uses

The immobilisation of coasts that are unable to respond to changes weakens their ability to accommodate predicted rise in sea levels due to global warming. The coast must be able to respond to changes in sea level if risks to our own infrastructure and to the natural environment are to be minimised. This natural response would be one of relocation of the coastal landform mosaic, a movement that would allow each landform component to take up its correct energy niche. Without such adjustment, landforms and habitats will disappear and flooding and erosion potential will increase.

The current short-term perspective on management would be to construct ever higher flood defences and develop more effective erosion control. Such a response has not worked in the past and will not work in the future: it is unsustainable. In the words of the Ministry of Agriculture, responsible for coastal defence in England it "locks future generations into a unsustainable spiral of increasing investment in coastal defence" (MAFF 1993). Two examples of the future implications of this immobilisation of our coasts are examined in the following paragraphs:

Landward transgression of estuaries

The first example is the landward transgression of estuaries as a response to sea level rise as was first suggested by Allen (1990) for the Severn Estuary and has subsequently been the subject of research in the Blackwater Estuary, Essex, (Pethick 1997) and the Humber estuary (Environment Agency 1999). Estuarine transgression appears to enable an estuary to maintain its relative position within the tidal and wave energy frame. It entails a vertical upward movement, keeping pace with sea level rise, as well as a landward movement that maintains its position in the longitudinal energy frame.

The process of transgression consists of sediment erosion from the outer estuary, principally from upper intertidal areas, and its transfer to the inner estuary where it is re-deposited on the upper intertidal marshes. The increased elevation in these inner estuarine areas keeps pace with sea level rise and allows the estuary margins to migrate landward over previously supra-tidal areas. Research in the Humber has shown that the rate of such migration is in the order of 10 m of horizontal transgression for every 1mm of sea level rise (Pethick 2000).

The problem in the Humber estuary however is that the inner estuary intertidal area has been reclaimed and flood embankments now prevent the re-deposition of sediment on these areas. As a result, the estuary is prevented from landward migration and thus energy levels increase leading to accelerated intertidal erosion and amplification of the tidal range. This means increased risk for the considerable infrastructure, industrial, urban and agricultural, along the estuary shores, an increase in risk which is always ascribed to sea level rise but which is in fact due to mis-management of the mobile coastline.

Open coasts

The second example is from Norfolk and illustrates the difficulties for future management of an open coast due to past mis-management. The location of landform units along an open coast is often determined by relatively robust cell systems determining for example, pocket beaches between rock headlands. In some areas however, where such geological controls are not present and where shallow nearshore conditions reduce wave energy sufficiently, landform units along an open coast may be both diverse and sensitively adjusted to wave energy gradients. Wave energy gradients are themselves the result of wave refraction patterns in the nearshore that can result in wave foci along the shore where wave height and energy are accentuated.

The North Norfolk shoreline is characterised by particularly complex wave refraction patterns, the result of nearshore bathymetric variations over the tidal deltaic deposits of the Wash. Here the shoreline lies at an oblique angle to the nearshore contours and the resultant wave refraction pattern shows a series of wave foci spaced at approximately 10km intervals. Figure 2 shows the pattern for a northerly wave with an 8-second period. The wave foci are areas where wave rays are compressed causing increased wave heights and energy levels while the areas between foci experience diminished wave height and energy levels as the wave rays are spread out over wider areas of shoreline. The landform response to this sequence of energy gradients is also shown in Figure 2. Areas of high wave energy are characterised by coarse sediment beaches, such as those at Hunstanton, Wells and Blakeney. Between these wave foci, areas of low energy are occupied by salt marshes such as those at Thornham and Stiffkey.

These landform sequences merge into one another and their boundaries are blurred by the variations in location of wave foci as wave approach angle changes during successive storm events. Nevertheless, the overall sequence is extremely sensitive to the imposed wave energy gradients and therefore to changes in sea level. The impact of sea level rise is demonstrated in Figure 3, in which sea level is assumed to have increased by 1 m (i.e. taking place over the next 150 years at an average rate of sea level rise of 6 mm per year). The wave foci are shown to have migrated along the shore by 8km, at a rate of 53 m per year, so that locations that previously experienced minimum now experience maximum wave energy. The response of the landform sequences along this coast is predicted to be equally dramatic. Sand beaches such as those at Wells are predicted to be replaced by mudflats and salt marshes, while sand beaches will replace

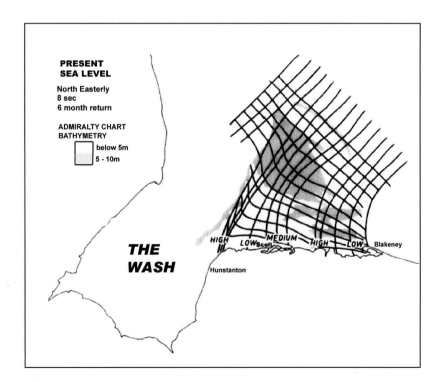

Fig. 2. Wave refraction and coastal landform units on the North Norfolk coast. Modelled wave is 8-sec period from north-east.

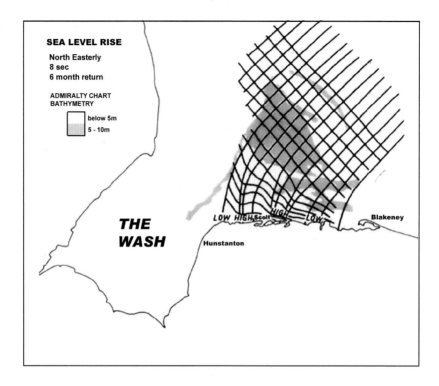

Fig. 3. Wave refraction and coastal landform units on the North Norfolk coast as in Fig. 2, but with a 1 m increase in sea level. Note transition of wave foci and corresponding landform units.

the designated salt marsh and sand dune areas of Thornham, Scolt and Stiffkey.

Two outcomes of this predicted migration of landforms must be of great concern to coastal management. First, the inertia of existing infrastructures means that development based upon coastal landforms, for example the recreational industry based on the proximity of sandy beaches or on bird watching habitats, will not be able to move along the coast in response to landform migration and will deteriorate. Second, the existence of coastal defences, even on this relatively undeveloped coast, means that the predicted migration of landforms cannot take place in its entirety. This will result in major losses of, for example, salt marshes, again threatening jobs and industries that depend on the coast and its landscape for their livelihood.

These two examples demonstrate that the physical legacy of reclamation and the attitude it engendered of treating the coast as a static line rather than a dynamic zone have reduced our options for the sustainable management of the Anglian coast. Where we have constructed towns or industries in high-risk coastal sites we may have no option in the future but to increase defence heights. The protection such defences appears to provide in turn encourages further development: a positive feedback process that cannot be sustained.

Solutions

Solutions to the problems of sustaining human uses of coastal areas and resources posed by the series of errors committed through past management practices become increasingly difficult to find, because we continue to compound the errors by inappropriate developments of all kinds. This drives us further and further away from strong sustainability.

At first sight, the obvious solution appears to be restoration of coastal ecosystem functions in order that natural self-sustaining processes might be resumed. For example, research on the Humber estuary and the natural protection of the coast afforded by the tidal delta (Pethick 2001) has indicated that the lateral extent of the ebb-tide delta may increase by as much as 386 m per year assuming an average rate of sea level rise of 6mm per year. This increase in the size of Humber ebb-tide delta in response to sea level rise appears, from the work of Pethick and others, to offer a natural mechanism for coastal defence that needs little or no management interference or financial input. Coastal managers have not appreciated the importance of this natural form of coastal defence. Since conventional hard defences may cost in the region of three thousand pounds per metre this could involve savings of one million pounds per year on the Lincolnshire coast. Extrapolating the figures for the Humber delta system to the coast of Britain as a whole is difficult. However, it may be that the reduction in the necessity for conventional defences provided by the extended tidal deltas might make more efficient use of natural capital and reduce the need for substitution by man-made capital. This would move us towards more strong forms of sustainability.

Large-scale restoration of the natural functions of coastlines is perhaps not practical in the short term due to the major concentrations of human settlements, industrial development and major infrastructure such as international airports. The designation of Special Areas of Conservation (SACs) and Special Protected Areas (SPAs) under the EU Habitat and Bird Directives also poses serious management problems in working toward strong sustainability. Under these Regulations, the competent authority and measures taken to prevent such changes must notify deterioration of designated habitats; changes in habitat due to natural processes, however, may be accepted. The key issue is whether natural habitat migration due to sea level rise can take place unimpeded by human interference. For example, existing flood defences, both for agricultural land and also for the urban development along the coast, may prevent landform migration and consequently the relative extent of different types of habitat may alter. If this is interpreted as contrary to the Regulations then

it will be necessary, under current provisions, to provide replacement habitat elsewhere. Replacement of habitat that is part of a functional coastal system in an alternative location is difficult if not impossible and it appears that a major impasse may be inevitable.

Conclusions

We are at a crossroads in respect to sustainable use of many of Europe's coastal regions. The spiral towards weak sustainability where coastal defences encourage development that, in turn, create the need for enhanced defences, must be broken, first by preventing future inappropriate development. The recent House of Commons Select Committee report on coastal management (House of Commons 1998) signalled the need for changing our current coastal management policies. The Report, states that 'We are of the opinion that flood and coastal defence policy cannot be sustained in the long term if it continues to be founded on the practice of substantial human intervention in the natural processes of flooding and erosion' 'We believe that a clear presumption should be made against future development in flood plain land'. Second, and most controversially, the introduction of a long-term policy to move those industrial or urban structures that would impede the natural coastal function. It urged '...the formulation of long term adaptive policies, for example encouraging the gradual managed abandonment of certain coastal areas, possibly over the course of many decades, and conferring residual life on defence works currently protecting assets which are untenable in the long term'.

The report continued by emphasising the need to allow the coast more space in order to carry out its primary function of energy dissipation, thus conferring benefits to all users. It was however, careful to point out that any programme for managed retreat should be carefully phased so as to avoid some of the problems outlined above: 'Greater priority should be given within national policy to managed realignment and sediment control. In each case, the total area of land which would be affected represents only a tiny fraction of the national land surface, and the associated costs could be diminished by implementing managed realignment of the coast over long time-scales.'

Perhaps most important, the report advised that the system for coastal management in the UK should be reviewed in order to provide larger scale administrative structures. This would overcome the problems of local scale pressures that ignore wider interactions at the coast and have been responsible for all of the problems of the Anglian coast that have been outlined in this paper.

The natural defence of coastal areas provided by wetlands, tidal deltas and other components of the coastal mosaic have yet to be fully appreciated by policy makers, planners and managers. If we adopt a more dynamic viewpoint, however, then it appears that these same coastal landforms remain intact, it is only their location that will change. If we persist in applying our static coastal management systems as sea levels rise then an increasing disparity will arise between our needs and the coastal resource. This will mean a further move away from "strong" sustainability and a loss of sustainable development opportunities. Instead we must begin to manage change at the coast in a more positive manner.

For this to happen we need a European Marine Policy that:
• Links terrestrial and marine management perspectives, and
• Encourages cross-disciplinary research to both create new knowledge that will help to inform coastal policy and strengthen management.

The key issue will be the exposition of options for coastal management that will move us away from weak sustainability and towards strong sustainability.

References

Allen JRL (1990) The Severn Estuary in Southwest Britain: Its retreat under marine transgression and fine sediment regime. Sediment Geol 66:13-28

Bijlsma L (1997) Climate change and the management of coastal resources. Climate Res 8:47-56

Bird ECF (1993) Submerging Coasts: The Effect of Rising Sea-Level on Coastal Environments. J Wiley & Sons Chichester, UK

Chappell J (1990) Some effects of sea level rise in marine and coastal Geol. Soc Aust SP 1:37-45

Crooks S and Turner RK(1999) Integrated coastal management: Sustaining estuarine natural resources. Adv Ecol Res 29:241-289

English Nature (1992) Coastal Zone Conservation: English Nature's Rationale, Objectives and Practical Recommendations. English Nature

Environment Agency (1999) Humber estuary: Geomorphological studies. Internal report

European Union (2000) A Communication from the Commission to the Council and the European Parliament on "Integrated Coastal Zone Management: A Strategy for Europe" (COM/00/547 of 17 Sept. 2000), Brussels

Fitzgerald DM and Penland S (1987) Backbarrier dynamics of the East Friesian Islands. J Sed Petrol 57:746-754

House of Commons (1998) Flood and coastal defence. House of Commons Agriculture Select Committee. HMSO

IPCC (Intergovernmental Panel on Climate Change) (1996) Second Assessment Report: Climate Change: The Science of Climate Change. Cambridge University Press, Cambridge, UK

MAFF (1993) National Flood and Coastal Defence Strategy for England and Wales. Ministry of Agriculture, Fisheries and Food, London, UK

Pethick J (1997) The Blackwater estuary: Geomorphological trends 1978 to 1994. Report to the Environment Agency London, UK

Pethick J (2000) Coastal management and sea level rise. Roy Soc Edinburgh Transactions, Edinburgh, UK

Pethick J (2001) The Anglian Coast. paper presented to the Dahlem Symposium on Science and Integrated Coastal Management (ICM) Berlin, December 12–17, 1999. In: Von Bodungen B and Turner K (eds) Proc Symp Science and Integrated Coastal Management (ICM). Dahlem University Press Berlin, Germany

Turner RK, Adger WN, Lorenzoni I (1998) Towards Integrated Modelling and Analysis in Coastal Zones: Principles and Practices. LOICZ Reports and Studies No. 11. IGBP/LOICZ, Texel, The Netherlands, iii+122 pp

WCED (1987) Our Common Future (The Brundtland Report) World Commission on Environmenta and Development. Oxford University Press. ISBN: 019282080.

Coastal and Shelf Processes, Science for Integrating Management

H.J. Lindeboom[1*], P.R. Burbridge[2], J.W. de Leeuw[3], A. Irmisch[4], V. Ittekkot[5], M. Kaiser[6], R. Laane[7], J. Legrand[8], D. Prandle[9], K. Reise[10], J. She[11]

[1]*Royal Netherlands Institute for Sea Research / ALTERRA, PO Box 167, 1790 AD Den Burg, The Netherlands*
[2]*Dept. of Marine Sciences and Coastal Management, Centre for Coastal Management, Ridley Building, University of Newcastle upon Tyne, Newcastle upon Tyne NE1 7RU, UK*
[3]*Royal NIOZ, PO Box 59, 1790 AB Den Burg, The Netherlands*
[4]*Projektträger BEO, Bereich Meeres- und Polarforschung, Geowissenschaften, Seestr. 15, 18119 Rostock, Germany*
[5]*ZMT, Fahrenheitstr. 6, 28359 Bremen, Germany*
[6]*School of Ocean Sciences, Menai Bridge, Gwynedd, LL59 5EY, UK*
[7]*RIKZ, Postbus 20907, 2500 EX Den Haag, The Netherlands*
[8]*IFREMER, TMSI/EC, BP 70, 29280 Plouzané, France*
[9]*Proudman Ocean Lab, Bidston Observatory, Birkenhead, CH43 7RA, UK*
[10]*AWI, Wattenmeerstation Sylt, 25992 List, Germany*
[11]*Danish Meteorological Institute, Lyngbyvej 100, 2100 Copenhagen, Denmark*
**corresponding author (e-mail): h.j.lindeboom@alterra.wag-ur.nl*

Abstract: A major challenge facing us is to sustain human use of coastal areas and their natural resources within the context of increasing human pressures and uncertainty over the scale and nature of climate changes. To address this challenge, there is an urgent need to more effectively use available scientific knowledge and to fill critical gaps in our understanding of coastal systems. This paper identifies user problems and corresponding scientific opportunities. Critical knowledge gaps which need more or new research include: natural variability in space and time, experimental management, potential role of habitats and species, effects of changing nutrient regimes, preturbations of food web dynamics, and ocean-atmosphere-sea-coast coupling. Safety risks, extreme events-including thinking the unthinkable, food security, sustainable dynamic land-sea boundaries, multiple use, and water quality are other items which demand more attention. Indispensable tools include: permanent integrated coastal oberservational systems, large scale research facilities like mesocosms, marine reintroduction sites and protected areas and a stronger framework to help integrate knowledge. Since in most cases we are dealing with global issues and problems, the need for more co-operation with developing countries and capacity building is stressed.

Introduction

The coastal regions of Europe have been the focus of human settlement and economic activity for many millennia. One of the most important challenges facing the Member States of the European Union (EU) is to maintain the continuity of human use of coastal areas and their natural resources

From WEFER G, LAMY F, MANTOURA F (eds), 2003, *Marine Science Frontiers for Europe.* Springer-Verlag Berlin Heidelberg New York Tokyo, pp 229-242

within the context of increasing uncertainty over the scale and nature of climatic change and the associated impacts on coastal systems. To address the management challenges facing us in the promotion of wise and durable use of our coasts, there is an urgent need to make more effective use of the available scientific knowledge. Such a strategy would inform and support the formulation and implementation of policy, investment strategies and planning and management systems governing the expansion and diversification of human activities. There is also a need for new scientific information to fill critical gaps in our understanding of how coastal systems will react in response to changes in environmental processes and human management interventions. This is reflected in the findings of the recent EU funded demonstration programme on Integrated Coastal Management and the subsequent formulation of the European Coastal Management Strategy. This is further highlighted in the recent statement on one of the major coastel activities by Margot Wallström, (EU Environmental Commissioner): "What still lies ahead of us is the arduous exercise of defining environmental objectives for fisheries together with a system of indicators for the future monitoring of policy performance, and the adoption of a long-term strategy and legal instruments to achieve policy objectives."

The coastal zone

The term the "Coastal Zone" has been widely adopted to denote a zone of transition between purely terrestrial and purely marine components of the global ecosystem. It contains the crucial boundaries between ocean, coastal systems and land. In the LOICZ (Land Ocean Interactions in the Coastel Zone) definition it is the area between 200 m above and 200 m below sea level. Although other studies may define these boundaries differently, all agree that to understand and manage the coastal zone both the catchment areas and the processes along the continental margins should be included. What makes the management of human activities in this "zone" different from inland or offshore areas is the high concentrations of energy, sediments, and nutrients that stimulate both high biological productivity and

a wide diversity of habitats and species. The richness and diversity of resources found in coastal areas has long been recognised by human society and there has been a corresponding concentration of human activities and settlement along shorelines and estuaries throughout Europe. These powerful and dynamic forces that continuously shape our coasts also pose risks to human activities. Poor perceptions of these powerful and dynamic processes increase man's vulnerability to natural hazards.

User problems

The history of the management of Europe's coastal areas and resources has often been characterised by major human interventions in coastal processes. Interventions, such as the reclamation of land from the sea, have been sustained through a massive investment of finance. There has also been a reduction of natural capital represented by the goods and services that would have been generated by the coastal systems if they had not been subjected to human induced changes (Fig. 1). Science has helped to clarify the nature and value of such natural capital. Science has also helped to demonstrate the adverse effects of past management practices in reducing options for longer-term economic and social development and the increasing vulnerability of human activities to sea level rise and other changes in large-scale natural systems affecting coastal regions.

We are faced with a dilemma if we are to maintain the continuity and positive evolution of the human use of coastal areas and resources (Table 1) . We either change the way we manage human interventions in coastal systems or we continue down a path that is leading to a decrease in opportunities for the sustainable economic and social uses of Europe's coastal assets. If we are to achieve a positive change and move towards a stronger form of sustainability and a more diverse and equitable base for economic and social development, we must recognise the powerful and dynamic forces that create coastal ecosystems and maintain the stream of environmental and economic goods and services that sustain human development.

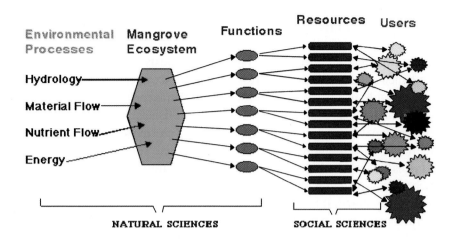

Fig. 1. Understanding the Linkages Among Coastel and Marine Ecosystems and Human Activities.

- To protect coastal marine biodiversity and ecosystems from damage arising from human activity
- To promote water quality in catchment and coastal areas, and control the impact of fisheries, in order to ensure the sustainability of resources including tourism and the dynamism of economy. More particularly, in the interest of human health, it is important to prevent the occurrence, or mitigate the adverse effects, of virus, bacteria and toxic algae in coastal waters
- To meet public concerns over the discharge and dispersion of various kinds of waste (effluents, oil spills, residues and pollution arising from the exploitation of sands and gravel
- To protect sensitive stretches of the coastline against erosion, excess sedimentation and the consequences of extreme events
- To promote the preservation and valorisation of a large part of the European cultural heritage

Table 1. Central objectives of research to provide a scientific background for Integrated Coastal Zone Management as proposed by the ESF Marine Board. These topics are addressed in the table.

Science opportunities

Advances in science have helped to explain how ill-advised human interventions in coastal systems and processes have led to less efficient use of both man-made capital and natural resources. These advances have also clearly demonstrated the need for a broader understanding of the linkages between ocean, atmospheric, terrestrial and coastal systems and large-scale environmental processes. For example, increased vulnerability of coastal settlements to flooding is a result of rising sea levels, subsidence, increased storm surges, changes in land cover in river catchments and continued expansion of the coastal settlements themselves. To address such issues of sustainable use of coastal areas, knowledge of the natural and man-induced changes in the surrounding regional marine and terrestrial systems is essential. While the concept of the Coastal Zone remains a useful tool in defining an area within which special planning and management arrangements are required, it is increasingly recognised that such plans and management arrangements must be based on sound scientific knowledge of dynamic environmental processes over a wider geographic scale and longer temporal scale (Lindeboom 2002a).

Sustainability

The concept of "Sustainable Development" was defined in the report of the World Commission on Environment and Development report "Our Common Future" as that which "meets the needs of the present without compromising the ability of future generations to meet their own needs" (WCED 1987 p8). This concept gained further public recognition as a result of the United Nations Conference on Environment and Development (UNCED) and the production of "Agenda 21" the global plan of action designed to promote more sustainable forms of development. Coastal management was given explicit recognition in Agenda 21 Section 17 dealing with the marine environment. In this section Integrated Coastal Management was given priority as the single most important means of reconciling a number of non-sustainable marine resource development issues. In a recent publication from the Land-Ocean Interchange in the Coastal Zone (LOICZ) programme sustainable coastal development was described as "the proper use and care of the coastal environment borrowed from future generations" (Turner et al. 1998).

The term "Sustainability" is as much an expression of social choice as it is an expression of the ability of the environment to continue to support human needs and aspirations. Throughout history, man has modified coastal systems to suit current

needs and aspirations. Reclamation of salt marshes to form sites for agriculture, industry and housing are common examples. However, such changes can seriously undervalue the resources generated by natural systems. For example, the reclamation of wetlands can reduce environmental services that support fisheries where wetlands act as spawning, feeding and nursery grounds for commercially important fish stocks.

Changing these systems to create short-term economic benefits gained by specific activities can foreclose future development options and can create long term costs that are poorly perceived or ignored. Degradation of coastal systems and consequent loss of natural functions such as biological production or stabilisation of sediments means that man has had to spend time, effort and monetary resources to compensate for the loss of environmental and economic goods and services. For example, the need to defend coastal infrastructure such as nuclear power stations located on eroding shorelines could be termed "weak sustainability" because of the need to allocate natural resources (shingle and sand = natural capital) and man made capital (money and machines to move the shingle and sand) to sustain the viability of the power plant. Weak sustainability is also associated with the assumption that there can be unlimited substitution possibilities between different forms of capital via technical progress. For example, the reclamation of intertidal mudflats and marshes has proceeded on the basis that there was little capital value represented by these coastal ecosystems and greater man-made capital could be achieved by their transformation into dry land. However, advances in scientific knowledge have identified highly valuable functions performed by wetlands as well as risks to the sustainable use of reclaimed areas resulting from both the destruction of the functions of wetlands and other coastal systems, and rising sea level. "Strong sustainability" is associated in economic thinking with the conservation of different forms of capital (man-made, human, natural, and social/moral) in respect to meeting the needs of human populations over time. Strong sustainability applied to marine and coastal systems would mean that their natural capital expressed in terms of biological diversity, generation or renewable resources, and maintenance of natural processes and functions would remain constant or increase.

Programmes for future scientific research:

In addressing the challenge of meeting the social and economic development needs of European society it is imperative that the Marine Science Plan for Europe addresses two critical coastal management and scientific investment issues, namely:
• filling critical gaps in available scientific information required to reduce uncertainty in dealing with predicted changes in large-scale environmental processes affecting coastal systems.
• investment to improve the utility of existing scientific knowledge from a wide array of social and natural science disciplines in the formulation of coastal policy, development strategies, investment and natural resources management practices.

Filling critical gaps

Natural variability of the coastal zone in space and time

Before discussing the sustainable use of coastal and shelf ecosystems implying acceptable, i.e. reversible human-induced perturbations of these ecosystems, we first have to establish the natural variability of these ecosystems.

Natural variability in (European) coastal zones (and to some extent in land-locked or inland seas like the Baltic, the Mediterranean and the Black Sea) is a long recognised phenomenon. However, the time scales, the amplitudes and the nature of this variability is far from understood.

Over the last decade it has become increasingly clear that the Northern Atlantic Oscillation (NAO) determines to a large extent the natural variability of the coastal ecosystems in Europe and that the time scales involved vary from seasonal to annual to decadal to centennial (Fig. 2). Moreover, the NAO as such interacts with other oceanic phenomena such as the THC variability, the Tropical Atlantic Variability (TAV) and the Arctic Variability (AV) and is possibly influenced by global warming. Apart from the NAO, other factors that influence natural variability have to be taken into consideration. For

example the variability in tidal current strengths is partly caused by the so-called moon cycle of 18.6 years (Oost et al.1993) and there are strong indications based on high-resolution lake sediment analyses that several decadal and centennial solar activity cycles are important as well.

Consequently, coastal ecosystems are confronted by cold or warm winters, frequent or sporadic storms, low or high average wave heights, low or high run off and by more or less erosion, sedimentation and resuspension. Such natural variability substantially affects the physical and morphological boundaries of coastal ecosystems and inland seas and, as a consequence of that, their chemistry and biology (Lindeboom 2002b).

Hence, before we can determine to what extent humans can explore coastal ecosystems in a sustainable way we first have to define the physical, chemical and biological bandwidths of the natural ecosystem.

This is a difficult task since presently there are few pristine European coastal ecosystems anymore. Thus, we have to rely on historical data, long-term data series and high-resolution analyses of sedimentary archives (to reconstruct the past and predict the potential future). To establish the natural variability of European coastal ecosystems is a major scientific task in itself, which can only be accomplished by intense European collaboration of (palaeo)-climatologists, meteorologists, oceanographers and "coastal scientists" critically analysing their historical data and long-term data series, pooling such data and contemporary data and careful interpretation of these huge data sets. Once the natural variability has been established the second major line of research can start focussing on the sustainable use of coastal ecosystems (and inland seas). Hence, what are the limits of physical, chemical and biological perturbations without irreversibly changing the variable natural ecosystem?

Experimental management

Applied research in general is retrospective, addressing issues that have arisen as a result of human activities in the marine environment. The outcome of such research provides the basis for formulating potential new management regimes. Of-

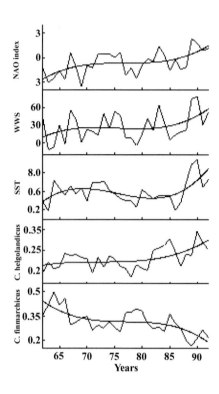

Fig. 2. Year-to-year fluctuations and long-term trends in the North Atlantic Oscillation index (NAO), West Wind Stress (WWS) in m²/s², Sea Surface Temperature (SST) in °C, and abundance of *Calanus finmarchicus* and *C. helgolandicus* (log-transformed) from 1962 to 1992 over the northeast Atlantic and the North Sea. Supplied by J.-M. Fromentin and B. Planque, adapted from Fromentin and Planque (1996). Polynomial fits were used to provide a visual indication of the long-term trends of these series. The NAO index is the difference in normalized sea level pressures between the Azores and Iceland calculated on the December-April period. WWS and SST values are the mean of December-April monthly data over the eastern-north Atlantic and the North Sea. Calanus values are the annual mean of log-abundance (log-number of organisms per sample) of copepodite V and adults of each species calculated over the eastern-north Atlantic and the North Sea. Calanus species are collected in near-surface water and each sample has a volume of about 3m³ (each annual value is calculated from about 2300 samples).

ten there is resistance to the initiation of novel management regimes because of the uncertainty associated with their outcome. Yet there is a need for research outcomes to be implemented, otherwise our endeavours are wasted. Consequently, there is an urgent need for managers to adopt experimental management regimes to enable us to test the predictions that have arisen from experimental studies. Generically, such regimes would include different levels of human interference ranging from no removal to partial removal and on to complete removal. An example of such a system may be the exclusion of bottom fishing activities from certain areas of the seabed to remove the effect of anthropogenic physical disturbance to the seabed and its biological components. An experimental management regime that permitted varying levels of disturbance would permit a rigorous examination of community responses to human disturbance. The outcome of such experiments may conclude that intermediate levels of disturbance result in a more productive system that may have benefits for the wider fishery and may result in a more diverse range of habitats and communities. In some cases, unexpected findings may occur. For example, 17 000 km² of the seabed off the eastern coast of the US was closed to all forms of towed bottom fishing gear to protect stocks of yellowtail flounder.

Not only did this closure protect this fish stock; it also permitted development of a more diverse seabed assemblage and resulted in a 14-fold increase in the biomass of commercially important scallops within the area (Fig. 3). Such findings enable us to continue to use the marine environment in a manner that is both sustainable but that is also not entirely prohibitive.

If parts of the coastal seas will be filled with fields of windturbines, the chance should be taken to explore experimentally new trajectories for a multiple and sustainable use in these area. 'Wind Parks' may be developed into 'Marine Parks' where fisheries with towed gear are excluded. On the other hand, there is a wide potential for other uses in addition to generating electricity. This particularly applies to aquacultures typically practiced inshore where they cause an array of environmental problems. Within offshore wind parks such

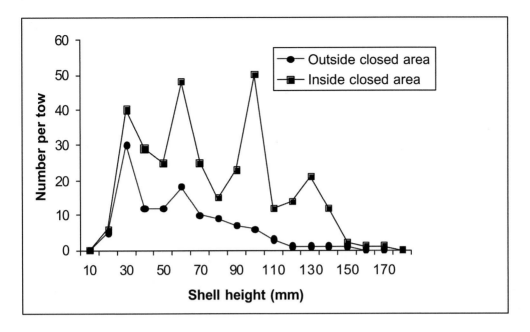

Fig. 3. Standardised abundance of sea scallops (numbers per dredge tow) by shell height, taken in the July National Marine Fisheries Service dredge survey on Georges Bank. Data are presented separately for the areas closed and those open to scallop dredging (adapted from Murawski et al. 2000).

problems may be mitigated by the larger water bodies passing through such an area. As examples, mussels may be raised on longlines, lobsters in crabs pots, attached oysters and macroalgae may be grown on rafts, etc. Performing these activities offshore needs partially new methods, and the consequences on the marine ecosystem need careful evaluation. However, coastal management may shift aqua-cultures from inshore to offshore to unburden shoreline habitats.

Another example where an experimental approach would have a wide potential for stimulating management practices are sand nourishments at coasts where erosion prevails. Where these coasts have been armoured by coastal protection to prevent shoreline retreat, natural habitats have been lost, there dynamics ceased and biodiversity decreased. Sand replenishments offer a chance to restore lost habitats and their dynamics. Particularly, experiments should be conducted with different designs of the supplied sand deposits. These may be arranged as sandy spits in front of seawals or dikes. In the shelter of such a sand bar, mud may accumulate and salt marshes may develop. Another option would be arcs of sandy isles and sand bars, providing shelter and at the same time offering coastal birds sites for undisturbed breeding colonies. Depending on sea level rise and the resulting sand hunger at a particular shoreline, such deposits may last for decades or centuries but continuously will change morphologically in size and shape, generating a diversity of coastal habitats otherwise in short supply. Because of the ephemeral nature of such sand deposits, future generations are not compromised in their use of the coastal zone.

Potential role of habitats and species

Habitat fragmentation and distribution of habitats of shoreline organisms entailed a loss of species along European coasts. Consequences are:
• Impaired ecosystem functions such as missing biotic filters between land and sea.
• A truncation of the coastal food web at its upper meshes.
• A degraded aesthetic and recreational value.

Species played and partially still play an important role in the biogenic morphodynamics of the shore. Other species directly or indirectly support essential coastal functions such as storage of organic carbon and nutrients, nurseries for offshore adult populations, food supply for flocks of birds, fisheries and human welfare. Based on deletion and addition experiments as well as observations on wider scales, science has to demonstrate the cascading effects species have on the integrity of the coastal system. The main emphasis will have to be on restoring sizeable habitats, mostly for spontaneous and self-organising recolonization of the coastal species assemblages. The target is not to achieve fixed species associations or habitat structures. Instead, the abiotic prerequisites for coastal life to unfold need to be supplied on European wide scales. To achieve this, research is requested to reconstruct the role of lost habitats and species in the coastal ecosystems at their time. In a second step, predictions are needed on the role of these elements in the context of the modern coast. In selected cases, individual species may be reintroduced which have an ecological flagship role such as sturgeons or the European flat oyster.

Effects of changing nutrient regimes

Human activities in the catchment area are often visible in higher nutrient concentrations, increased plankton growth and blooms and oxygen deficient conditions. However, more subtle long term changes are also occurring in the catchment area and coastal zones which are related to changes in the relative contribution of different nutrient elements. Nutrient concentrations in the coastal zones are mainly affected by the riverine inputs reflecting changing human activities in the catchment area. In addition to an increase in the concentrations and fluxes of nutrient elements such as N, P and Si, there is also a change in their ratios between the different elements compared to those in pristine or natural coastal systems. This change in ratio has been reinforced in the industrialised western countries by a sharp drop in phosphate but nearly no reduction in nitrogen. Retention of nutrients in lakes and reservoirs behind dams has been found to retain besides N and P also Si.

The changing nutrient ratios in turn has an effect on the species composition of phytoplankton and consequently on the food web structure in the catchment area and in the coastal zone.

Perturbations of food web dynamics

Physical interventions (for example, constructions, land reclamation, dikes, beam trawling), withdrawal of material (fishing, sand and gravel mining, gas and oil exploration), changes in the input of matter and energy (oil spills, sediments, changes in storm surges, wave action, currents and the introduction of alien species) affect the coastal marine food web structure both directly and indirectly by habitat modification and selective removal or addition of biota. In many cases these changes are taking place unnoticed during a certain time span, since they can also be attributed to the natural variability before the actual causes suddenly become manifest.

There have been several instances in the past where scientists were not equipped to respond or were not able to respond rapidly to environmental events, perceived by the public and policy makers as catastrophes, which occurred in European Seas such as algal blooms or a decline in populations of several species. For instance, the decline in the Wadden Sea of the Eider duck populations in the past few years which appeared to be related to events that had their beginnings at least 10 years earlier. These and other events such as oil spills require that adequate institutional structures with a continuity in basic research are in place to respond to them.

To anticipate food web changes due to the above reasons at an early stage much more detailed insight into the microbial and macro food web structure is required.

Optimal design of observing networks

To establish a permanent coastal ocean observing system, optimal design of the observing system is required to reach cost-effective performance. This means a fordable observing system with the capability to provide data for the diagnosis and prognosis of coastal/shelf oceans and to support decision-making in coastal zone management and marine industry.

The scientific issues concerned include quality of existing and optimal design of future coastal ocean monitoring networks. The first objective is to evaluate scientific quality of the current existing coastal observing systems by developing suitable indicators and methods to assess the quality of existing systems. Methods to estimate reconstruction error and effective information need to be developed and applied to identify data redundancy and insufficiency. Observing System Simulation Experiments via data assimilation should be used to evaluate the significance of the observing system. The second objective is to optimally integrate existing observing systems and state-of- the-art monitoring technology and determine a cost-effective monitoring strategy for European coastal oceans, i.e. where to put our limited resources. Cost-benefit analysis and optimisation methods should be combined with approaches used in observing system evaluation. Pilot field experiments may be needed to demonstrate the quality of optimally designed coastal observing systems.

Ocean-atmosphere-sea-coast coupling

Global Climate Change concerns focus attention directly on this issue. How might changes in Atlantic circulation impact on coastal climate and related dynamics to ecology? How will this impact be conveyed through shelf-edge exchanges, internal circulations, atmospheric coupling? Noting the vastly reduced inertial lag of the shelf seas – are there first signs of Global Climate Change in shelf seas. Such scientific questions are foremost world wide, the integrated communications provided within the Operational Oceanographic approach allows individual, specialist scientists to interact with and impact on these issues.

The models by which such integration can be facilitated need to be explored along with the needs for infrastructure investment in Very High Performance Computers, High Performance Data Networks etc. The success of the ECMWF in stimulating European research into meteorological, climate and oceanographic is noted.

Improve the utility of existing scientific knowledge

There is an urgent need to invest time, effort and funds in gaining additional value from the scientific knowledge that exists. To do this we propose to identify critical management issues affecting the sustainable use of European coastal regions and then to undertake a series of meta-analyses of existing studies. Such an analysis should more clearly represent the mosaic of coastal ecosystems, the key environmental processes that govern the health and productivity of individual coastal ecosystems, and to assess the effect of current management policies on the sustainable use of those systems and their renewable resources. The major issues identified so far are:
• Risk to life, property public and private investment
• Food security relating to both fisheries and aquaculture.
• Sustainability of the land-sea boundary.
• Optimisation of the use of the coastal zone which would include multiple uses.
• Water quality in catchment and coastal areas.

A meta-analysis approach adds value to existing science by identifying where we appear to have reasonably complete scientific information. Perhaps more importantly this approach identifies critical gaps in our knowledge permitting us to identify those areas of research that need to be addressed. Finally, meta-analysis may yield new knowledge enabling us to better understand how our coastal systems function.

Risks

To address the challenge to protect coastal marine biodiversity and ecosystems from damage caused by human activity, it is a pre-requisite that we must appreciate the relative influence of human activities in comparison with natural processes. For example, human sources of physical disturbance to the marine environment must be assessed against the background of natural physical perturbations (wave action, seabed currents) that occur. It is important to realise that even without the influence of human activities coastal ecosystems and biodiversity will evolve through time. Hence we would caution the use of the word "protect" as this suggests that we would attempt to maintain a system in a condition that may not be appropriate (i.e. it may have changed even without human influences acting upon it). Such efforts would be scientifically flawed and extremely costly. In order to assess the natural variability within the coastal system, we urgently require areas that remain free from interference. Such a goal is achievable when considering activities that directly impinge upon coastal systems (e.g. fishing, dredging, land-reclamation) but is perhaps unattainable with respect to the chemical composition of the seawater.

Extreme events – thinking the unthinkable

Earlier EU studies have both advanced individual aspects of marine science (models, technology, process studies) and linked some components of these in COHERENS, PROMISE, ERSEM etc. Comprehensive 'risk analysis' makes it now opportune to examine extreme, extraordinary events that may occur due to:

• Unsustainable, accelerating rates of change in meteorological forcing associated global response (mean sea level) and localised conditions (supply of sediments, river flow etc).
• Exception events - seismic, accidental spillage.
• Peculiar interactions between physical, sedimentary, chemical and biological parameters.
    The scientific issues concerned include:
• Extension from linear to non-linear systems and from 'primitive equations' to 'group dynamics'.
• Cross-spectral exchanges in physical – biological parameters, across temporal and spatial scales under exceptional circumstances.
• Time variability of extreme events in different time scales and its reasons.
• Value-added prediction for extreme events to reduce the risk.

Food security relating to both fisheries and aquaculture

The decline of marine fish resources in the European coastal zone is primarily caused by over-fishing and the lack of political will to implement the measures required to halt this decline. This is a fisheries management issue. However, fishing activities have wider ecosystem effects on both the biota and habitat, these include:
• Disturbance of the upper 1 cm – 20 cm of the substratum causing short-term resuspension of sediments, remineralisation of nutrients and contaminants, and re-sorting of sediment particles.
• Direct removal, damage, displacement or death of a proportion of the infaunal and epifaunal biota.
• A short-term (0-72 hrs) aggregation of scavenging species.
• The alteration of habitat structure (e.g. flattening of wave forms, removal of rock and biogenic structures).

From an ecological perspective, fishing removes specific components of the marine community that perform different functions within that community (e.g. fish may be important predators whilst bivalves are important filter feeders transporting energy from the water column to the seabed). Because of the complex number of interactions within the fish community of the North Sea, it is unlikely that even the total extirpation of one species (e.g. cod) would lead to dramatic 'cascading' effects within the system. However, the removal of species that are important habitat engineers (e.g. bivalves that form reefs, bioturbating sea urchins etc) may alter the habitat such that it becomes less suitable for the maintenance of some species, whilst other species may benefit. Thus the degradation of seabed surface topography by bottom fishing may reduce the suitability of the habitat for roundfish (e.g. cod) but may favour flatfish populations (e.g. plaice and sole). While the physical disturbance of soft-sediment communities by fishing gears may elevate productivity and promote the smaller body-sized biota (e.g. worms that provide food for fishes), excessive levels of disturbance will ultimately reduce overall production. Whilst the benefits of extra food might be easy to measure in terms of population benefits (e.g. increased growth rate of plaice in the North Sea) the unknown benefits of habitat composition and structure require further research.

Aquaculture is an effective method for increasing the provision of dietary protein when it is based on species that convert organisms at low trophic levels (e.g. plankton and plants) into protein. Hence, bivalve mariculture and the production of herbivorous and omnivorous fishes such as carps and tilapias is one of the most efficient methods of generating protein for human consumption. The cultivation of piscivorous species typically consumed by western societies is less efficient because of the use of other fish species in the production of fish-feed. Research to identify plant-derived feeds that can be used to rear piscivorous fishes is of the utmost priority. In addition, the activities associated with intensive aquaculture can have deleterious effects upon the marine environment, these include:

• The use of anti-biotics to suppress disease may lead to disease resistance and changes in the microbial community in the vicinity of fish cages.

• The build up of organic matter in proximity to aquaculture sites alters benthic community structure.

• Poor management may lead to the production of toxic algal blooms.

• Escaping fish may lead to loss of genetic diversity.

Towards more sustainability of dynamic land-sea boundaries in Europe

Most developed coastlines in Europe lost their morphodynamic flexibility and biotic diversity. Adaptability to a gradual rise in sea level and to exceptional meteorological events such as storm tides ceased. Maintaining these shorelines in their present form will require an increasing effort in structural as well as financial terms. There is a lack of sustainability which asks for scientific innovations and guidelines to initiate an integrated management allowing self-sustaining processes to unfold to the benefit of continued human use of the coast.

The dynamics of the land-sea boundary must be given more room. To achieve this, geomorphological and ecological scenarios need to be elaborated. These may be worked out along a continuum. At one end there would be managed retreat by deconstructing and abandoning unsustainable portions of the coast. At the other are attempts to supplement structurally defended coastlines with a buffer of sand borrowed from offshore bottoms. The aim is to find solutions, which offer a wide potential for multiple uses but no longer are based on a permanency of infrastructure compromising future generations.

Optimisation of the use of the coastal zone which would include multiple uses

A feature of complex coastal ecosystems is the wide array of physical, biological and chemical functions they perform. Primary and secondary biological production are examples. These functions generate environmental resources such as fish, which have tangible social and economic value. Other valuable functions and resources are common to coastal ecosystems. For example, seagrass beds help to reduce wave energy that otherwise would erode soft coastlines. The value of such environmental services is often poorly perceived by policy makers, planners and coastal managers with the result that they are afforded little protection and can be degraded. With mounting population pressures on coastal areas and increasing demands for resources, we can ill afford to degrade or lose either the economic or environmental goods and services provided by coastal systems (Fig. 4).

A solution is to promote multiple-use of complex coastal systems and to seek to optimise the stream of economic, social and environmental benefits to society. To do this requires sound knowledge of what form of human activity are dependent on the functions and resources created by coastal systems and how such activities can be integrated in time or space as part of coastal management.

This is an example of where science needs to be used to better inform management and is also an example where cross-disciplinary science can help to define management options for wise and sustainable coastal development.

Water quality in catchment and coastal areas

From hazard to risk assessment Billions of EUROs are spent each year during the last 30years to assess the water quality in rivers and coastal zones. Office shelfs, books and data bases are piled up with data with concentrations of certain priority substances. They are assessed in national annual reports and international Quality Status Reports of different coastal seas. However, it is still difficult to say if water quality is getting better or worse. We are data rich and information poor.

To assess the water quality, objectives or quality values, based on toxicological data, have been set by different national authorities. To be absolute sure to prevent organisms from hazardous effects, all kind of safety factors were introduced, lowering the quality objectives, sometimes below the natural background values.

However, more and more it is recognised that this substance-oriented assessment with these quality values is not possible anymore. First of all this assessment method is just an identification of a possible potential effect, not a real effect in the environment it self. Also, it is not possible to assess the toxic effects of all 150000 substances separately and in combination with each other, because it will take too long time and is too expensive. The ultimate effect of the last 10 years is that the quality values differ enormously (factor of at least 1000) between different countries. It is beginning to be recognised that the monitoring programmes do not always include the substances, which cause the real effects.

So, more and more, real effects in the lab and field are determined with bioassays and rapid screening methods.

With these bio-effect tools it is possible with the Toxicity Identification Evaluation (TIE) procedures, to determine the substances causing the real effects.

These developments will have an impact on the monitoring strategy to assess water quality. Policy measures to reduce the input of substances and to meet their quality values, are taken nowadays at an international level. However, the monitoring strategies are still at a local or national level. International harmonisation and standardisation in monitoring strategies in catchment areas and coastal zones will be necessary in the near future to fulfil the information needs of policy makers and the general public.

Tools

Permanent Integrated Coastal Observational System

European requirements for developing and assessing science and informing management include permanent *in situ* integrated observation systems in the coastal areas and the Atlantic equipped with automated physical, chemical and biological sensors.

There is a need to optimise investments by:
• co-ordinating existing observational networks.
• adding (European value) to on-going/proposed. observational missions (sea/ground truth; proxy relationships etc.
• maximizing synergy between satellite-aircraft-ship-buoy-radar-lander platforms.
• determining key indices, critical positions, appropriate sampling intervals.
• standardisation/harmonisation of instrumentation/analytical procedures.
• delegated responsibilities for data analyses, dissemination and archiving.
• establishing links with terrestrial, atmospheric, oceanic monitoring programmes.
• interfacing with modelling requirements (set up bathymetry), initialisation, forcing, assimilation, assessment) and utilising developing modelling capacities.

The scientific issues concerned include quality of existing and optimal design of future coastal ocean monitoring networks.

The objective is to evaluate scientific quality of the current existing coastal observing system. Developing suitable indicators and methods to assess the quality of the existing coastal observing system; Developing and applying theories estimate reconstruction error and effective information in order to identify data redundancy and insufficiency. Observing System Simulation Experiment via data assimilation should be used to evaluate the significance of the observing system on improving prediction.

Another objective is to optimally integrate existing observing system and state-of-art monitoring technology and determine a cost-effective monitoring strategy for European coastal ocean, i.e. where to put our limited resources. Cost-benefit analysis and optimisation methods are combined with approaches used in observing system evaluation. Pilot field experiments may be needed to demonstrate the quality of optimally designed coastal observing systems.

Large scale facilities, e.g. mesocosms

To understand the different processes in the complex coastal systems and to execute experiments with different management options, both laboratory and outdoor experiments are necessary. Experiments in which the impact of different boundary conditions or different management regimes can be tested need large-scale facilities like mesocosms, both to ensure realistic scaling and to enable sufficient replicates. Although a co-operation among several existing facilities has been established within the EU, we urge for a further organisational effort. This should lead to a better co-ordination of experiments, prevent duplication and provide additional funding for these expensive facilities.

Marine reintroduction sites

Habitat loss, fisheries and pollution have led to local extinction's of many species. The urge for protection of biodiversity and sustainability has started the discussion on reintroduction of several species like sturgeons, flat oysters, sabellaria reefs, seagrasses and rays.

We recommend to establish marine reintroduction sites for species that were common in the past but have now disappeared from large parts of the coastal zones. Apart from sites in the field, which will need special protection such that the original causes of disappearance have been taken away, facilities were viable specimen for reintroduction can be bred are urgently needed.

Need for stronger framework to help integrate knowledge

There is a need for a stronger framework to bridge the gap between science and management. The dynamic tension between real user needs and the perception of scientists often leads to miscommunication and frustration on both sides.

To overcome this problem we should identify user needs and balance fundamental and applied science taking into account integration and application. Scientist more often have to ask the question WHY?, and will have to indicate and propagate the relevance of science to management.

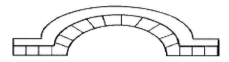

Fig. 4. a) Coastel Land and Ocean systems are Interdependent like the stones in a bridge: Bridge is strong because elements are linked and mutually supporting.

Fig. 4. b) If you damage 1 or more ecosystems you weaken the carrying capacity of the coast and coastal seas.

Fig. 4. c) Extreme damage can severely reduce capacity of coastal systems to sustain human needs.

On the other hand, managers have to set priorities of what is really important and what options are realistic, also in a social and economical context. Furthermore, there will always be a mismatch between the managers needs of to-day, and the results of freshly started research.

Partly this may be overcome by a better integration of the natural and social sciences, and also elaborate GIS and MSS systems may help to bridge the gap. But to do so, there is a need to back up science with investment in infrastructure. Europe has a task to influence broader international science and management agendas.

A major problem in coastal and shelf seas management is the invisibility of the underwater world. Out of sight, out of mind. For protection of habitats and species, public and political awareness and support is indispensable. In order to attain this, we will have to develop ways and tools so that the public can enjoy and appreciate the underwater world.

Co-operation with developing countries

The cooperation with developing countries in the field of marine sciences should take place in the form of long-term co-operative exercises with a strong capacity building component. These co-operative efforts will also entail the transfer of European expertise and tools in the form of environmental packages for application in the coastal zones of developing countries. Suggested are additional initiatives to establish regional centres of excellence in developing countries, which will further contribute to ensuring the international role of European science. From the scientific point of view Co-operation with developing countries in several regions such as S and SE Asia will help to obtain valuable data sets on marine and atmospheric processes that are interlinked and of a global nature (global teleconnections).

References

Lindeboom HJ (2002a) The coastal Zone: An ecosystem under pressure. In: Field JG, Hempel G, Summerhayes CP (eds) Oceans 2020; Science, Trends and the Challenge of Sustainability. Island Press Washington, London pp 49-84

Lindeboom HJ (2002b) Changes in coastal zone ecosystems. In: Wefer G, Berger WH, Behre K-E, Jansen E (eds) Climate Development and History of the North Atlantic Realm. Springer Berlin pp 447-455

Oost AP, de Haas H, Ijnsen F, van den Boogert JM and de Boer PL (1993) The 18.6 year nodal cycle and its impact on tidal sedimentation. Sediment Geol 87: 1-11

Turner RK, Adger NW and Lorenzoni I (1998) Towards integrated modelling and analysis in coastal zones: Principles and practices. LOICZ Reports and Studies No.11. LOICZ, Texel, The Netherlands 122 p

Ecosystem Functioning and Biodiversity: Bioengineering

G. Graf

Universität Rostock, Institut für Aquatische Ökologie, Freiligrathstr. 7/8, D-18055 Rostock
corresponding author (e-mail): Gerd.graf@biologie.uni-rostock.de

Abstract: In addition to the characterisation of ecosystems by the use of their biomass and fluxes, biodiversity is introduced as an important state variable describing the development of marine ecosystems. It is discussed in relation to the Redundant Species Hypothesis, the Rivet Hypothesis and the Diversity-Stability Hypothesis. The relation between diversity and fluxes in marine systems is discussed using processes such as sedimentation and bioturbation as examples. It is concluded that pelagic and benthic systems may respond differently to losses in biodiversity. In pelagic systems, modifications of the food web structure represent the major problem for the future ecosystem development, while it is the loss of ecosystem engineers in benthic systems.

Introduction

An alarming decline in species richness has been reported for terrestrial ecosystems such as tropical forests. Based on species-area relationships the loss of habitat causes a world wide extinction rate of up to several thousand species per year, whereas the evolution of new species is a very slow process which may result in only a few newcomers per year. Due to human impacts the natural process of the extinction of species has been accelerated by a factor of 100 to 1000. It is a topic of highest priority in modern ecosystem research to investigate whether or not this dramatic change to our ecosystems will also affect major functions, e.g. primary and secondary production, fluxes between different compartments such as pelagial and benthal, but also aspects such as ecosystem stability. Understanding these effects is of the same fundamental relevance for the future development of the biosphere as understanding present day anthropogenically induced climate changes.

In marine ecosystems to our knowledge, comparable losses of species are so far restricted to coastal systems. The most significant changes over the last decades are found especially in areas with coral reefs, mangroves, and wetland, where the loss of habitat is the main reason for species extinction, as it is in terrestrial systems (see GESAMP 1997; Carlton et al. 1999). In addition marine species diversity is affected by overexploitation, pollution and climate changes, which leads to increased UV-radiation or may cause pH-decreases in the future. The latter two effects would cause a global change and thus also threaten the open ocean (Fasham et al. 2001).

Species diversity is determined by species richness and the proportional distribution of individuals among the species. This so-called heterogeneity diversity occurs at different spatial scales, i.e. a single sample, habitats, large ecosystems and biogeographical provinces (Gray 2000). In addition genetic diversity will be relevant for certain functions in an ecosystem. The genetic diversity within a population is important as it provides the basic mechanisms for adaptation to environmental changes and will be relevant for the diversity-stability debate, e.g. more diverse ecosystems may decrease the frequency of successful invaders (McCann 2000). Finally, the functional diversity of an ecosystem describes the existence of different functions such as feeding guilds (predators, grazers, etc.), or guilds of primary producers (algae, bacteria).

This paper will exclude the question why species diversity is high in certain systems and low in others. One of the few attempts to explain this phenomenon is the intermediate disturbance hypothesis (IDH) (Connell 1978) which predicts the highest species diversity at medium size and medium frequency of disturbance. For marine systems this hypothesis has only been tested for some shallow water systems and it is rather unclear what an intermediate disturbance would be, e.g. in the deepsea. The aim of this paper is instead to discuss the effect of a differing species composition on ecosystem functions, especially on the fluxes of energy and matter. Whereas in the pelagic realm a more direct interaction of species within food chains and the microbial loop will be relevant, weaker links between species are proposed for benthic systems and the greater importance of ecosystem engineers will be stressed.

Is biodiversity a state variable of marine ecosystems?

Energy flow is the basic prerequisite for ecosystems to stay at a certain distance away from thermodynamic equilibrium. Any fluctuation in the supply of light energy, or in chemically bound energy for aphotic systems, will modify the state of the system, i.e. the distance from the thermodynamic equilibrium. This process was described during many field experiments e.g. by measuring the primary production or oxygen consumption, biomass development or other state variables of the system. An example for benthic systems was given in Graf et al. (1982), showing the tight link between fluxes from the pelagic to the benthic zone and the elucidated benthic responses. Figure 1 gives a schematic view of such fluctuations in energy flow.

For a given system at state A, a change in energy flow (e.g. a settling bloom) as indicated by the black arrow, may either cause short fluctuations of the system which may then return to state A (resilience), or as indicated by the open arrow may be shifted to a different state B_1 or B_2. The next event may then shift the system to state C_1 or C_2 or D_1 and D_2 respectively, indicating that the history of the system is relevant for the state which is reached after certain fluctuations in energy flow. This example is defined on a time scale of days to weeks, but for major events such as an oil spill or climate changes e.g. the North Atlantic Oscillation, the

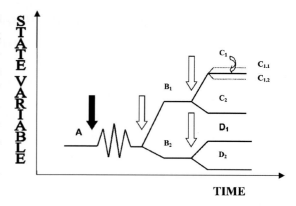

Fig. 1. Hypothetical development of ecosystem state variables such as fluxes of energy and matter, primary and secondary production, and possibly biodiversity. The black arrow indicates a disturbance, which was not sufficient for pushing the system into a different stable state.

scheme may be relevant on longer time scales as well.

This purely energetical view of ecosystem functioning would generate the provocative Null Hypothesis of diversity having no effect on the state of an ecosystem and on energy flow. On the other hand it is well known that certain species may be more important than others, e.g. a top predator acting as a keystone species may significantly change the food web and thus energy flow (Paine 1994). The hypotheses discussed presently are excellently summarised by Lawton (1994) as illustrated in Figure 2.

The redundant species hypothesis states that only very few species are needed to maintain the major ecosystem functions and that most species are redundant and lost of species will be replaced without significant modification of ecosystem function. An increase in species number would have no effect. The existence of rare species might be explained by the benefit of "back-up species" that can replace the role of species lost after a certain disturbance in the system. The rivet hypothesis predicts that every species has a certain effect on ecosystem functions, mainly because of the interspecies connections. However, it admits that some species are more significant than others. Finally the idiosyncratic response hypothesis states that loss or addition of a species will result in un-

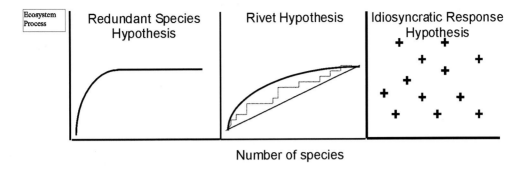

Fig. 2. Hypothesis on the relation of diversity to ecosystem functions (modified after Lawton 1994)

predictable changes of ecosystem functions. For the two latter hypotheses the sequence in which species are added or lost is also of great importance.

Referring to the energy flow scheme above (Fig. 1) the question is, whether the change in heterogeneity diversity is a state variable that affects the absolute value of states A, B and C? Will the addition or loss of a keystone species be able to cause a major change as indicated by the arrows, or would other species such as invaders just slightly modify the levels A,B, and C into sublevels e.g. $C_{1.1}$, $C_{1.2}$ etc. which, however, would be still significantly different from C_2. It is obvious that the absolute values of these levels will strongly depend on the ecosystem process that was measured. More generally speaking it is the question of physical forcing on ecosystems versus endogenic factors represented by species and their genetic adaptation, that determine the development of ecosystems

Diversity-stability hypothesis

An ecosystem which is not disturbed may stay at a certain attractor state where the energy flow is relatively stable and a constant distance from the thermodynamic equilibrium can be maintained (see A in Fig. 1). One criterion for stability is the ability to return to this attactor state after perturbation, a phenomenon called resilience. To date, few experiments in the marine sciences have examined whether or not diversity - and first of all species richness - positively affects stability. McCann (2000) reviews the present state of the art and concludes that there is a positive correlation between species richness and stability. This is explained by the probability that more species produce or result in greater redundancy in the system, i.e. similar species with only slight differences can response differently to the perturbation. This argument can also be used on the level of genetical diversity, which may stabilise a population, but also on the level of functional diversity, which may stabilise the ecosystem.

To date, the most convincing experiments have been presented by terrestrial ecologists. Tilman and Dowing (1994) tested the drought resistance of experimentally constructed grassland plots and demonstrated that a higher species number increased the resistance of the system against this important abiotic factor. It was, however, a criticism of this approach that the composition of functional groups was important in this experiment.

In recent literature on food web theory the importance of weak interactions between consumer and resources is stressed. According to McCann (2000) weak interactions guarantee that a resource is not overexploited, which would destabilise the system and create huge fluctuations in both, consumer and resource. The scheme in figure 3 illustrates this phenomenon. If a consumer **C** has a strong link to resource **R1** representing its major food, but changes to resource **R2**, an inferior competitor to **R1**, a process called negative covariance occurs. Whenever **R1** is strongly consumed **R2** benefits, because it is released from competitive limitations. High densities of **R2** will reduce the consumption of **R1**, whereas for low densities of **R1** a higher consumption of **R2** is expected.

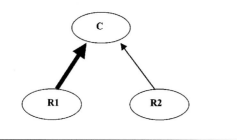

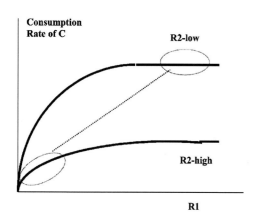

Fig. 3. The coupling of a strong consumer-resource interaction (**C - R1**) with a weak consumer-resource interaction (**C - R2**) (from McCann, reprinted with permission from Nature, Copyright (2000), Macmillan Magazines Limited).

For marine systems the diversity stability discussion is relevant for younger systems such as the Baltic Sea which depicts a significantly reduced species richness or such as those coastal systems that have already lost certain habitats and thus species. These marine ecosystems are more sensitive to disturbances, especially for species invasion. The lesser species and functional groups are available the stronger the links created in the food webs will be and the stronger fluctuations in species composition have to be expected.

Diversity and fluxes

The most convincing experiments on the relation between species richness and fluxes were once again carried out by terrestrial ecologists. Tilman et al. (1996) demonstrated in 147 mesocosm plots with 7 species-richness treatments from 1 to 24 different species, that the uptake of nitrate within the rooting zone and below it was significantly higher with increasing species richness. In more complex systems with several trophic levels Naeem at al. (1994) showed that the remineralisation of organic matter in soil was significantly increased with a higher complexity of the food chain.

Recently Naeem et al. (2000) ran an experiment to test the co-dependency of producers and decomposers. The effect of 12 different bacteria cultures, that were added in different combinations, on the productivity of 8 different alga species was investigated in 118 freshwater microcosms. The results depict that the production rose with increasing numbers of alga species, but it also is dependent on the decomposer diversity. A manipulation of decomposer species richness could almost double the primary production. The highest algae biomasses were reached in combination with a medium number of additorial decomposer species. The bacteria managed to build up a significant proportion of biomass only in the microcosms with highest decomposer species richness. Such a systematic investigation is not yet available for marine ecosystems. However, there are some considerations on the effect of different functional groups on fluxes.

A nice example for the marine pelagic system was presented by Wassmann (1998) as depicted in figure 4. He describes the export from the euphotic zone following a phytoplankton bloom, in terms of cell, aggregate and faecal pellet sedimentation, to be significantly dependent on the amount of meso-zooplankton and in addition, dependent on the feeding guild, i.e. herbivory, omnivory, and carnivory. These different feeding guilds will change the partitioning of fluxes and the amount of matter which settles to deeper waters and to the seafloor varies by more than a factor of two.

For benthic systems Elmgren (1984) and Elmgren and Hill (1997) describe the effect of a loss of a functional group, when they compared the energy flow in the North Sea with 3 subsystems of the Baltic Sea, namely the Baltic Proper, the Bothnian Sea and the Bothnian Bay, and came to the conclusion that in spite of a decreasing primary production along the salinity gradient in the Baltic, the energy flow remained unchanged with the exception of the Bothnian Bay. Due to the reduced salinity in this subsystem the loss of benthic suspension feeders, mainly bivalves, causes a signifi-

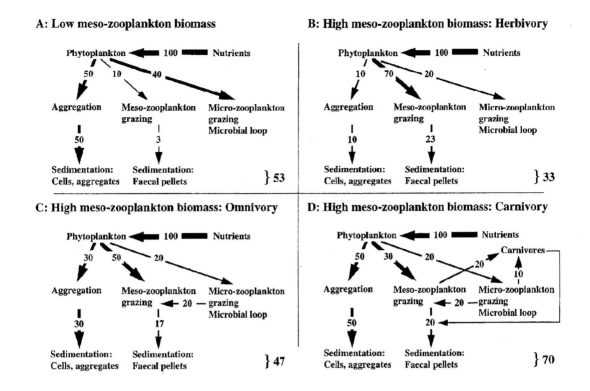

Fig. 4. Partitioning of the sedimentation of a phytoplankton bloom in relation to the amount of mesozooplankton and its feeding guild. (from Wassmann 1998, with kind permission from Kluwer Academic Publishers)

cant change with significantly reduced benthic fluxes and the unusual situation that meiofauna-biomass dominates macrofauna-biomass.

Additional benthic examples refer to the effect of different species on bioturbation. This process leads to a multiplication of the diffusive transport mechanisms by a factor of 2 to 10 and to a distribution of particulate organic matter down to the bioturbation depth and is thus a major factor that determines the thickness of the sediment layer where significant microbial decomposition takes place. Ritzrau et al. (2001) described that the bioturbation depths in the Norwegian-Greenland Sea varied between 3 to 10 cm, although the fluxes to the sea floor are very similar, i.e. the regional different species composition determines this process and not fluxes. A similar result was given for the Indian Ocean by Turnewitsch et al. (2000), who also demonstrated the independence of bioturbation rate and depth on fluxes of organic matter to the sea floor. In addition they depicted that surface liv-

ing deep-sea fauna in a eutrophic situation can show the same bioturbation intensity as deep-burrowing animals in an oligitrophic system. The deepest bioturbation depth, however was achieved when both functional groups overlapped.

Food web versus ecosystem engineering

The investigations on species interactions and food web structures mainly focus on predator -prey interactions and on competition and, as discussed above, it is difficult to explain the existence of so many species and why better adapted species do not extirpate the inferior ones. In benthic ecology it is long known and very obvious that additional effects such as animal - sediment interactions determine the habitats and thus the existence of additional niches and species. In the classical work by Reise and Ax (1979) they demonstrated how the burrow structure of *Arenicola marina* creates habitats for several groups of meiofauna, i.e. large benthic animals determine the extent and existence of the habitats for smaller organisms.

A general approach to this sort of species interaction was defined by Jones et al. (1994). They termed these organisms as ecosystem engineers and their first example was the beaver, who constructed a dam, that cause the existence of a completely new ecosystem, furthermore it is plain that there is no food web relationship between the beaver and most of the organisms living in the newly created pond.

Jones et al. (1994) defined an ecosystem engineer as an organism that directly or indirectly modulate the availability of resources (other than themselves) to other species by causing physical state changes in biotic and abiotic materials. They distinguished between autigenic engineers, who change the environment by their own physical structure, e.g. corals, and allogenic engineers, who transform living or non-living materials from one physical state to another, via mechanical means or other means (e.g. the above mentioned burrow construction by *Arenicola marina*).

Using these broad definitions it is not surprising that ecosystem engineers can be identified in all marine habitats. In pelagic systems the nitrogen fixation of blue-green algae and the production of pellets modify important fluxes and may be regarded as engineering. Nevertheless, ecosystem engineering in the benthic systems is much more significant. All bioturbation processes change the sediment habitats significantly, determine the extension of this habitat (bioturbation depth) but also the fluxes in and out of the sediment. Graf (1999) summarised the effects of animal constructions on deposition and resuspension (Fig. 5). He concluded that the passive effect of these features alone, e.g. pits, mounds tubes and tube lawns, increased the particle fluxes by a factor of 2 to 10. Soft sediment communities create and live in a 3-dimensionally engineered world, with many microniches, which may contribute to explain the higher diversity of these systems as compared to pelagic systems.

Of course there are also many food web relationships in benthic communities and some of the best existing examples derive from benthic ecologists, e.g. the sea otter, sea urchin and kelp relationship by Paine (1986). Most examples, however, derive from sediment surface and hard substratum systems, i.e. the animals can interact directly in a similar way as in the pelagic systems. In sediments it is much more difficult to find prey and it is extremely energy consuming. Many infauna species and surface feeders directly feed on detritus or even sediment itself and are thus part of the decomposer compartment in the system. In addition many benthic species can vary their feeding mode, e.g. the polychaete *Nereis diversicolor*, and therefore

indirect deposition and resuspension

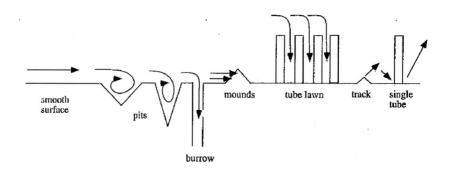

Fig. 5. Schematic effect of biogenic structures on particle deposition and resuspension (after Graf 1999)

create only a very weak link in the food web. Especially in the deep-sea most benthic species are omnivorous.

Table 1 summarises the major differences between pelagic and benthic soft bottom communities, as far as they are relevant for diversity considerations. Besides the difference in the degree of food web structuring versus ecosystem engineering, also the microbial loop may be different in both systems. While in pelagic systems exudation by pimary producers is a major dissolved organic carbon (DOC) source, in the sediments the high DOC pool is mainly created by exoenzymes. The resultant bacteria are particle associated and less accessible for grazers as compared to those in pelagic systems. A further major difference is the fact that relative abundance of particulate organic matter in the sediments is dramatically higher as in pelagic systems, i.e. only less than 5 % mostly less than 1% of the organic matter in sediments is living biomass, whereas in pelagic systems it is up to 90 %.

Concluding it can be stated that these basic differences between pelagic and benthic systems have consequences for the sensitivity of these two marine habitats to changes in biodiversity. In pelagic systems extinction or invasion of a key predator may completely change the system, in benthic sys-

tems a loss of a major ecosystem engineer is the greatest danger for decreasing biodiversity.

References

Carlton JT, Geller JB, Reaka-Kudla ML, Norse EA (1999) Historical extinctions in the sea. Annu Rev Ecol Syst 30:515-438

Connell JH (1978) Diversity in tropical rainforests and coral reefs. Science 199:1302-1310

Elmgren R (1984) Trophic dynamics in the enclosed, brackish Baltic Sea. Rapports et Procés-verbaux des Réunions. Conseil International pour l'Exploration de la mer 183:152-169

Elmgren R, Hill C (1998) Ecosystem function at low biodiversity - the baltic example. In: Ormond RFG, Gage JD, Angel MV (eds) Marine Biodiversity. Press Syndicate of the University of Cambridge, Cambridge pp 319-336

Fasham MJ, Balino BM, Bowles CM (2001) A new vision of ocean biogeochemistry after a decade of the Joint Global Ocean Flux Study (JGOFS). Ambio special report, No 10 30p

GESAMP (1997) Marine biodiversity: Patterns, threads and conservation needs. Rep Stud GESAMP 62:1-27

Graf G (1999) Do benthic animals control the exchange between bioturbated sediments and benthic turbidity zones? In: Gray et al. (eds) Biogeochemical Cycling and Sediment Ecology. Kluwer Acad Press pp 153-159

Pelagic Systems	Benthic Systems
Mainly food web structered	Mainly ecosystem engineered
Microbial loop based on exudated DOC	Microbial loop based on DOC provided by exoenzymes
Less bacteria particle associated	Almost all bacteria particle associated
Ratio of living OM/dead OM < 1:100	Ratio of living OM/dead OM >50:100

Table 1. Summary of the major differences between pelagic and benthic soft bottom communities

Graf G, Bengtsson W, Faubel A, Meyer-Reil L-A, Schulz R, Theede H, Thiel H (1982) The importance of the spring phytoplankton bloom for the benthic system of the Kiel Bight. Rapp P-v Reun Cons int Explor. Mer 183:136-143

Gray JS (2000) The measurement of marine species diversity, with an application to the benthic fauna of the norwegian continental shelf. J Experiment Mar Biol Ecol 250:23-49

Jones CC, Lawton JH, Shachak M (1994) Organisms as ecosystem engineers. OIKOS 69:373-386

Lawton JH (1994) What do species do in an ecosystem? OIKOS 71:367-374

McCann KS (2000) The diversity-stability debate. Nature 405:228-233

Naeem S, Thompson LJ, Lawier SP, Lawton JH, Woodfin M (1994) Declining biodiversity can alter the performance of ecosystems. Nature 368:734-737

Naeem S, Hahn DR, Schuurman G (2000) Producer-decomposer co-dependency influences biodiversity effects. Nature 403:762-764

Paine RT (1986) Benthic community-water column coupling during the 1982-83 el nino. Are community changes at high latitudes attributable to cause or co-incidence? Limnol Oceanogr 31:351-360

Paine RT (1994) Marine rocky shores and community ecology: An experimentalist's perspective. Ecology Institute Nordbünte Oldendorf/Luhe Germany pp 94

Reise K, Ax P (1979) A meiofaunal "Thiobios" limited to the anaerobic sulfide system of marine sand does not exist. Mar Biol 54:225-237

Ritzrau W, Graf G, Scheltz A, Queisser W (2001) Bentho-pelagic coupling and carbon dynamics in the northern North Atlantic. In: Schaefer P, Ritzrau W, Schlueter M, Thiede J (eds) The Northern North Atlantic: A chamging environment. Springer, Heidelberg pp 207-224

Tilman D, Dowing JA (1994) Biodiversity and stability in grasslands. Nature 367:363 - 365

Tilman D, Wedin D, Knops J (1996) Productivity and sustainability influenced by biodiversity in grassland ecosystems. Nature 379:718–720

Turnewitsch R, Witte U, Graf G (2000) Bioturbation in the abyssal Arabian sea: Influence of fauna and food supply. Deep-Sea Res II 4:2877-2911

Wassmann P (1998) Retention versus export food chains: processes controlling sinking loss from marine pelagic systems. Hydrobiologia 363:29 - 57

What Structures Marine Biodiversity and why does it vary?

C. Heip

Centre for Estuarine and Coastal Ecology, Netherlands Institute of Ecology,
NL-4400 AC Yerseke, The Netherlands
corresponding author (e-mail): heip@cemo.nioo.knaw.nl

Abstract: Marine ecological biodiversity research is a scientific field with few observational data to support a weak theory largely borrowed from terrestrial ecology and lacking in experimental verification. The relative lack of scientific interest and effort until recently was a consequence of the general feeling that marine biodiversity is far less threatened than terrestrial biodiversity. This view is not sustainable. There is now ample evidence of widespread changes in most coastal habitats in populated areas around the world (coral reefs, mangroves, seagrass fields, intertidal rocky shores and subtidal sediments on the continental shelf and margin) due to exploitation of marine resources, introduction of exotic species and the increased pressure from mariculture and fisheries. The sustainable exploitation of the seas requires development of a sound theoretical framework for marine biodiversity, including genetic, species and habitat diversity and especially the relationships between them. At the present state of knowledge such a general theory is still far from being reality. In this paper an overview is given of the main elements that an ecological theory of species diversity should include and what aspects of human pressure on the biodiversity of marine ecosystems should be given priority attention.

Introduction

The very rapid and widespread changes due to human activities in the abundance and distribution of biological species that have been documented over the last decade have raised the question whether these changes in biodiversity have important consequences for ecosystem functioning. Ecosystems perform a series of vital functions for human society, many of which depend on a variety of organisms present in a variable environment. The relationships between biodiversity and ecosystem functioning are far from clear and before this issue can be properly researched a number of more basic questions have still to be answered.

In current thinking biodiversity equals biological variability, from genes to species and habitats within ecosystems. The mechanisms that generate and maintain variation at these different levels in the organisation of the biosphere are very different. A single coherent theory of biodiversity change and maintenance will have as a major challenge the linking of these mechanisms over the spatial and temporal scales at which they operate. This is still

a far and perhaps an elusive goal. Even the basic job of providing an adequate description of biodiversity is far from done. On the one hand there is still a major effort required in scientifically describing the species that inhabit the seas and much concern has been expressed about the shear number of undescribed species and the decline of taxonomic expertise. On the other hand we are at the dawn of unprecedented progress in charting and explaining genetic diversity in the oceans. For the first time the prospect of a complete inventory of marine microbial populations is no longer utopic. The research that will be required to couple knowledge on genetic structure to the functioning of marine systems will take many years and even decades to be accomplished. The genomes of about 300 species will be characterized in the near future, but only very few of them are relevant to marine ecology; so, although the start has been taken, the work is only beginning.

It is clear that many basic facts are missing and that it will require many years of scientific effort to obtain them. This task can only be accomplished in

From WEFER G, LAMY F, MANTOURA F (eds), 2003, *Marine Science Frontiers for Europe.* Springer-Verlag Berlin Heidelberg New York Tokyo, pp 251-264

a fruitful way if a coherent and testable theory on marine biodiversity will be constructed over the next few years. Whether a unified theory on biodiversity should be based on characteristics of individual organisms, including their genetic composition, or on species or communities in landscapes remains an open question. The most easily recognized operational building block in the biodiversity hierarchy is the individual organism, which is often ignored in marine ecology. All individual organisms are different to some extent, even in clonal species. The potential genetic diversity present in populations is much larger than the number of individuals in which it is expressed and each generation in a sexually reproducing population is a new and unpredictable sample from a very large gene pool. In a recent paper, Pachepsky et al. (2001) present a framework for studying the dynamics of communities which generalises the prevailing species-based approach to one based on individuals that are characterised by their physiological traits.

At the habitat or ecosystem level as well, there is an important random element due to the unique and unpredictable geological setting in which ecological processes take place. When one discusses characteristics and changes in habitats, a review would have to address the complex physical, biogeo-chemical and geological processes and especially their interactions that structure landscapes in the sea, including the pelagic environment, and that can operate at very different time and space scales. The newly coined term 'biogeomorphology' probably comes closest to what kind of questions we have to formulate, at least for the sea floor. The interactions within biofilms and microbial mats, the biological impact on sediment characteristics such as bottom roughness, the structure of the benthic boundary layer and benthic-pelagic coupling are examples of research areas where a close link between biology and geology exists. Coral reefs in the tropics are the most prominent case of an environment nearly completely created by biotic processes, and several less spectacular reef forming organisms are active in temperate waters, including maerl forming red algae, molluscs and polychaetes, deep water corals and others. At the other extreme we have environments such as sandy beaches which are nearly completely built by physical forces and to

which organisms have to adapt without being able to modify them. To include habitats in the biodiversity concept is therefore a far from simple task, although the debate is part of the marine ecological literature at least since Sanders (1968) in the frame of the stability-time hypothesis discussed the difference between biologically and physically dominated communities.

Although it would be fascinating to summarize the processes that structure and change genetic, species and habitat variability, and to try and link them, the task would be far outside the scope of this paper. Even restriction to the classical use of the term biodiversity - species diversity - requires addressing nearly the whole field of ecology. This is clearly impossible and I will restrict myself to a series of headings indicating what we should consider to know in order to describe and explain marine biodiversity and a short summary of what is known under that heading. I have heavily borrowed from a number of review papers that appeared during recent years and that show the explosive growth of interest in the subject. I also refer to the book of Ormond et al. (1997) for an important collection of papers that cover many aspects of marine biodiversity that are ignored here, e.g. phylogenetics and evolution.

Is marine biodiversity special?

The theoretical foundations as well as the experimental approach required to understand marine biodiversity are very poorly developed, in general and also when compared to terrestrial ecology. In fact, the whole literature is so much dominated by theory developed for terrestrial ecosystems that one can scarcely speak of a field of marine biodiversity. An important part of this theory has been developed by plant ecologists (e.g. Loreau 2000 for an example) and its application to the oceanic environments is of course problematic. One basic question is whether terrestrial and marine systems are similar enough to allow theory from one domain to be used for the other. Most probably this is not the case. Marine systems have a series of characteristics which distinguish them from terrestrial systems, as explained in the following box (from Heip et al. 1999).

Box 1. The distinctive features of marine biodiversity

1. Life has originated in the sea and is much older in the sea than on land. As a consequence, the diversity at higher taxonomic levels is much greater in the sea where there are 14 endemic (unique) animal phyla whereas only 1 phylum is endemic to land. There is also a remarkable diversity of life-history strategies in marine organisms. The sum total of genetic resources in the sea is therefore expected to be much more diverse than on land.

2. The physical environment in the oceans and on land is totally different. Marine organisms live in water, terrestrial organisms live in air. Environmental change in the sea has a much lower frequency than on land, both in time and in space.

3. Marine systems are more open than terrestrial and dispersal of species may occur over much broader ranges than on land. Although most species in the ocean are benthic and live attached to or buried in a substratum, in coastal seas a very large proportion of them has larvae that remain floating in the water for days to months. These high dispersal capacities are often associated with very high fecundities and this has important consequences for their genetic structure and their evolution.

4. The main marine primary producers are very small and often mobile, whereas on land primary producers are large and static. The standing stock of grazers is higher than that of primary producers in the sea, the opposite to the situation on land. Ocean productivity is on average far lower than land productivity. In the largest part of the ocean, beneath the shallow surface layers, no photosynthesis occurs at all.

5. High level carnivores often play key roles in structuring marine biodiversity and yet are exploited heavily with unquantified but cascading effects on biodiversity and on ecosystem functions. This does not occur on land, where the ecosystems are dominated by large herbivores and, of course, increasingly by humans which monopolise about 40 % of the total world primary production.

6. A greater variety of species at a higher trophic level is exploited in the sea than on land: man exploits over 400 species as food resources from the marine environment; whereas on land only tens of species are harvested for commercial use. Exploitation of marine biodiversity is also far less managed than on land and amounts to the hunter-gatherers stage that humans abandoned on land over 10,000 years ago, yet exploitation technology is becoming so advanced that many marine species are threatened to extinction. Insufficient consideration has been given to the unexpected and unpredictable long-term effects that such primitive foodgathering practices engender.

7. All pollution (air, land and freshwater) ultimately enters the sea. Marine biodiversity is thus most exposed to and critically influences the fate of pollutants in the world. Yet marine species are probably least resistant to toxicants. The spread of pollutants in marine food chains and therefore the quality of marine food is uncontrollable by man.

There is (probably) less species diversity and more genetic diversity in the marine environment than on land. If one looks at the arthropods, the insects and chelicerates on land and the crustaceans in the oceans, the difference is striking. A single tree in the tropical forest may harbour over a thousand species of insects. The entire planet harbours eighty species of euphausiids. This indicates that the mechanisms of speciation are very different in the sea and that resource competition does not constitute a dominant selective pressure, although in fine-grained environments such as marine sediments there are - as expected - more species than in the water column. The upper water column has a very dominant vertical gradient in light availability and nutrient concentration and, at least in some groups mostly from the micro- and picoplankton, more species may exist in the plankton than one may expect. This was called the paradox of the plankton by limnologist G.E. Hutchinson (1959) and was later applied to the marine environment by Margalef (1968). However, no studies have attempted to define resources in the sea at the same level of detail as is customary in the terrestrial environment. The distinction between different food species during different phases of the life cycle, or the view that different parts of food species may offer different resources allowing for specialization and speciation has not been tested. Overall, the smaller number of marine species, when confirmed, make it reasonable to assume that the mechanisms of diversity generation and maintenance are (very?) different.

One area where this generalization was not confirmed by reality was the deep-sea, where until 1968 the general consensus was that species numbers were low. In that year Sanders (1968) showed that, on the contrary, the deep-sea is very species rich when compared to shallower waters and estuaries. This he attributed to the constant and very old deep-sea environment which would allow for large diversification and specialization and consequently narrow niches, a concept he called the stability-time hypothesis.

Another generalisation that was derived from terrestrial studies is the lower diversity among the smallest animals described by May (1975). Again, it is unclear whether this pattern holds for the marine environment where two groups of small to very small marine animals exist that are claimed to be the most numerous animals on earth, the copepods in the plankton and the nematodes in the benthos. Especially in the nematodes and other groups of the meiofauna, the tendency to mineraturisation allows the exploitation of niches within finely grained environments and consequently an increase in local biodiversity.

A striking characteristic of marine species diversity is that, whatever the number of species, at the higher taxon level the marine environment is certainly much richer than the land. Of the 33 animal phyla, only five do not occur in the seas and 13 are endemic to it (Grassle et al. 1991). This implies that genetic, biochemical and physiological diversity is also much higher in the oceans than on land (Lasserre 1992). The recent discoveries by using molecular markers of many new archaea and bacteria, and three new entire classes of marine algae prove that the same probably holds for microbiota and plants. What the consequences of this high genetic diversity are for ecosystem functioning and whether this larger genetic pool provides a better buffering capacity for change is not known.

Coastal environments at the largest spatial scales are discrete units separated from similar units by other types of environment. Even the pelagic environment can be termed discrete on the global scale. At the other side of the spectrum are coastal lagoons and salt marshes that are very small and sometimes very far away from each other but often contain very similar faunas and floras. Small islands and estuaries fall somewhere in between. It is clear that fragmentation has to be considered as a function of the genetic and ecological dispersal capacity of species. Marine plant and animal species have developed efficient strategies to be able to colonize such habitats. Knowledge on dispersal and life-history in general of many marine plants and animals remains vert scarce. Even knowledge on distribution of almost all marine species is fragmented and has not received much attention in the last decades.

What is marine biodiversity?

As the study of genetic and habitat diversity in the sea is only just starting, I will restrict this discussion to species diversity. Species diversity can be studied globally (i.e. the total number of existing species), regionally or locally (the number of species occurring in a region, habitat, patch etc.). Globally, at our present state of knowledge, marine species diversity appears to be low. There are probably about 7-8 million eukaryote species on Earth, although some authors claim higher numbers of 30 and even 100 million species. Most of those species (85 %) are terrestrial and most of them are animals. About two thirds of all species occur in the tropics, largely in humid forests. Out of these 7-8 million species, about 2 million have been scientifically described.

It is not well known how many marine species have been described. In the literature figures vary between 200,000 and 500,000 marine animal species. Described species of marine viruses and micro-organisms (bacteria, fungi, protozoans, micro-algae) number in the thousands and perhaps 20,000 marine plant species are described; the marine plant inventory must be more complete than that of the animals, since there are no deep-sea plants.

One of the most striking characteristics of marine biodiversity is the difference between the benthic and the pelagic environment. The pelagic oceanic environment is the largest habitat on earth; it is immense. It has very strong vertical gradients, in light and temperature, in nutrients and in food, typically in scales of tens to hundreds of meters, but sometimes much less. It has also important horizontal gradients existing over very large scales, hundreds to thousands of kilometres. The pelagic environment has been called coarse grained and uniform, but this has been contested by phyto-plankton and microbial ecologists, who at the one hand have described the paradox of the (phyto) plankton already mentioned and on the other have recently discovered and discussed the very fine scaled gradients around bacterial and phytoplankton cells (Azam et al. 1994). Still, pelagic animal species are grouped in clearly evident assemblages and all show

vaste geographical distributions. There are distinct oceanic spatial patterns of species diversity and these are large and few in number. These patterns are correlated with the size and shape of the great oceanic gyres and the equatorial currents and counter currents. Smaller scale circulation patters, such as mesoscale eddies, rings and the circulation in small basins, do not seem to have provided the same degree of persistence of habitat features required for the coevolution of the diversity of species that is required for the establishment of functional ecosystems (McGowan and Walker 1993).

As a consequence, species diversity in the oceanic pelagic environment is extremely low. The number of species in the upper 200m of the pelagic oceanic environment is well known for four groups of animals, the Euphausicacea, Chaetognatha, Pteropoda and Copepoda, which dominate the biomass everywhere. There are only 80 species of euphausiids, 50 of chaetognaths, about 40 of pteropods and less than 2,000 for the most diverse group, the calanoid copepods. These data are based on more than 20,000 net tows and, although new species will certainly continue to be discovered, it is obvious that pelagic biodiversity is of another order than both terrestrial and marine benthic diversity (McGowan and Walker 1993).

This low number of (animal) species is in striking contrast with the diversity of animals in sediments. About 200,000 species are currently known from benthic environments, most of them have been described from coral reefs, and only about 60,000 are known from soft bottom habitats that cover most of the earth's surface. Benthic species from the temperate shallow waters of Europe are reasonable well known, especially in the larger macro- and megafauna. The smaller meiofauna (mm sized animals) is less well described and a survey of the benthos in the North Sea in 1986 yielded about 40 % of benthic copepod species new to science (Huys et al. 1992).

For the continental margin and the deep-sea, the state of knowledge is poor. As an example, in the very extensive EU-OMEX programme on the Goban Spur continental margin it was often not possible to name even dominant species of the macrofauna (Flach et al. in press). Consequently, there has been much debate recently on the number of marine benthic species that is to be expected. Numbers such as 10^8 nematodes species globally have been published in the literature, although without proper justification. The main argument for a high number of undescribed species in the macro-benthos comes from a survey of Grassle and Maciolek (1992). The survey covered ten stations sampled with a grab along 176 km of the 2100m isobath (depth contour) off the east coast of the US and four additional stations at depths of 1500m and 2500m. In the ten stations a total of 798 species representing 171 families and 14 phyla were identified on a total sampled surface of 21 m². Of these species, 460 (58 %) were new to science. About 20 % of the species were found at all ten 2100m stations and 34 % occurred at only one station. Of the total soft-sediment fauna 28 % of species occurred only once and 11 % only twice. The number of species found increased continuously as more samples and individuals were collected. At a single station species were added at a rate of about 25 per 0.5 m². When samples were added along the 176 km transect the rate of increase was about 100 species per 100 km. The rate of increase across depth contours is even greater. Since the deep-sea at depths greater than 1000m occupies about 3.10^8 km² of the earth's surface, if a linear addition of one species per km is extrapolated to 1 species per km², the global deep-sea macrofauna would contain on the order of 10^8 species. Taking into account the fact that the deepest, oligotrophic areas of the ocean floor have densities of macro-fauna of about one order of magnitude lower, this may be reduced to 10^7 species. This estimate depends on the hypothesis that rare species are different in different areas and May (1992) pointed out that, since half of the species found in the Grassle and Maciolek (1992) study were known to science, the implication that the total number of species is double the number of species described also provides some kind of (minimum) estimate, that is two orders of magnitude lower. Again its accuracy depends on whether the rare species are the same ones in different localities on earth.

In view of all these uncertainties, it is still premature to give reliable estimates of the species number in the marine benthos, but there may be several millions. Marine sediments are fine grained environments with gradients in the order of millimeters and centimeters in the vertical and centimetres to metres in the horizontal. It is to be expected that the mechanisms of speciation are very different in marine sediments than in the pelagic environment. This is well illustrated for instance by marine nematodes, where many congeneric species may be found together in a single sediment core, each living at a certain depth horizon only millimetres apart (Soetaert and Heip 1995; Heip 1996).

Patterns of species diversity

Before we can answer the question why biodiversity varies, we need to know the basic patterns of its distribution in space and time. The most fundamental data of diversity are the numbers of species in different places. This is a fundamental problem for marine biodiversity studies because this is largely unknown. There are some exceptions, such as some animal groups from the zooplankton, a number of plant and animal species from intertidal and shallow subtidal zones, and increasingly the microbiota and fauna from hydrothermal vents. But we know next to nothing about the distribution and the dynamics of the large majority of species living in the sediments covering millions of square kilometers of the deep-sea floor.

Terrestrial ecologists have used geographic distributions of species extensively and have discovered relationships between these data and latitude, climate, biological productivity, habitat heterogeneity, habitat complexity, disturbance, and the sizes and distances of islands. Several of these relationships have suggested mechanisms that might regulate diversity but in the terrestrial environment as well a general and comprehensive theory of diversity accounting for most or all of these relationships does not exist.

Spatial scale is the overriding variable that needs to be considered when discussing the changes in diversity and what are their causes.

Definition of scales is not straightforward, neither in terrestrial nor in aquatic environments. Scales are often defined from the perception of the human observer and less as a function of the species or communities considered. It is customary to distinguish between local, regional and global spatial scales. Locally, species diversity in any locality is seen as a balance between two opposing forces. On the one hand local abiotic processes, interactions between species and chance tend to reduce diversity; on the other hand immigration from outside the locality tends to increase diversity. Each local population is seen as a sample from a larger species pool. Theories on larger, mesoscale patterns take migration and dispersion explicitly into account. The metapopulation concept and connectivity of land (sea)scapes are central in this approach. Global patterns are for instance latitudinal gradients. Within most groups of terrestrial organisms, the number of species reaches its maximum in tropical latitudes and decreases both northward and southward toward the poles. In many cases, the latitudinal gradient in diversity is very steep. Tropical forests, for example, may support ten times as many species of trees as forests with similar biomass in temperate regions (Latham and Ricklefs 1993).

Since many factors vary in parallel with latitude the causal mechanisms that explain such patterns are difficult to distinguish and, moreover, nearly all studies are from terrestrial environments. In marine communities, the existence of such patterns over large geographical scales has only rarely been studied (Rex et al. 1993). Whether they are as widespread as in the terrestrial environment is questionable, but even in terrestrial environments the general trend in diversity is sometimes reversed, as it is for shorebirds, parasitoid wasps, and freshwater zooplankton, of which more species occur at high and moderate latitudes than in the tropics. These counterexamples may reflect the latitudinal distribution of particular habitat types, the history of the evolution of a taxon, or ecological circumstances peculiar to a particular group.

What structures species diversity?

The number of species co-occurring in a certain area depend on the characteristics of that area: its history,

its size, the physical, geological and chemical environment, other species occurring in the area, its distance to other areas etc. There is thus a large number of very different factors that determine whether a species can exist and maintain itself in a certain habitat. Because theory has often concentrated on only one or a few of those factors, it is very difficult to present a comprehensive picture of what generates and maintains marine biodiver-sity.

Most of the ideas on this topic again have their origin in empirically observed patterns. Species richness appears to be related generally to climate (Terborgh 1973). In particular, conditions that favor biological production - warm temperatures and abundant precipitation in terrestrial ecosystems - are often associated with high diversity. In aquatic systems this generalization probably does not hold. Highly productive ecosystems, such as occur in upwelling areas or as a consequence of eutrophication, are not linked to temperature and they often hold a limited number of species.

A striking pattern is the connection between diversity and habitat complexity. Salt marshes are extremely productive but harbor few species of plants and animals; deserts occupy the other end of the productivity gradient but may support a diverse flora and fauna. Regional or landscape complexity has also been implicated in patterns of diversity. Ecologists have long been aware of the greater diversity of mountainous regions compared with flatlands (Simpson 1964), or of rocky versus sandy shores. This pattern may arise because of the increased numbers of species distributed allopatrically on isolated mountains or in isolated valleys on the land, or in sandy pockets, tidal pools, crevices etc. of rocky shores.

The influence of geography on diversity has long been apparent in island settings, where the number of species tends to increase with island size and to decrease with distance from sources of colonists (Hamilton et al. 1963; Connor and McCor 1979; Abbott 1980). MacArthur and Wilson's (1963, 1967) theory portraying diversity on islands as a balance between colonization and extinction emphasized the influence of both local and regional processes on the diversity of local communities. The theory of island biogeography can readily be applied to coastal environments (estuaries, lagoons, beaches, rocky shores) but as well to isolated deep-sea environments such as deep-sea mountains and hydrothermal vents.

Other processes implicated in the regulation of diversity have occurred to ecologists as a result of more theoretical considerations. These include density-dependent predation (Paine 1966; Janzen 1970), and temperal and spatial stochasticity (Levins 1979; Chesson and Warner 1981; Thiery 1982; Clarke 1988). Theories based on these processes obtain little empirical support from geographical patterns of diversity because ecologists lack a general understanding of the geographical distribution of these factors. That ecologists generally have a higher regard for empirically based hypotheses, emphasizes our difficulty in distinguishing between phenomena as inspirations for hypotheses and as tests of the mechanisms proposed to explain them (Schluter and Ricklefs 1993)

Early in the previous century, species diversity was regarded primarily as reflecting the accumulation of species over time, and thus largely an evolutionary subject. By the early 1960s, diversity had come to be perceived as the outcome of ecological interactions, particularly competition, that were resolved within small areas and habitats and over short time scales (MacArthur 1972). This is reflected in the niche theory. This theory predicts that stable coexistence between species depends on each being a superior competitor in its own niche. The more niches there are in a certain habitat, the more species can co-exist. Complex habitats will provide more niches and therefore more room for more species. Niches may also change temporally, allowing for seasonal succession of species. The problem is that many species depend on the same resources and acquire them in a limited number of ways and niche overlap is consequently high so that exclusion of all species except the superior competitor should be the result (see however Huisman and Weising 1999). To overcome this difficulty a series of hypotheses have been advanced relying e.g. on niche separation in time and space with consequent smaller overlap of species on relevant niche axes.

Such separation may be due to disturbance and the so-called intermediate disturbance theory has made an important contribution to explain observed distribution patterns of species, one of the main contributions from marine sciences to general ecological theory. The greatest effect of disturbance is at intermediate levels because a high level of disturbance precludes many species and a low level cannot prevent competitive exclusion by the superior competitor. The theory is based on work in rocky intertidal communities by Connell (1975) and Paine (1966), who showed that when these communities are left undisturbed they become eventually dominated by a single dominant competitor. Disturbance, either by a physical event such as a storm, or by the appearance of a predator, allows more species to co-exist with the dominant competitor. The intermediate disturbance hypothesis is closely linked with the succession theory of Clements (1935). Landscapes typically include a mosaic of patches of disturbance of different extent and intensity such that each patch exists in some stage of succession, out of equilibrium with the prevailing climax in its species composition (Watt 1947). Thus, the entire mosaic, which is part of a larger regional equilibrium, contains more species than any individual patch.

Such species may form metapopulations, inhabiting suitable patches in fragmented landscapes and linked by migration. Most of metapopulation theory has been developed for the terrestrial environment. The most basic aspects have been analysed with simple spatially implicit models such as the Levins model in which a fraction h of the habitat patches is suitable for occupancy, the equilibrium fraction of patches occupied out of suitable patches is given by $p^* = 1 - \delta/h$, where δ is the ratio of the extinction to the colonization rate. A spatially realistic model for a finite number of habitat patches of known areas and spatial locations can be constructed by modelling the rate of change in the probability that a patch is colonized. Extinction rates are considered to be inversely proportional to the area of the patch and colonization rates depend on the distance between patches and the migration distance of a species (Hanski and Ovaskainen 2000).

If landscapes are mosaics and populations disperse over them in a random matter, habitat loss will inevitably lead to species loss. To convert habitat loss to species loss the principles of island biogeography are applied. The relationship between species and island area is nonlinear and from this one can predict how many species should become extinct as the size of the islands shrinks. These doomed species do not disappear immediately however. The term extinction debt has been coined to demonstrate the idea that many of the species present nowadays are doomed for immediate (in geological terms) extinction, even if conservation measures are taken. In this category are 12 % of all plants and 11 % of all bird species on earth (Hanski and Ovaskainen 2000).

In terrestrial environments such calculations show that on the order of thousand species per decade per million species go extinct. However, this process is not spread evenly over the globe. Myers et al. (2000) have shown recently that roughly 30-50 % of plant, amphibian, reptile, bird and mammal species occur in 25 biodiversity hotspots that occupy no more than 2 % of the ice-free land surface. The oceans are not included in this analysis, but species dependent on coral reefs are similarly concentrated. It is clearly one of the urgent tasks for marine biologists to come up with reliable estimates of the danger of species extinctions in the marine environment

One factor that clearly influences extinction rate is body size. The vast majority of the earth's animals have relatively small body sizes; there are far fewer species with large body sizes. The reason for the smaller number of large-bodied animals may be the higher extinction rates of larger species associated with their lower population density. Although smaller and more numerous organisms should experience lower extinction rates, they also may experience greater population fluctuations for several reasons. First, on average, smaller organisms have greater maximal population growth rates, which may give them a greater tendency to oscillate (e.g. May 1975). Second, smaller organisms also have shorter lifespans and a lower ability to store resources, and thus live in a more unpredictable environment than do larger organisms. Greater population fluctuations should increase the chance of extinction.

May (1986) offered several explanations for the larger number of smaller species, including the possibility that it might be caused by the fractal nature of spatial heterogeneity. A habitat may be more heterogeneous to a small organism than to a large one. May suggested that the quantitative relationship between organism size and habitat heterogeneity, which may be represented by the fractal dimension of the habitat, might help to explain the large number of small-sized animal species. This idea was recently taken one step further by Ritchie and Olff (1999) who discuss how many patterns of biodiversity can arise from simple constraints on how organisms acquire resources in space. These patterns include well-known responses of biological diversity to different factors such as the number of available niches in space, productivity, area, body size and habitat fragmentation. Ritchie and Olff (1999) use spatial scaling laws to describe how species of different sizes find food in patches of varying size and resource concentration. From this they derive a mathematical rule for the minimum similarity in size of species that share these resources. This packing rule yields a theory of species diversity that predicts relations between diversity and productivity. The theory also predicts relations between diversity and area and between diversity and habitat fragmentation. Thus, spatial scaling laws provide potentially unifying first principles that may explain many important patterns of species diversity.

The influence of humans: is marine biodiversity threatened?

Loss of biodiversity is now occurring on a scale which is without precedent in the last 65 million years. Most concern about elimination of genetic, species, and ecosystem diversity has focused on the terrestrial realm. It is now clear that marine ecosystems under human pressure are also at risk, including estuaries, coral reefs, seagrass beds, mangrove forests, intertidal shores, and continental shelves and slopes worldwide.

At first sight, it seems reasonable to assume that the sea's extent makes it less vulnerable than forests. The world ocean, covering 361×10^6 km² to an average depth of nearly 4 kilometers, constitutes more than 99% of the biosphere permanently inhabited by animals and plants. In contrast, forests and woodlands cover 38×10^6 km², or 7.5% of the

Earth's surface (Carlton et al. 1999). However, the most productive marine ecosystems, the world's continental shelves, cover 28 x 10^6 km², less than the area covered by forests. The most charismatic marine ecosystems, including coral reefs, kelp forests, seagrass beds, and mangrove forests, constitute a very small portion of the sea. Coral reefs, which are considered the marine equivalents of tropical forests, occupy only 0.6 x 10^6 km², or 0.1 %, of the Earth's surface, a small fraction of the 12 x 10^6 km² of closed tropical forests (Carlton et al. 1999).

That the sea is not immune to extinctions is clear from the geological past. At several times in the history of the planet more than 80 % of the marine biota got extinct in a relatively short time. Marine organisms are often thought to be less prone to extinction because of having widespread popu-lations and wide dispersal by ocean currents, or because they live in refugia from human predation. Only a few marine extinctions are documented, but this is hardly surprising. Few scientists work on marine extinction issues. As with the once abundant and widespread eelgrass limpet, the extinction of which went unremarked for more than five decades, the fact that these extinctions have not been documented is not evidence that they are not occurring. "One should not be surprised that the marine equivalent of the passenger pigeon will soon be found as being one of the larger and/or slow growing elasmobranch or bony deep-sea fish species that are being exploited just in the recent past (Carlton et al. 1999)".

Trawling of the Sea Floor

Ecological extinction caused by overfishing precedes all other pervasive human disturbance to coastal ecosystems (Jackson et al. 2001). Overfishing of large vertebrates and shellfish was the first major human disturbance to all coastal ecosystems that were recently examined. The magnitude of losses was enormous and large animals are now absent from most coastal ecosystems in the world. All other changes to coastal ecosystems came much later. Jackson et al. (2001) even conclude that overfishing may be a necessary precondition for eutrophication, species introductions and

disease to occur and the microbization of the coastal ocean as well as the consequences of climate change should not ignore the disappearance of the larger species.

A large proportion of the world's fish catch comes from continental shelves. Many fishing methods are harmful because, in addition to killing target species and those incidentally brought on deck (by catch), they severely disturb the seabed and organisms that provide food and hiding places. Trawling and dredging have been used for a long time on smooth, shallow bottoms near industrialized nations. With the hugely increased engine power of modern fishing vessels, trawling can now be practiced on nearly any bottom type, with the exception of stony areas and around obstacles such as ship wrecks, not only in shallow water on the continental shelf but increasingly in deeper waters, including the upper continental slope, and seamounts from subpolar to tropical waters (Carlton et al. 1999).

The effects of mobile gear on seabed biota resemble those of clearcutting in forests, removing the complex structures that are hiding and feeding places for many species (Carlton et al. 1999). Bottom trawling affects a greater area than any other benthic disturbance: Watling and Norse (1998) estimated that an area equivalent to 14.8 x 10^6 km² is trawled annually. Worldwide, an area equalling the world's continental shelf is trawled every two years. The North Sea floor on average is trawled twice a year (Lindeboom and De Groot 1998), but some areas (the best ones for fishing) are trawled much more frequently. Small spatial scale studies in the North Sea have shown that fishermen can very selectively look for target species (Rijnsdorp et al. 1998).

For longer-lived species, repeated removal and physical disturbance on so large a scale can make extinction all but inevitable. In Europe one of the oldest documented local extinctions is that of the flat oyster *Ostrea edulis* in the Wadden Sea, more than a century ago, due to bottom trawling (Reise 1982; Wolff 2000b). The long-lived, large (1.5 m long) barndoor skate (*Raja laevis*), a Northwest Atlantic shelf -dwelling fish that was once common in trawl bycatch, is now nearing extinction (Carlton et al. 1999). The same happened to all skate species in the southern North Sea (Wolff 2000a). Concern in Europe is rapidly growing on the widespread destruction of banks of the deep-sea cold water coral *Lophelia* with its associated fauna, occurring in a narrow band along the continental slope. As in tropical forests, the disappearance of the most observable species is likely a strong indication of greater extinction.

Mariculture

The impact of aquaculture on aquatic ecosystems has been recently reviewed by Naylor et al. (2000). More than 220 species of freshwater and marine finfish and shellfish are farmed; the range includes marine species such as giant clams, mussels and salmon. These activities impact on marine biodiversity in different ways. They share or compete for many coastal ecosystem services, including the provision of habitat and nursery areas, feed and seed supplies, and assimilation of waste products. Production practices and their impacts on marine ecosystems vary widely. Molluscs are generally farmed along coastlines where wild or hatchery-reared seed are grown on the seabed or in suspended nets, ropes or other structures. The animals rely entirely on ambient supplies of plankton and organic particles for food. Several systems – ponds, tanks or cages – are used in farming finfish. Most marine and diadromous finfish are reared in floating net cages nearshore, and all their nutrition is supplied by formulated feeds. In such intense cultivitation the density of individuals is increased, which requires greater use and management of inputs, greater generation of waste products and increased potential for the spread of pathogens. Shrimps dominate crustacean farming and are grown in coastal ponds which are often constructed in former mangrove areas.

Cultured species imported from elsewhere can escape and establish in the wild. A spectacular example of this is the dramatic increase of the Japanese oyster *Crassostrea gigas* in the Ooster-schelde area in the Netherlands, where this species now outcompetes the native cockles and oysters for space and food. Also well documented is the escape from net pens of over 255,000 Atlantic salmon in the Pacific Ocean, where they now interbreed with local stocks. Movement of feed and stocks also increases the risk of spreading pathogens.

Global production of farmed fish and shellfish has more than doubled in the past 15 years. Many people believe that such growth relieves pressure on ocean fisheries, but the opposite is true for some types of aquaculture. Farming carnivorous species reguires large inputs of wild fish for feed. Some aquaculture systems also reduce wild fish supplies through habitat modification and wild seedstock collection. Fry of wanted species are often obtained from the wild at the expense of fry of non-wanted species (Naylor et al. 2000).

Habitat modification

The coastal environment in Europe has undergone profound modifications in many areas. Subtidal sediments, sandy beaches and dune systems have been transformed through the construction of harbours, dykes and breakwaters; coastal inlets and rivers have been dammed with decreased export of sediments to the coastal zone as a consequence; sand and gravel are exploitated for the building industry, sand is suppleted to beaches to prevent erosion etc. The many forms of individual human use (tourism, collection, food gathering etc.) change the physical environment as well, e.g. by turning stones on the beach, cutting algae that provide shelter etc.

In the tropics the situation is perhaps even worse. Hundreds of thousands of hectares of mangroves and coastal wetlands have been transformed into milkfish and shrimp ponds. This transformation results in loss of essential ecosystem services generated by mangroves, including the provision of nursery habitat, coastal protection, flood control, sediment trapping and water treatment. Mangrove forests serve as nurseries that provide food and shelter to many juvenile finfish and shellfish caught as adults in coastal and offshore fisheries. In southeast Asia, mangrove-dependent species account for roughly one-third of yearly wild fish landings excluding trash fish. A positive relationship between finfish and shrimp landings and mangrove area has been documented in Indonesia, Malaysia and the Philippines. Mangroves are also linked closely to habitat conditions of coral reefs and seagrass beds (Naylor et al. 2000).

Loss of mangrove forests results in increased sediment transport onto downstream coral reefs. Fisheries capture from reefs contributes about 10% of human fish consumption globally and much more in developing countries. The loss in wild fisheries stocks due to habitat conversion associated with shrimp farming is large. Naylor et al. (2000) estimate that a total of 400g of fish and shrimp are lost from capture fisheries per kilogram of shrimp farmed in Thai shrimp ponds developed in mangroves. If the full range of ecological effects associated with mangrove conversion is accounted for, including reduced mollusc productivity in mangroves and losses to seagrass beds and coral reefs, the net yield from these shrimp farms is low.

Invading species

The risk of the establishment of exotic species in the seas is not yet widely acknowledged outside biological circles, but there are now many spectacular, documented examples. Species are transported in high numbers and high densities over vaste distances, mainly through the ballast water of ships. The sheer number of invasions is staggering and has been calculated at one species of macrofauna and flora per week in areas with heavy shipping traffic. There is nearly no information on microbes. Most invasions probably even do not take off, but a large number does. This can lead to qualitative changes in ecosystem functioning when empty niches are filled or the global erosion of biodiversity with the loss of valuable genetic material. There are many case studies that report on invading species. There are now many spectacular examples of animal (e.g. *Marenzelleria viridis*) and plant (e.g. *Caulerpa taxifolia*) outbreaks, but not many of these cases have been followed up sufficiently long to be able to say something about the long term success and consequences of these invasions. An excellent summary, focusing on North American coastal communities, is provided by Ruiz (2000). This is an active and important area of research with many practical consequences, e.g. in mariculture, for which codes of practice have been established (ICES).

Conclusion

It is clear that marine biodiversity is already heavily changed by human exploitation, especially of the large mammals, reptiles, fish, crustaceans and mollusks that have been greatly reduced in coastal waters worldwide. Besides other local problems with eutrophication, pollution, invaders and disease, one may expect changes in the distribution and abundance of marine plants and animals due to effects of global climate change, including sea level rise and changes in temperature, and perhaps more subtle effects such as changing pH or Fe of the ocean surface waters. All these changes take place against a background that is hardly understood. It is not known how many marine species exist, where they exist and why. The evolutionary history and the scope for adaptation of most species is unknown. There is hardly any information on microbes and only scattered information on almost all other species, including the large vertebrate species. It is not known what the extinction of known and unknown marine species would mean for ecosystem functioning, including the delivery of a large range of ecosystem goods and services that are essential to human well-being. An important research effort aimed at underpinning sustainable use of marine resources is therefore a first requirement in the near future.

Acknowledgments

This is paper nr. 2891 of the Centre for Estuarine and Coastal Ecology, Netherlands Institute of Ecology. I thank the organizers of the Hanse Meeting for inviting me and Angelika Brandt and Gerd Graf for constructive comments on an earlier draft of the manuscript.

References

Abbott J (1980) Theories dealing with the ecology of landbirds on islands. Adv Ecol Res 11:329-371

Azam F, Smith DC, Steward GF, Hagstrom A (1994) Bacteria - organic-matter coupling and its significance for oceanic carbon cycling. Micro Ecol 28:167-179

Carlton JT, Geller JG, Reaka-Kudla ML, Norse EA (1999) Historical extinctions in the sea. Ann Rev Ecol Syst 30:515-538

Chesson PL, Warner RR (1981) Environmental variability promotes coexistence in lottery competitive systems. Amer Nat 117:923-943

Clarke RD (1988) Chance and order in determining fish-species composition on small coral patches. J Exp Mar Biol Ecol 115:197-212

Clements FE (1935) Nature and structure of the climax. J Ecol 24:252-289

Connell JH (1975) Some mechanisms producing structure in natural communities: A model and evidence form field experiments. In: Cody ML and Diamond J (eds) Ecology and Evolution of Communities. Harvard University Press, Cambridge, Mass 460-490

Connell JH (1979) Intermediate-Disturbance Hypothesis. Science 204:1345

Connor EF, McCoy (1979) The statistics and biology of the species-area relationship. Amer Nat 113:791-833

Duarte CM (2000) Marine biodiversity and ecosystem services: An elusive link.

Flach E, Muthumbi A and Heip C (in press) Meiofauna And Macrofauna Community Structure In Relation To Sediment Composition At The Iberian Margin Compared To The Goban Spur (Ne Atlantic). Prog Oceanogr

Grassle JF, Lasserre P, McIntyre AD and Ray GC (1991) Marine biodiversity and ecosystem function. Biology International Special Issue 23; 19p

Grassle JF, Maciolek NJ (1992) Deep-sea species richness - regional and local diversity estimates from quantitative bottom samples. Amer Nat 139:313-341

Hamilton TH, Rubinoff I, Barth RH, Bush GL (1963) Species abundance: Natural regulation of insular variation. Science 142:1575-1577

Hanski I (1999) Habitat connectivity, habitat continuity, and metapopulations in dynamic landscapes. Oikos 87:209-219

Hanski I, Ovaskainen O (2000) The metapopulation capacity of a fragmented landscape. Nature 404:755-758

Heip C (1996) Biodiversity of Marine Sediments. In: di Castri F and Younes T (eds) Biodiversity, Science and Development. Towards a New Partnership. pp 139-148

Heip C, Warwick RM, d'Ozouville L (1999) A European Science Plan on Marine Biodiversity. European Science Foundation, Stasbourg.

Huisman J, Weissing FJ (2000) Biodiversity of plankton by species oscillations and chaos. Nature 402:407-410

Huys R, Herman PMJ, Heip CHR, Soetaert K (1992) The meiobenthos of the north sea: Density, biomass trends and distribution of copepod communities. Ices J Mar Sci 49:23-44

Hutchinson GE (1959) Homage to Santa Rosalia or Why are there so many kinds of animals? Amer Nat 93:145-159

Jackson JBC, Kirby MX, Berger WH, Bjorndal KA, Botsford LW, Bourque BJ, Bradbury RH, Cooke R, Erlandson J, Estes JA, Hughes TP, Kidwell S, Lange CB, Lenihan HS, Pandolfi JM, Peterson CH, Steneck RS, Tegner MJ, Warner RR (2001) Historical overfishing and the recent collaps of coastal ecosystems. Science 293:629-638

Janzen DH (1970) Herbivores and the number of tree species in tropical forests. Amer Nat 104:501-528

Lasserre P (1992) The role of biodiversity in marine ecosystems. In: Solbrig, van Emden and van Oordt (eds) Biodiversity and Global Chang. IUBS Press, Paris, France pp 105-130

Latham RE, Ricklefs RE (1993) Global patterns of tree species richness in moist forests. Energy-diversity theory does not account for variation in species richness. Oikos 67

Levins R (1979) Coexistence in a variable environment. Amer Nat 114:765-783

Lindeboom HJ, De Groot SJ (1998) The effects of different types of fisheries on the North Sea and Irish Sea benthic ecosystem. NIOZ Report nr. 1. ISSN 0923-3210. 404 p

Loreau M (2000) Biodiversity and ecosystem functioning: Recent theoretical advances. Oikos 91:3-17

MacArthur RH (1972) Geographical Ecology: Patterns in the distribution of species. Harper and Row, New York

MacArthur RH, Wilson EO (1963) An equilibrium theory of insular zoogeography

MacArthur RH, Wilson EO (1967) The theory of island biogeography. Princeton Univ Press, Princeton NJ

Margalef R (1968) Perspectives in Ecological theory. Chicago Univ Press

May RM (1975) Biological Populations Obeying Difference Equations - Stable Points, Stable Cycles, And Chaos. J Theor Biol 51:511-524

May RM (1986) Biological Diversity - How Many Species Are There. Nature 324:514-515

May RM (1992) Biodiversity - Bottoms Up for the Oceans. Nature 357:278-279

Mc Gowan JA, Walker PW (1993) Pelagic Diversity Patterns. In: Ricklefs RE and Schluter D (eds) Species Diversity in Ecological Communities. Univ of Chicago Press, Chicago pp 203-214

Myers N, Mittermeier RA, Mittermeier CG, de Fonseca GAB, Kent J (2000) Biodiversity hotspots for conservation priorities. Nature 403:853-858

Naylor RL, Goldburg RJ, Primavera JH, Kautsky N, Beveridge MCM, Clay J, Folke C, Lubchenco J, Mooney H, Troell M (2000) Effect of aquaculture on world fish supplies. Nature 405:1017-1024

Ormond RFG, Gage JD, Angel MV (1997) Marine Biodiversity: Patterns and Processes. Cambridge University Press

Pachepsky E, Crawford JW, Bown JL, Squire G (2001) Towards a general theory of biodiversity. Nature 410:923-926

Paine RT (1966) Food web complexity and and species diversity. Amer Nat 100:65-75

Reise K (1982) Long-term changes in the macrobenthic invertebrate fauna of the wadden sea - are polychaetes about to take over? Neth J Sea Res 16:29-36

Rex MA, Stuart CL, Hessler RR, Allen JA, Sanders HL, Wilson GDF (1993) Global-scale latitudinal patterns of species diversity in the deep-sea benthos. Nature 365:636-639

RijnsdorpAD, Buys AM, Storbeck F, Visser EG (1998) Micro-scale distribution of beam trawl effort in the southern North Sea between 1993 and 1996 in relation to the trawling frequency of the sea bed and the impact on benthic organisms. ICES J Mar Sci 55:403-419

Ritchie ME, Olff H (1999) Spatial scaling laws yield a synthetic theory of biodiversity. Nature 400:557-560

Ruiz G.M, Fofonoff PW, Carlton JT, Wonham MJ and Hines AH (2000) Invasion of coastal marine communities in North America: Apparent patterns, processes, and biases. Ann Rev Ecol Syst 31:481-531

Sanders HL (1968) Benthic marine diversity: A comparative study. Amer Nat 102:243-282

Schluter D, Ricklefs RE (1993) Species diversity: an introduction to the problem. In: Ricklefs and Schluter (eds) Species Diversity in Ecological Communities. Historical and Geographical Perspectives. The Univ of Chicago Press pp 1-10

Simpson GG (1964) Species densities of North American mammals. Syst Zool 13:361-389

Soetaert K, Heip C (1995) Nematode assemblages of deep-sea and shelf break sites in the North Atlantic and Mediterranean Sea. Mar Ecol Prog Ser 125:171-183

Terborgh J (1973) On the notion of favorableness in plant ecology. Amer Nat 107:481-501

Thiery RG (1982) Environmental instability and community diversity. Biol Rev 57:671-710

Watling L, Norse EA (1998) Disturbance of the seabed by mobile fishing gear: A comparison to forest clearcutting. Conserv Biol 12:1180-1197

Watt AS (1947) Pattern and process in the plant community. J Ecol 35:1-22

Wolff WJ (2000a) Causes of extirpations in the Wadden Sea, an estuarine area in the Netherlands. Conserv Biol 14:876-885

Wolff WJ (2000b) The south-eastern North Sea: Losses of vertebrate fauna during the past 2000 years. Conserv Biol 95:209-217

Marine Microbial Food Web Structure and Function

G.J. Herndl[*] and M.G. Weinbauer

*Dept. of Biological Oceanography, Netherlands Institute for Sea Research (NIOZ),
P.O. Box 59, 1790 AB Den Burg, The Netherlands
* corresponding author (e-mail): Herndl@nioz.nl*

Abstract: The major challenge for future research on the biodiversity and function in microbial communities is to develop approaches that allow identification of key species among the microbial community and to determine their abundance and function accurately. Furthermore, it is essential to provide information on the number of species capable of fulfilling the same function. Thus, we have to advance from a simple assessment of the species richness of a microbial community to linking species richness with other parameters determining diversity such as evenness and also addressing functional aspects. Only if we are able to define the specific microhabitats for a given microbial species will we be able to determine and understand the life strategies of key species and consortia responsible for the biogeochemical cycling of the major elements.

Introduction

Advances in the field of marine microbial ecology have revolutionized our knowledge on the carbon and energy flow through marine food webs. It is now well-established that bacterioplankton are the only significant transformers of dissolved organic carbon (DOC) in the ocean which represents, together with soil humus, the main biologically available organic carbon source on earth (Hedges 1992). On average, about 50 % of the organic carbon produced by phytoplankton, the main marine primary producers, is channeled through the so-called 'microbial loop' of the food web and bacteria represent the largest living surface area in the ocean (Whitman et al. 1998). This microbial loop consists of phytoplankton, bacteria, hetero- and mixotrophic protists and viruses, the latter 2 compartments controlling bacterio- and phytoplankton abundance. All these compartments together comprise the biotic components of the microbial loop (Fig. 1). The biomass of the functional groups acting in the microbial loop, as well as the fluxes of carbon and energy mediated by them, have been determined for a variety of marine systems, ranging from coastal to open ocean environments (Fuhrman 1999; Ducklow 2000).

The approach used to study these interactions was, until recently, essentially a system or a "black box approach" where input into and output from the different biological and abiotic compartments (Fig. 1) have been quantified. Due to the lack of distinct morphological features especially of bacteria and viruses but also of the bulk of the protists (particularly the flagellates), almost no insight into dynamics at a population- or species-level has been obtained with this system approach. The availability of molecular techniques for microbial ecology now allows us to open these 'black boxes' and to gain insights into the dynamics and regulation mechanisms of the microbial food web. The ultimate goal is to advance from a description of microbial processes to a mechanistic understanding of the functioning of microbial communities. Understanding interactions between microorganisms on a scale relevant to them, the microscale, will advance our knowledge on regional and basin-scale variability in biogeochemical cycles.

What do we know about microbial diversity and its relation to general ecosystem function?

Diversity can be measured in various ways which can be conceptually divided into three approaches: Richness, evenness and difference (Pedros-Alio

From WEFER G, LAMY F, MANTOURA F (eds), 2003, *Marine Science Frontiers for Europe.* Springer-Verlag Berlin Heidelberg New York Tokyo, pp 265-277

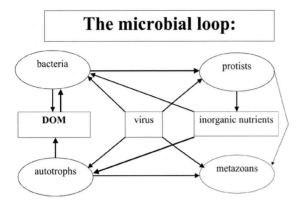

The microbial loop:

Fig. 1. Scheme of the main material flux through the microbial food web. Bacteria take up dissolved organic matter (DOM) and are preyed upon by hetero- and mixotrophic protists providing remineralized nutrients for the autotrophic component (phytoplankton) which releases DOM during photosynthesis and, more importantly, during decay. Viruses are an important component in the microbial food web as well exerting control on all the biotic components of the microbial loop.

1993). Species richness is the number of species in a habitat. For bacteria and even more so for viruses, the species concept poses problems (see below). Higher-taxon richness is often used in diversity studies, particularly if species determination is not feasible. Species evenness is a measure of the number of individuals per species and reveals whether or not species dominate numerically in a habitat. Species richness and evenness are frequently combined in indices such as the Shannon index for diversity. Difference accounts for the fact that some species in a habitat are very similar, whereas others are very different. This feature can be seen as phylogenetic difference which is the sum of the branch lengths in the evolutionary tree linking them.

The morphological diversity of prokaryotes is low compared to (multicellular) eukaryotes. However, prokaryotes exhibit a vast metabolic variability which is the result of more than 3 billion years of evolution. Ecosystem functions of prokaryotes in pelagic marine systems are production and decomposition of organic matter, regeneration of nutrients and building-up of biomass from 'dead' organic matter which would otherwise be lost from

the food web. The role of prokaryotes in the functioning of marine systems has been intensively studied over the last 2-3 decades. For example, the biomass of phototrophic prokaryotes in the open ocean can be as high as that of eukaryotic phytoplankton (Falkowski 1994) and the central role of the microbial food web for nutrient and carbon cycling has been documented (Azam and Cho 1987). Protists contribute to photosynthetic carbon fixation as well as to controlling bacterial activity and nutrient regeneration. A function of viral infection is cell lysis which retains the lysed organic matter in the microbial food web and renders bacterial production unavailable to higher trophic levels. This latter function may also retain elements such as N, P and Fe in the layers of the water column where lysis occurred by producing non-sinking dissolved organic matter. However, the interaction between the diversity of microbes and ecosystem functioning is not well investigated. For example, it is not clear how microbial diversity contributes to the stability of ecosystems.

How do we measure microbial diversity?

One of the species concepts for bacteria and viruses integrates different kinds of information using a polyphasic taxonomy (bacteria) or polythetic classes (viruses). According to a more genotypic approach, two bacterial strains belong to the same species at DNA-DNA hybridization values of >70% (for a recent review of the bacterial species concept (see Rossello-Mora and Amann 2001). This concept has been heavily criticized. For example, by this definition all primates would belong to one species. Recent methodological progress using sequence information of ribosomal RNA has provided novel information on the phylogeny of life. Cellular organisms are now separated into the three domains *Bacteria, Archaea* and *Eukarya.* Two prokaryotic populations differing by >97% in the sequence of their 16S rRNA gene are generally considered as two different species. This difference corresponds to a cross hybridization of DNA of > 70%. It goes without saying that this concept is highly arbitrary. Indeed, it has been argued that genome data shake the tree of life (Pennisi 1998), since lateral gene transfer jeopardizes the use of hierarchical universal classification. However, based on sequence data, phylogenetic distances can be calculated which can be used to estimate diversity even if there is some subjectivity left as to where to set the line that defines species in the phylogenetic tree.

For diversity studies 16S rRNA or its gene are extracted from environmental samples and amplified by the polymerase chain reaction (PCR) using universal or more specific primers. The 16S rDNA is then separated by sequence-specific methods methods such as denaturing gradient gel electrophoresis (DGGE) or single strand conformation polymorphism (SSCP) electrophoresis or by creating clone libraries (for more details on methods see Fuhrman 1996; Stahl 1996). These techniques also enable circumventing the 'great plate count anomaly', i.e. the fact that most marine bacteria (typically >99%) cannot be grown and thus isolated on culture plates as a prerequisite for species identification (Amann et al. 1995). Culture-independent approaches also permit sequencing of the separated 16S rDNA and affiliating these 'phylotypes' in the tree of life. Thus, these data can provide information on the species richness as well as phylogenetic diversity. However, due to inherent problems with PCR (which are not fully understood), not all sequences are amplified, as e.g., sequences of rare species, and the amplification of sequences itself is not homogeneous. It has been shown for example that the *Cytophaga-Flavobacter-Bacterioides*-(CFB) group is underrepresented, whereas α-Proteobacteria are overrepresented in clone libraries (Cottrell and Kirchman 2000b). Thus, the density of bands of phylotypes on a gel or their representation in clone libraries is at best a poor indicator of the significance of species and cannot be used to evaluate species evenness. Another problem with the 16S rDNA approach is that the number of RNA operons (groups of genes encoding for ribosomes) differs between species (from 1 to ca. 15) and may differ with respect to sequence within one single clone. Other molecular chronometers such as the RNA polymerase beta subunit (*rpoB*) (Dahllof et al. 2000) have the advantage that they occur in only one copy number. The database for such molecules is still small, however.

From known sequence data, oligonucleotide probes (small DNA sequences) can be constructed which are specific for different taxonomic levels, from species to domain level (Amann et al. 1995). These probes target rRNA and are conjugated with a fluorochrome allowing detection and identification on a single cell basis using epifluorescence microscopy, confocal scanning laser microscopy or flow cytometry. Using fluorescent *in situ* hybridization (FISH), cell numbers and taxonomic information are obtained, typically at the level of higher taxonomic groups such as the subgroups of *Proteobacteria*. A disadvantage of FISH is that only a limited number of samples can be processed. Moreover, the specificity of the probes for some taxonomic groups might not be sufficient. Quantitative PCR is a promising new tool to circumvent these problems (Suzuki et al. 2000), but it has not been tested sufficiently on complex environmental samples.

Protist species richness has also been studied using 18S rDNA sequences. Novel sequences from eukaryotic picoplankton were retrieved by using clone libraries (Lopez-Garcia et al. 2001; Moon-van der Staay et al. 2001) suggesting that the species richness of small eukaryotic picoplankton was also underestimated.

The assessment of viral species richness is even harder to accomplish than that of prokaryotes. This is a consequence of the 'great plate count anomaly', since bacterial (or algal) host strains have to be isolated in culture before attempts can be made to isolate viruses. This is done by using plaques that are formed on bacterial lawns in agar plates due to viral lysis of cells. Thus, this 'great plaque count anomaly' results in an underestimation of viral abundance and diversity. The most frequent type of marine viruses, tailed phages with a double-strand DNA, may have a common ancestry and may be as old as 3.5 to 3.7 billion years (Ackermann 1999). Since there seem to be no common sequences for these phages, it is unlikely that molecular chronometers can be used such as the 16S rDNA for cellular organisms. However, some approaches have been developed to amplify previously unknown sequences of bacterial and algal viruses from marine systems. These sequences also allow a phylogenetic affiliation of viruses. Thus, it might be at least possible to develop molecular chronometers for some taxonomic groups such as algal viruses (Chen and Suttle 1995). Another approach is based on size-specific separation of viral genomes obtained from environmental samples using pulsed field gel electrophoresis (PFGE) (Wommack et al. 1999). While an estimation of the abundance of viruses corresponding to single genome sizes is possible with this technique, it is possible that viral genomes from different species have the same genome size.

Overall, despite the progress that has been made, there is still an urgent need to develop techniques for reliably and economically investigating the three aspects of diversity in prokaryotes and viruses, species richness, evenness and difference.

How is microbial function measured?

We can reasonably well estimate production of entire bacterial communities, e.g. as incorporation of radiolabeled thymidine and leucine in bacterial DNA and protein, respectively (Fuhrman and Azam 1982; Kirchman et al. 1985; Simon and Azam 1989). Other specific radiolabeled compounds can be used to determine the potential of the microbial community to transform these compounds. Also, bacterial respiration can be assessed as oxygen consumption, although major problems are inherent to the method such as its sensitivity and the incomplete separation of bacterial from eukaryotic respiration (Del Giorgio et al. 1997; Del Giorgio and Cole 1998). Other parameters that can be quantified are specific activities such as nitrogenase and ectoenzyme activity (Hoppe 1991). The use of stable isotopes offers the potential to determine the flux of specific compounds through specific members of the bacterial community (Boschker et al. 1999; Radajewski et al. 2000). Recently developed methods have combined FISH with microautoradiography (STARFISH and MICRO-FISH) (Cottrell and Kirchman 2000a; Ouverney and Fuhrman 1999). Using these methods, the uptake of radiolabeled substrates such as thymidine, amino acids or phytoplankton-derived organic matter by different groups of prokaryotes including *Archaea* can be studied. Analysis of gene expression by quantification and identification of mRNA encoding for enzymes may be a useful technique to combine functional and diversity studies.

The measurement of viral function is less elaborated than that for bacteria. Methods have been developed to estimate lysogeny (the occurrence of 'dormant' viral genomes in the host genome), viral production and viral mortality of bacteria and microalgae (Fuhrman 1999; Wilhelm and Suttle 1999; Wommack and Colwell 2000). Some attempts have been made to assess virus-mediated gene transfer (Paul 1999). Another potential function of viral lysis is the control of species diversity. However, there is an urgent need to develop and refine techniques to assess the functional aspects of viruses.

For measurements on the functioning of protists, we have to distinguish between autotrophic and heterotrophic activity. The autotrophic activity can be readily measured via the incorporation of bicarbonate into the autotrophic compartment of the microbial food web or, more specifically, into the antenna pigments of specific autotrophs (Cailliau et al. 1996). Also the Rubisco gene expression can be assessed. For heterotrophic protists, activity is most frequently determined as consumption of labeled prey (Sherr et al. 1989).

With the advent of molecular techniques it was hoped that the phylogeny of microorganisms could also provide insight into metabolic functions. However, although it might be true for some groups, this hope has been shattered by several studies and by looking in more detail at already available data. Based on recently published studies which show that phylogenetically closely linked species behave very differently in the natural environment (occupy different niches), we can conclude that sequence data are of limited value with respect to estimating metabolic rates or function (Fuhrman and Campbell 1998). Thus, there is still a need to determine metabolic functions by separate techniques and link them to diversity.

Relationship between diversity, stability and productivity in microbial communities

The enormous biodiversity found on earth has puzzled scientists for a long time. Indeed, global patterns of biodiversity are difficult to explain. The finding that species richness generally increases towards the equator has attracted considerable attention. Attempts have been made to explain such latitudinal gradients of species richness by the amount of energy available or productivity in a system. However, the reason for species richness on large scales remains frustratingly uncertain, although there is an attempt to provide a unifying principle that explains diversity patterns by spatial scaling laws (Ritchie and Olff 1999). Some studies indicate that an increase in the number of species in crucial functional groups increases productivity and ecosystem reliability (Naeem and Li 1997). Such findings are in accordance with the 'insurance effect' of biodiversity meaning that a high species richness with a high redundancy of species contributes to the buffering capacity of ecosystems as perturbations occur (Yachi and Loreau 1999). Also, it has been argued that the Hutchinson's 'paradox of the plankton', addressing the question why there are so many (phytoplankton) species in the presence of only a few limiting resources, can be resolved by the 'intermediate disturbance hypothesis' (IDH) (Flöder and Sommer 1999; Huisman and Weissing 1999). IDH predicts a peak in species richness at intermediate disturbance frequencies and intensities, allowing for the co-existence of competitive dominants as well as pioneer species. Current evidence suggests that the number of dominant bacterial species is low in marine pelagic systems (see below). Considering the effects of viral lysis and protist grazing might suggest that for microorganisms, Hutchinson's paradox should be reformulated as: Why are there so few dominating species when simple steady-state models can predict so many (Thingstad 2000)? Other than for phytoplankton where the growth-limiting factors are availability of solar radiation and major and trace nutrients, heterotrophic microorganisms and especially bacteria might be limited by an almost indefinite number of factors. It is well-known, at least for bacterial strains, that most bacteria have specific substrate requirements in terms of quantity and quality.

There is a heated discussion on the linkage between ecosystem function and stability and diversity (Woodward 1994; Tilman 1996; Loreau and Behera 1999; Sankaran and McNaughton 1999).

This diversity-stability debate has strong implications for mankind, since ecosystems are our ultimate life-support systems. The bottom line of results obtained from studying multicellular organisms is that the declining diversity observed in many ecosystems may result in an acceleration of the simplification of these ecosystems. However, many of these studies consider only species richness but not evenness and difference. Since unicellular organisms such as most phyto- and bacterioplankton species turn over rapidly, they are ideal for studying the causes of biodiversity. For example, it has been shown using monoclonal antibodies that two bacterial populations were present in a eutrophic lake for 6 years corresponding to 900 and 3,400 generations, respectively. This would correspond to an investigation period of ca. 27,000 to 100,000 years for trees assuming a generation time of 30 years (Weinbauer and Höfle 1998).

The use of culture-independent approaches has revealed numerous new bacterial species and clades in the ocean which are not cultivable (Giovannoni et al. 1990; Amann et al. 1995). In the ocean, there are 11 major groups of prokaryotic plankton with nine groups of *Bacteria* and 2 groups of *Archaea* (Giovannoni and Rappé 2000). A recent estimation indicates that ca. 70% of prokaryotic cells in the ocean belong to *Bacteria* and ca. 30% to *Archaea* with the latter dominating in deep water (Karner et al. 2001). Interestingly, the β-subgroup of *Proteobacteria* is only of minor importance in offshore marine systems but is dominant in freshwater systems. The first cluster of genes that was recovered from clone libraries was SAR11 which showed only a distant relationship to sequences from isolates in databases (Giovannoni et al. 1990). This group was found in coastal and offshore environments, in deep ocean waters and in permanently anaerobic marine sediments. Thus, this cosmopolitan cluster may be the most abundant group of marine *Bacteria*. Another remarkable feature revealed by clone libraries is that phylogenetic groups form clusters of related genes rather than single lineages. This uneven distribution of phylogenetic difference may indicate common advantages that favor the speciation of these groups. Although coastal and offshore communities seem to be similar, at least regarding major phylogenetic groups, there is a distinct depth distribution discernable among bacteria. Some groups (not only phototrophs) dominate in the photic zone, whereas others such as *Archaea* dominate in the aphotic zone. Also, bacterial communities living on organic particles differ strongly from their free-living counterparts. Organic particles are enriched with cells belonging to the *Cytophaga-Flavobacter-Bacterioides* (CFB) group and the order Planctomycetales. Using MICRO-FISH, it has been demonstrated in coastal marine environments, that the CFB group takes up preferentially chitin, *N*-acetyl glucosamine, and protein, α-*Proteobacteria* take up preferentially *N*-acetyl glucosamine and amino acids, and γ-*Proteobac-teria* amino acids (Cottrell and Kirchman 2000b). Other studies have shown that the bacterial community composition changed depending on the origin of organic matter, i.e. from cyanobacteria or diatoms. This indicates that organic matter composition influences bacterial diversity, however, it also suggests that the composition of bacterio-plankton determines their ecosystem function as decomposers of organic matter.

The number of known dominant bacterial species in the ocean ranges from 2 to 15 (note, however, the problem of species identification for bacteria as outlined above) (Pinhassi and Hagström 2000). This number is 1-2 orders of magnitude lower than the number of species detected in soil. Bacterioplankton species richness is still unknown; the number of species retrieved from environmental samples is highly correlated with the effort to screen for new phylotypes, e.g. with the size of genome libraries. A low number of dominant species in the presence of high species richness would indicate a low species evenness for marine bacterioplankton. Bacterial successions have been shown to occur in marine systems (Pinhassi and Hagström 2000). While some species showed a strong seasonal variability, others were present at rather stable and low 'background' numbers. The assessment of seasonal variability of bacterial diversity (or its absence) is necessary for a full characterization of bacterial diversity in a system. Such an information is also a prerequisite for assessing topics such as latitudinal gradients of bacterial diversity and its link to system productivity.

A cosmopolitan occurrence of viruses was demonstrated for a virus infecting *Micromonas pusilla*, a picophytoplankter, with a world-wide distribution (Cottrell and Suttle 1991). Using pulsed field gel electrophoresis (PFGE), changes in the viral communities on the scale of days and months have been shown, as well as geographic and depth-related differences in community structure. The number of bands from individual PFGE fingerprints and thus the potential number of dominant viral species ranged from 7 to 16 which is in the range of estimated dominant bacterial species in the ocean. Viral diversity likely depends on the presence of the number of host species as well as on the production and survival rates of the viruses released during cell lysis. Research on the diversity of marine viruses is still in its infancy, and we do not know how viral diversity relates to ecosystem functioning.

The microbial food web in action: Interaction between functional groups and mechanisms maintaining diversity and function

The different functional groups of microorganisms are tightly linked to other components of the microbial food web due to their similar generation times. Considerable emphasis has been given over the past 1-2 decades to the quantifying of grazing rates of protists and lysis rates of viruses and to relating these rates to the abundance and size of the prey and its predator. Efforts have been put into calculations of contact rates at varying prey concentrations and to relating them to the energy demand of the prey. This type of calculation has allowed the determination of threshold levels of food concentrations for the major functional groups interacting in the microbial loop. There is now evidence accumulating that the abundance of a specific functional group and the size of the organism involved in this prey-predator relationship are not the only factors controlling the grazing efficiency. Microorganisms are more selective than hitherto assumed with respect to food selection. Currently, we have probably resolved just the tip of the iceberg in our efforts to shed light into the "behavior" of microorganisms which ultimately regulates diversity and function of the microbial food web.

In the following, some selected examples of microorganism behavior will be given for all functional groups of the microbial food web (bacteria, phytoplankton, viruses and protists). The aim is not to present an extensive review here but rather to stress the need for a better understanding of the multifaceted possibilities of the reactions of microorganisms to environmental stimuli. These responses will ultimately have impact on the diversity and function of marine microbial food web in particular and on the overall productivity of the system in general.

Bacterial behavioral repertoire to sustain diversity

It is well recognized that not all the bacteria present in a given environment are metabolically active. It is assumed that bacteria will become dormant if their specific substrate requirements are not met. Thus one might envisage a scenario where a few bacterial species in a community are highly active because of the availability of suitable substrate while a large number of other species is in a dormant stage but can potentially readily resume growth if the appropriate environmental conditions are established again. Investigating depth distributions suggests that some of the bacterial species might be more specialized than others. Analysis of the structure of the bacterioplankton community in the Mediterranean Sea revealed that about 25 % of the bacterial species occurred throughout the water column while more than 50 % of the bacterial species were present either in the surface or the deep water (below 300 m) (Fig. 2). Also large differences between the community analysis on the DNA- and the RNA-level have been observed indicating that the bacterial community present does not resemble well that which is metabolically active (hence detectable on a RNA-level) (Moeseneder et al. 2001). This suggests that for a large number of bacterial species the substrate requirements is not or only insufficiently met. Identifying the main bacterial species playing a key role in the transformation of organic matter under specific conditions is a major challenge for future research. In model ecosystems it was shown recently that, while the composition of methane-producing bacterial species changed significantly over time, the rate of methane formation remained unchanged (Fernandez et al. 1999). Thus while diversity changed, the specific function (i.e. the rate of methane formation) remained stable. This example highlights also the importance of addressing the question of how many bacterial species are present to potentially perform the same function. Also the mechanism leading to the dominance of one species over another species with the same or similar function needs to be elucidated (see also below).

Co-metabolism between different bacterial species might also be a more common phenomenon than assumed hitherto. Thus, one might speculate that whenever a key bacterial species becomes limited in substrate supply or diminished by viruses or grazers, the entire bacterial consortium acting in concert to break down specific compounds might be affected. At present there are only a few reports on bacterial consortia responsible for specific biogeochemical processes. One example has been recently described in gas hydrates. There, anaerobic methane degradation has been found to be mediated by the concerted action of methane producers (*Archaea*) and sulfate reducing bacteria (*Bacteria*) living in close proximity (Boetius et al. 2000). It is unclear whether similarly stable consortia exist in the water column of the ocean. There is indirect evidence, however, that the water column is more structured on a microspatial scale than we have assumed (see below).

Potential mechanisms of grazing-selectivity of flagellates

Protists are known to feed preferentially on actively growing bacteria and some examples are documented of species selectivity (Del Giorgio et al. 1996). The mechanisms responsible for this selectivity, however, remain obscure. At present we do not know whether there is a link between metabolic activity and motility of bacteria. There is evidence that motility in marine bacteria is widespread and that bacteria are capable of using this motility to seek nutrient-rich microenvironments (Blackburn et al. 1998). Studies on the orientation of bacteria in a nutrient gradient have all been performed under laboratory conditions (Mitchell et al.

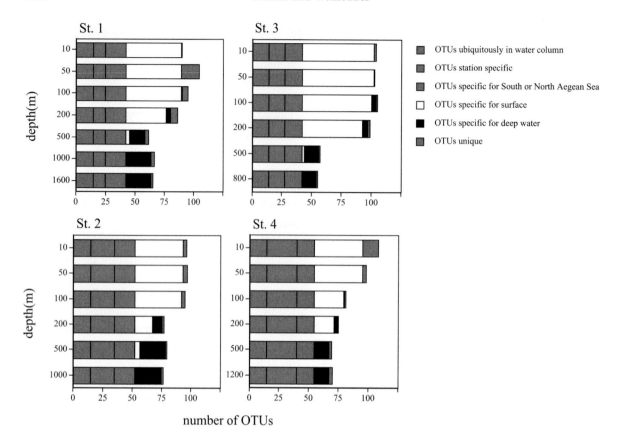

Fig. 2. Analysis of the spatial distribution of OTUs for free-living Bacteria using 16S rDNA and T-RFLP from 4 different stations in the Eastern Mediterranean Sea (Sts 1 and 2 are located in the Southern Aegean Sea, Sts. 3 and 4 in the Northern Aegean Sea). The number of OTUs obtained from the 16S rDNA is grouped in 6 different categories of occurrence: OTUs found at every station at every depth (ubiquitously occurring); OTUs found only at a single station but throughout the water column (station specific); OTUs found at either in the South or North Aegean Sea but throughout the water column (Specific for South or North Aegean Sea); surface-specific OTUs found only at one station; deep water specific OTUs found only at one station; OTUs unique for a specific depth layer (reprint from Moesender et al. 2001 with permission from American Society of Limnology and Oceanography, Inc.).

1995; Mitchell and Barbara 1999), but we may assume that such an orientation in a nutrient field might occur also in nature. Motility of bacteria could be one mechanism allowing protists to distinguish between metabolically active and dormant bacteria.

Another mechanism allowing protists to distinguish between active and dormant bacterioplankton might be the presence or absence of signal molecules anchored to the surface of bacteria. It has been shown that active bacteria express a polysaccharidic capsule which has to be continuously renewed (Stoderegger and Herndl 1998). Once

bacteria become inactive, they rapidly loose this capsule (Heissenberger et al. 1996). In laboratory experiments, it has been shown that capsulated bacteria are grazed 2-4 times more efficiently than non-capsulated, inactive or dormant bacteria (Stoderegger and Herndl in press). The preferential feeding on metabolically active, capsulated bacteria over non-growing bacteria ensures ultimately the preservation of the richness of bacterial species commonly found in the oceanic water column (Giovannoni and Rappé 2000). Dormancy can only be a successful survival strategy if the grazing pressure is significantly lower on dormant bacteria than on the actively growing fraction of the bacterial community. If this were not so, the non-growing (dormant) bacterial community would be rapidly eliminated and consequently the species richness greatly reduced.

The surface characteristics and charge of bacteria influences the efficiency of the main predators, the protists, as well. It has been shown that with increasing hydrophobicity of the cell surface the uptake rates of specific nutrients increase and also the probability of being grazed (Monger et al. 1999). Thus there is a trade off for bacterial cells between nutrient uptake efficiency and grazing risk. The extent to which bacteria are capable of regulating the hydrophobicity of their cell surface needs to be investigated.

The role of viruses in sustaining microbial diversity

Viruses, together with protists and metazoans, are known to biologically control phytoplankton, protists and bacteria (Fuhrman 1999). Viral infection has also been linked to the extinction of species. Viruses are predominantly species-specific as far as we know. Thus, they might exert control on the evenness distribution of prey species as does every species-specific predation. It has been suggested that viral lysis can sustain the diversity of bacteria by 'killing the winner' and thus keeping in check competitive dominants (Thingstad 2000). This hypothesis is based on the idea that host densities need to exceed a threshold before viruses can exert a significant control. A lower threshold value of 10^4 cells ml^{-1} has been reported for successful replication of viruses infecting bacteria. As the contact rate between hosts and viruses and thus infection increases with their abundance, viral lysis may finally result in a collapse of the host population which, in turn, allows the survival of competitively weaker species.

Phytoplankton bacteria interaction

Bacterioplankton abundance is related to phytoplankton biomass and it is generally assumed that phytoplankton exudation is the major carbon source for bacteria in open ocean systems where allochthonous carbon sources are scarce. Heterotrophic bacterioplankton may compete with phytoplankton for inorganic nutrient sources (Kirchman and Wheeler 1998). Phytoplankton, in turn, receive inorganic nutrients via the remineralization of organic matter in the microbial food web. Thus, there is a direct dependency between phyto- and bacterioplankton. The question now is: how does this dependency influence diversity and function of the 3 functional groups involved, phyto-bacterio-plankton and protists. Naeem et al. (2000) found in lab experiments, that the productivity of the system depends on the diversity of both producers and decomposers (i.e. bacteria). The results, however, were surprisingly complex and do not support a simple relationship between diversity and productivity. Systems containing a single algal species and many bacterial species displayed a low combined productivity of algae and bacteria and most of the total biomass consisted of bacteria. Systems with moderate to high phytoplankton diversity tended to have a higher combined productivity with high algal biomass and comparatively low bacterial biomass, regardless of bacterial diversity. Bacteria in systems with high productivity could use a greater variety of carbon sources than those from systems with lower productivity. The ability of bacteria to use a variety of carbon sources did not depend solely on the number of bacterial species present, but also increased with increasing algal diversity.

Biodiversity may be seen as a form of biological insurance against the loss or poor performance of selected species (Naeem and Li 1997). Using multitrophic lab systems including several species of phytoplankton, decomposers (bacteria) and grazers (heterotrophic protists), it has been shown that the predictability of model systems increased with the number of species per functional group supporting the biological insurance hypothesis for biodiversity (Naeem and Li 1997).

What is needed to better understand microbial diversity and function?

From the previous chapters it is evident that the diversity and function relationship is a central problem in ecology which is not easily solved. What is generally needed are approaches that resemble more closely natural situations. A number of studies addressing diversity - function relationships have used microbial communities because microorganisms respond quickly to changing environmental conditions due to their short generation time. As mentioned above, microbial communities in the marine environment are the essential driving forces for biogeochemical cycles with established impacts on overall marine productivity and also on climate via the formation and consumption of greenhouse gases or volatile compounds (Kiene 1990) which are thought to influence climate such as dimethylsulfide. Thus, in order to understand biogeochemical fluxes and their variability over time and space it is essential to know how microbial systems react to disturbances and how stable they are in terms of function.

The two major types of mechanisms by which biodiversity influences function or productivity have been recognized: 1) functional niche complementarity and 2) selection of extreme trait values (Loreau 1998). In the complementarity effect, trait variation forms the basis for a permanent association of species that enhances collective performance. In the selection effect, trait variation comes into play as an initial condition, and a selective process then promotes dominance by species with particular trait values (Loreau 1998). Currently, we still do not know how many bacterial species drive biogeochemical cycles at a given ecological situation, i.e. the selection of trait values. We also have only extremely rudimentary knowledge on the complementarity effect. One might assume that bacterial consortia in which different species interact synergistically might be common in sediments and also in the water column on colloidal aggregations. Also unknown is the number of microbial species coexisting with the same function, i.e. the redundancy of species. Thus, our overall knowledge on the diversity and function of microbial systems is still rather poor.

A plea to study the microbial microenvironment

Other than in macrobial systems, the interactions between microbes are at a spatial scale not easily accessible by humans. The oceanic water column appears to us as a rather homogeneous environment. We measure nutrient concentrations using hundreds of ml volumes and determine in this way the nutrient concentrations in different water masses. Is this the relevant scale for microbial communities? As far as we know now, probably not. There is serendipical evidence that the nutrient field even in the oceanic water column should be heterogeneous on a microspatial scale. Decaying phytoplankton cells might be centers of enhanced nutrient concentrations present in a "nutritive dessert". Motile bacteria might be capable to detect such nutrient halos around decaying cells or dead particulate organic matter and actively seek proximity to these nutrient micropatches as shown elegantly by Blackburn et al. (1998). Different bacterial species might have different affinities to nutrients. This would ultimately lead to a structuring of the microbial community in a seemingly homogeneous water column. Our approaches to determine microbial species richness are barely suited to shed light onto such microspatial structures of the microbial community in the water column. Thus, the environmental heterogeneity of the water column is most likely larger than we commonly anticipate. As shown by Kassen et al. (2000), heterogeneity even in a small volume of water supports larger bacterial diversity than a homogenous environment.

It follows that we should take into account the microspatial heterogeneity of the water column. Dissolved organic matter might spontaneously coagulate to form nanoparticles potentially aggregating further to form larger particles (Chin et al. 1998), recently described as transparent exopolymer particles. Colloidal particles occur at an abundance of up to 10^{10} ml^{-1} and transparent exopolymer particles are present in the euphotic zone at an abundance of around 10^3 ml^{-1} (Alldredge et al. 1993). Thus, there is sufficient potentially available microstructure influencing the nutrient field for microbial communities and therefore, ultimately supporting biodiversity of microbes if we extrapolate from Kassen's et al. (2000) simple experiments.

Even if we accept that the microspatial habitats of the microbial world in the oceanic water column are largely unknown and that these micro-habitats support microbial biodiversity, we still can question whether it really matters if we do not know all these microspatial details. Would it not be suf-ficient to assess the richness of the microbial com-munity in a given water mass and characterize this water mass with our traditional oceanographic methods and determine some overall activity pa-rameters to assess its function? We argue that in-formation on the abundance and ecological niches of the key microbial organisms is as essential as information on the microstructure of microbial communities since such microstructures determine ultimately the complexity of microbial systems, probably even in the water column. Knowledge of the microstructure of microbial habitats is essen-tial for a mechanistic understanding of the role of microbes in biogeochemical cycling in the sea.

The challenge for the future

The major challenge for future research on bio-diversity and function in microbial communities is to develop approaches that allow identification of key species and the accurate determination of their abundance and function. Furthermore, it is essen-tial to provide information on the number of spe-cies present capable of fulfilling the same function. Thus, we have to advance from a simple assessment of the species richness of the microbial community to truly linking species richness with other parameters determining diversity such as evenness and addressing functional aspects. All this has to be done by taking into account the specific micro-habitats. Only if we are able to define the specific microhabitats for a given microbial species will we be able to determine and understand life strategies of key species and consortia responsible for the biogeochemcial cycling of the major elements. Microbial communities have been used in the past to address general ecological questions on the biodiversity – function problem but observations are needed using experimental approaches more closely resembling natural situations. Microsensors to determine the microhabitat structure of specific microbial species and consortia have to be devel-oped and combined with methods allowing the de-termination of the species distribution (richness

and evenness) and metabolic activity without dis-rupting the potential microspatial arrangement of the community. For macrobial ecology this is com-mon practice but for microbial ecology this is still a major challenge. With innovative technology this ultimate goal is now within reach.

Acknowledgments

We want to thank Victor Smetacek and John Patch-ing for their helpful comments on a former version of the manuscript and the members of the Biodiversity working group at the Marine Science Frontier workshop for stimulating discussions. This is publication # 3684 of the NIOZ.

References

Ackermann H-W (1999) Tailed bacteriophages: The or-der *Caudovirales*. Adv Virus Res 51:135-201

Alldredge AL, Passow U, Logan, BE (1993) The exist-ence, abundance, and significance of large transpar-ent exopolymer particles in the ocean. Deep-Sea Res I 40:1131-1140

Amann RI, Ludwig W, Schleifer K-H (1995) Phylo-ge-netic identification and *in situ* detection of indi-vidual microbial cells without cultivation. Micro Rev 59:143-169

Azam F, Cho BC (1987) Bacterial utilization of organic matter in the sea. In: Fletcher M (ed) Ecology of Microbial Communities. Cambridge University Press, Cambridge, pp 261-281

Blackburn N, Fenchel T, Mitchell J (1998) Microscale nutrient patches in planktonic habitats shown by chemotactic bacteria. Science 282:2254-2256

Boetius A, Ravenschlag K, Schubert CJ, Rickert D, Widdel F, Gieseke A, Amann R, Jørgensen BB, Witte U, Pfannkuche O (2000) A marine microbial consortium apparently mediating anaerobic oxidation of methane. Nature 407:623-626

Boschker HTS, Brouwer JFCd, Cappenberg TE (1999) The contribution of macrophyte-derived organic matter to microbial biomass in salt-marsh sediments: Stable carbon isotope analysis of micro-bial biomark-ers. Limnol Oceanogr 44:309-319

Cailliau C, Claustre H, Vidussi F, Marie D, Vaulot D (1996) Carbon biomass, and gross growth rates as estimated from ^{14}C pigment labelling, during photo-acclimation in *Prochlorococcus* CCMP 1378. Mar Ecol Prog Ser 145:209-221

Chen F, Suttle CA (1995) Amplification of DNA polymerase gene fragments from viruses infecting microal-gae. Appl Environ Microbiol 61:1274-1278

Chin WC, Orellana MV, Verdugo P (1998) Spontaneous assembly of marine dissolved organic matter into polymer gels. Nature 395:568-572

Cottrell M, Kirchman D (2000a) Natural assemblages of marine proteobacteria and members of the *Cytophaga-Flavobacter* cluster consuming low- and high-molecular-weight dissolved organic matter. Appl Environ Microbiol 66:1692-1697

Cottrell MT, Kirchman DL (2000b) Community composition of marine bacterioplankton determined by 16S rRNA gene clone libraries and fluorescence *In situ* hybridization. Appl Environ Microbiol 66:5116-5122

Cottrell MT, Suttle CA (1991) Wide-spread occurrence and clonal variation in viruses which cause lysis of a cosmopolitan, eukaryotic marine phytoplankter, *Micromonas pusilla*. Mar Ecol Prog Ser 78:1-9

Dahllof I, Baillie H, Kjelleberg S (2000) rpoB-based microbial community analysis avoids limitations inherent in 16S rRNA gene intraspecies heterogeneity. Appl Environ Microbiol 66:3376-3380

Del Giorgio PA, Cole JJ (1998) Bacterial growth yield efficiency in natural aquatic systems. Annu Rev Ecol Syst 29:503-541

Del Giorgio PA, Gasol JM, Vaqué D, Mura P, Agusti S, Duarte CM (1996) Bacterioplankton community structure: Protists control net production and the proportion of active bacteria in a coastal marine community. Limnol Oceanogr 41:1169-1179

Del Giorgio PA, Cole JJ, Cimberis A (1997) Respiration rates of bacteria exceed phytoplankton in unproductive aquatic systems. Nature 385:148-151

Ducklow H (2000) Bacterial production and biomass in the oceans. In: Kirchman DL (ed) Microbial Ecology of the Oceans. Wiley-Liss, New York pp 85-120

Falkowski PG (1994) The role of phytoplankton photosynthesis in global biogeochemical cycles. Photosynthesis Res 39:235-258

Fernandez A, Huang S, Seston S, Xing J, Hickey R, Criddle C, Tiedje J (1999) How stable is stable: Function versus community composition. Appl Environ Microbiol 65:3697-3704

Flöder S, Sommer U (1999) Diversity in planktonic communities: An experimental test of the intermediate disturbance hypothesis. Limnol Oceanogr 44:1114-1119

Fuhrman JA (1996) Community structure: bacteria and archaea. In: Hurst CJ, Knudson GR, McInerey MJ, Stezenbach LD, Walter MV (eds) Manual of Environmental Microbiology. ASM Press, Washington pp 278-283

Fuhrman JA (1999) Marine viruses and their biogeo-

chemical and ecological effects. Nature 399:541-548

Fuhrman JA, Azam F (1982) Thymidine incorporation as a measure of heterotrophic bacterioplankton production in marine surface waters: Evaluation and field results. Mar Biol 66:109-120

Fuhrman JA, Campbell L (1998) Microbial microdiversity. Nature 393:410-411

Giovannoni SJ, Britschgi TB, Moyer CL, Field KG (1990) Genetic diversity in Sargasso Sea bacterioplankton. Nature 345:60-63

Giovannoni S, Rappé M (2000) Evolution, diversity, and molecular ecology of marine prokaryotes. In: Kirchman DL (ed) Microbial Ecology of the Oceans. Wiley-Liss, New York pp 47-84

Huisman J, Weissing F (1999) Biodiversity of plankton by species oscillations and chaos. Nature 402:407-410

Hedges JI (1992) Global biogeochemical cycles: Progress and problems. Mar Chem 39:67-93

Heissenberger A, Leppard GG, Herndl GJ (1996) Relationship between the intracellular integrity and the morphology of the capsular envelope in attached and free-living marine bacteria. Appl Environ Microbiol 62:4521-4528

Hoppe H-G (1991) Microbial extracellular enzyme activity: A new key parameter in aquatic ecology. In: Chrost RJ (ed) Microbial Enzymes in Aquatic Environments. Springer Verlag, New York pp 60-83

Karner MB, DeLong EF, Karl DM (2001) Archaeal dominance in the mesopelagic zone of the Pacific Ocean. Nature 409:507-510

Kassen R, Buckling A, Bell G, Rainey P (2000) Diversity peaks at intermediate productivity in a laboratory microcosm. Nature 508-512

Kiene RP (1990) Dimethyl sulfide production from dimethyl-sulfonioproprionate in coastal seawater samples and bacterial cultures. Appl Environ Microbiol 56:3292-3297

Kirchman D, K'Ness E, Hodson R (1985) Leucine incorporation and its potential as a measure of protein synthesis by bacteria in natural aquatic systems. Appl Environ Microbiol 49:599-607

Kirchman DL, Wheeler PA (1998) Uptake of ammonium and nitrate by heterotrophic bacteria and phyto-plankton in the sub-Arctic Pacific. Deep-Sea Res I 45:347-365

Lopez-Garcia P, Rodriguez-Valera F, Pedros-Alio C, Moreira D (2001) Unexpected diversity of small eukaryotes in deep-sea Antarctic plankton. Nature 409:603-607

Loreau M (1998) Biodiversity and ecosystem functioning: a mechanistic model. Proc Natl Acad Sci USA

95:5632-5636

Loreau M, Behera N (1999) Phenotypic diversity and stability of ecosystem processes. Theor Popul Biol 56:29-47

Mitchell JG, Barbara GM (1999) High-speed marine bacteria use sodium-ion and proton driven motors. Aquat Microb Ecol 18:227-233

Mitchell JG, Pearson L, Bonazinga A S, Dillon Khouri H, Paxinos R (1995) Long lag phase and high velocities in the motility of natural assemblages of marine bacteria. Appl Environ Microbiol 61:877-882

Moeseneder MM, Winter C, Herndl GJ (2001) Horizontal and vertical complexity of attached and free-living bacteria of the eastern Mediterranean Sea determined by 16S rDNA and 16S rRNA fingerprints. Limnol Oceanogr 46:95-107

Monger BC, Landry MR, Brown SL (1999) Feeding selection of heterotrophic marine nanoflagellates based on the surface hydrophobicity of their picoplankton prey. Limnol Oceanogr 44:1917-1927

Moon-van der Staay SY, De Wachter R, Vaulot D (2001) Oceanic 18S rDNA sequences from picoplankton reveal unsuspected eukaryotic diversity. Nature 409:607-610

Naeem S, Li S (1997) Biodiversity enhances ecosystem reliability. Nature 390:507-509

Naeem S, Hahn DR, Schuurman G (2000) Producer-decomposer co-dependency influences biodiversity effects. Nature 403:762

Ouverney CC, Fuhrman JA (1999) Combined microauto-radiography-16S rRNA probe technique for determination of radioisotope uptake by specific microbial cell types in situ. Appl Environ Microbiol 65:1746-1752

Paul, JH (1999) Microbial Gene Transfer. J Mol Microbiol Biotechnol 1:45-50

Pedros-Alio C (1993) Diversity of bacterioplankton. TREE 8:86-90

Pennisi E (1998) Genome data shake tree of life. Science 280:672-674

Pinhassi J, Hagström Å (2000) Seasonal successions. Aquat Microb Ecol 21:245-256

Radajewski S, Ineson P, Parekh NR (2000) Stable-isotope probing as a tool in microbial ecology. Nature 403:646-649

Ritchie ME, Olff H (1999) Spatial scaling laws yield a synthetic theory of biodiversity. Nature 400:557-560

Rossello-Mora R, Amann R (2001) The species concept for prokaryotes. FEMS Microbiol Rev 25:39-67

Sankaran M, McNaughton S (1999) Determinants of biodiversity regulate compositional stability of communities. Nature 401:691-693

Sherr BF, Sherr EB, Pedros-Alio C (1989) Simultaneous measurement of bacterioplankton production and protozoan bacterivory in estuarine water. Mar Ecol Prog Ser 54:209-219

Simon M, Azam F (1989) Protein content and protein synthesis rates of planktonic marine bacteria. Mar Ecol Prog Ser 51:201-213

Stahl DA (1996) Molecular approaches for the measurement of density, diversity and phylogeny. In: Hurst CJ, Knudson GR, McInerey MJ, Stezenbach LD, Walter MV (eds) Manual of Environmental Microbiology. ASM Press, Washington pp 102-114

Stoderegger K, Herndl GJ (1998) Production and release of bacterial capsular material and its subsequent utilization by marine bacterioplankton. Limnol Oceanogr 43:877-884

Stoderegger KE, Herndl GJ (in press) Distribution of capsulated bacterioplankton in the North Atlantic and North Sea. Microbiol Ecol

Suzuki MT, Taylor LT, DeLong EF (2000) Quantitative analysis of small-subunit rRNA genes in mixed microbial populations via 5'-nuclease assays. Appl Environ Microbiol 66:4605-4614

Thingstad TF (2000) Elements of a theory for the mechanisms controlling abundance, diversity, and biogeo-chemical role of lytic viruses in aquatic systems. Limnol Oceanogr 45:1320-1328

Tilman D (1996) Biodiversity: Population versus ecosystem stability. Ecol 77:350-363

Weinbauer MG, Höfle MG (1998) Distribution and life strategies of two bacterial populations in a eutrophic lake. Appl Environ Microbiol 64:3776-3783

Whitman WB, Coleman DC, Wiebe WJ (1998) Prokaryotes: the unseen majority. Proc Natl Acad Sci USA 95:6578-6583

Wilhelm SW, Suttle CA (1999) Viruses and nutrient cycles in the Sea. Biosci 49:781-788

Wommack K, Colwell R (2000) Virioplankton: Viruses in aquatic ecosystems. Microbiol Mol Biol Rev 64:69-114

Wommack KE, Ravel J, Holl RT, Chun J, Colwell RR (1999) Population dynamics of Chesapeake Bay virioplankton: Total-community analysis by pulse-field gel electrophoresis. Appl Environ Microbiol 65:231-240

Woodward FI (1994) How many species are required for a functional ecosystem? In: Schulze E-D, Mooney HA (eds) Biodiversity and Ecosystem Function. Springer, Berlin pp 271-291

Yachi S, Loreau M (1999) Biodiversity and ecosystem productivity in a fluctuating environment: The insurance hypothesis. Proc Natl Acad Sci USA 96:1463-1468

Ecosystem Function, Biodiversity and Vertical Flux Regulation in the Twilight Zone

P. Wassmann[1*], K. Olli[2], C. Wexels Riser[1] and C. Svensen[1]

[1] *Norwegian College of Fishery Science, University of Tromsø, N-9037 Tromsø, Norway*
[2] *Institute of Botany and Ecology, University of Tartu, Lai 40, 51005 Tartu, Estonia*
* corresponding author (e-mail): paulw@nfh.uit.no

Abstract: The current lack of adequate investigations of the vertical export above the depth of 200-500 m where the majority of long-term sediment traps have been deployed, results in difficulties to understand and model the carbon flux. There exists a black box of several hundred metres between the surface layers where measurements and algorithms of primary production exists and where data on the carbon export to the ocean interior are available. In this black box, the twilight zone, we face a lack of basic understanding on how vertical export of biogenic matter into the oceans interior is regulated. Essential for this regulation are planktonic key organisms and the structure and dynamics of the pelagic food web. To better comprehend the pelagic carbon cycle and sequestration of CO_2, it is instrumental to obtain a basic understanding how the biota determines and transforms the export production in the twilight zone. Here we discuss some of the key organisms involved in vertical flux regulation, present an idealised, conceptual model of vertical carbon export and focus upon the "pelagic mill" and vertical flux regulation in the upper 200 m. An adequate understanding of carbon cycling demands not only adequate investigations of primary production, but also concomitant research on the functional biodiversity of the pelagic zone, plankton dynamics, vertical flux and its regulation in the twilight zone.

Introduction

The export of biogenic matter from surface layers to the oceans interior has been a matter of global interest for almost two decades, as reflected by international programmes such as JGOFS. Prior to this the earliest vertical flux data derived primarily from the coastal zone (e.g. Hargrave and Taguchi 1978; Smetacek 1980; Burrel 1988; Wassmann 1991) and were used to formulate algorithms to predict the vertical export as a function of primary production in the surface layer and depth (e.g. Suess 1980; Betzer et al. 1984; Pace et al. 1987; Berger et al. 1989; Banse 1994). Here vertical export was extrapolated to the upper layers although data often were missing or difficult to interprete. During subsequent attempts to estimate the vertical export of carbon into the oceans interior large, automatic sediment traps have been generally deployed at depths below 200 m. This is in part caused by methodological difficulties to de-

ploy large-sized, poisoned traps in the upper layers (e.g. Gardner 2000). As a consequence, few high-quality data are available from the upper 200-500 m, the twilight zone. Recent investigations, however, emphasise that only a minor fraction of the export production leaving the euphotic zone is injected into the layers deeper than 200 m (e.g. Noji et al. 1999). In the Barents Sea, north Norwegian fjords and off the shelf of NW Spain the decrease of vertical POC export was largest in the upper 100 m (Andreassen and Wassmann 1998; Reigstad et al. 2000 and Olli et al. 2001, respectively). The quantitatively important processes of vertical flux regulation occur probably in the uppermost layers, often over short vertical distances.

Together with the lack of adequate algorithms for export production (e.g. Wassmann 1990, 1998), the inadequacy of our current understanding continues when particles penetrate through the twilight zone (e.g. Gardner 2000) and deeper into the ocean on their way to the sediment. It has been suggested that the dynamics of retention and export food chains determine the quality and quantity of the vertical flux of biogenic matter (Wassmann 1998). Accordingly, there is an apparent need to investigate more profoundly the "pelagic mill", i. e. the upper 500 m of the ocean with its food chains, mineralization rates and biogenic fluxes. Inside the substantial diversity of organisms constituting the "pelagic mill" only a few play a key role.

Artefacts induced by moored, voluminous and poisoned long-term sediment traps prevent satisfactory information from the twilight zone. However, short-term, non-poisoned and drifting sediment traps provide an adequate approach to investigate vertical flux in the twilight zone on a diurnal basis (e.g. Andreassen et al. 1999; Reigstad et al. 2000; Olli et al. 2001). In an attempt to initiate a more thorough analysis of vertical carbon flux attenuation in the twilight zone we suggest more profound investigations of key stone species and an idealised, conceptual model of vertical flux attenuation. We discuss also how the gap in knowledge in the twilight zone can be overcome in order to generate more realistic carbon models. We describe the attenuation of POC sedimentation below the productive or mixed layer as a function of depth, discuss some of the processes involved and indicate the composition of vertical POC export. It is argued that the general form of the vertical flux curvature is strongly influenced by pelagic heterotrophs and vertical mixing. The focus here is not a balanced, comprehensive study of vertical carbon attenuation and its regulation by key species. Rather we present glimpses from the subject matter in an effort to assemble some first pieces of evidence in an attempt to describe mathematically the pathway from export production (i.e. carbon flux out of the euphotic zone) to vertical flux throughout the twilight zone.

Key organisms and vertical flux regulation

Vertical flux of biogenic matter is tightly coupled to the functioning of the ecosystem. For instance, it is possible to characterise pelagic systems as retention- and export-systems, depending on the as-sociated flux of carbon. Both nutrient loading and grazers may regulate the pattern of algal sedimentation. It is also possible to distinguish organisms that play important roles in ecosystem functioning and hence vertical flux regulation.

Diatoms constitute an important phytoplankton group and are generally associated with the export system. Although a diatom cell with diameter 10 mm will sink < 1 m day^{-1} (according to Stoke's law), sinking rates of > 100 m day^{-1} have been reported. The reason for the enhanced sinking rates is that diatoms may collide and build up large (> 0.5 mm, MacIntyre et al. 1995) aggregates. Aggregation is dependent on the collision rate (through differential settling and shear), and the rate at which they stick together after collision (Jackson 1990). The stickiness of the cell is species dependent (Kiørboe et al. 1990), as well as variable in time (Dam and Drapeau 1995)

A contrasting phytoplankton community would consist of flagellates (pico- and nanoplankton < 20 µm) as this group is often associated with retention- rather than export-systems (Heiskanen 1998). However, flagellates may contribute to the vertical flux by settling associated with aggregates (Reigstad 2000; Olli et al. 2001). *Phaeocystis pouchetii* is a key species in boreal and sub-arctic pelagic communities (Verity and Smetacek 1996), but their role in vertical flux is still a matter of debate (Reigstad 2000).

Microzooplankton (20-200 µm protists) may exert a strong grazing impact on phytoplankton because of their short generation time. The faecal pellets of microzooplankton are small with low sinking velocities, and therefore the microzooplankton will retard rather than accelerate vertical flux.

The influence of zooplankton on the vertical flux of biogenic matter may be both positive and negative, depending on abundance and composition. An important key species in pelagic food chain of Northeast Atlantic waters and in the Norwegian Coastal zone is *Calanus finmarchicus* (Tande 1991). When appearing in large numbers, they may graze a large part of the phytoplankton. This process involves not only a reduction of phytoplankton available for sinking, but also packaging of small particles with low sinking velocities into larger, rapidly sinking faecal pellets.

Faecal pellets may sink out of the euphotic zone, or they may be grazed by other mesozooplankton and thus be remineralised within the twilight zone. It has been demonstrated that faecal pellets and detritus may constitute a great part of copepod diet (Dagg 1993; González and Smetacek 1994).

However, the retention capacity of faecal pellets seems to be highly dependent on the composition of the zooplankton community (Viitasalo et al. 1999; Wassmann et al. 1999; Wassmann et al. 2000) and hence variation in feeding behaviour. The cyclopoid copepod *Oithona sp.* is a key species in this respect. Its importance in vertical flux regulation is coupled to its raptorial feeding behaviour, and thus its ability to detect sinking and swimming prey particles (Svensen and Kiørboe 2000). Increasing evidence from both field and laboratory studies demonstrating negative relationship between abundance of *Oithona sp.* and faecal pellet flux further confirm their importance (e.g. González et al. 1994; González and Smetacek 1994; Wassmann et al. 1999; González et al. 2000). In many coastal areas only a small fraction of the copepod faecal pellets sink out of the upper layers. Wexels Riser et al. (2001) describes this phenomenon as the "copro-phagous filter".

Krill, salps and appendicularians are also key species for vertical flux of faecal pellets (Peinert 1986; Reigstad 2000). These organisms may appear in large swarms, and their pellets are probably too large to be grazed upon by flux feeding copepods (Reigstad 2000). They short-circuit the coprophagous filter and give rise to an instantaneous pelagic-benthic coupling.

Particulate vertical carbon export under idealised conditions

To envisage an idealised downward component of POC in the water column we assume an initial increase in the upper surface layer where primary production by algae exceeds the community respiration rate. The initial increase of the vertical flux slows down as a result of decreasing primary production when light becomes limiting with increasing depth, and increase of microbial degradation of the settling material. We can assume that the vertical flux peaks at the compensation point, where the primary production is balanced by community respiration rate. Below the compensation point the vertical flux starts to decrease and below the euphotic zone when no primary production is possible, the decrease is generally agreed to follow a power function (Fig. 1).

The decrease, though, can be quite variable, depending among other factors on bacterial degradation, the zooplankton grazing, coprophagy, aggregate formation and size/sinking velocity of sinking particles. In previous inquiries, the usual procedure to obtain the parameters of the power curve is to calculate a linear least-square regression between log-transformed depth and vertical flux estimates (e.g. Banse 1994). As the decrease of the vertical flux with depth depends on the structure of the planktonic community, so should the model parameters. Thus it should be possible to compare the vertical flux scenarios at different locations and during different seasons by looking at these model parameters.

In the conventional settings the depth scale (x axis) is measured from the surface and the y asymtote of the power curve is fixed at 0 m. This, however, produces a conceptual error because it assumes that the logarithmic decrease of vertical flux starts immediately below the surface layer and contradicts to the above idealised model where the decrease is logarithmic only below the euphotic zone. This error is probably not excessive when the sediment traps are at mid-water and greater depths. However, the model becomes a poor representation of reality when the traps are in the upper 200 m including the lower part of the euphotic layer. To account for this discrepancy we have proposed earlier (Olli et al. 2001) to adjust the y asymtote to a depth which minimises the error function of the curve fit (Fig. 1).

$$y = a \times (x - z)^b \qquad (1)$$

where z is the the specific depth (m) from the surface to the y-asymtote. In close to idealised conditions the asymtote will locate somewhere in the lower part to the euphotic zone (Fig. 1), but this can vary considerably depending on the actual estimated profile of the vertical flux and the actual depths of the sediment traps.

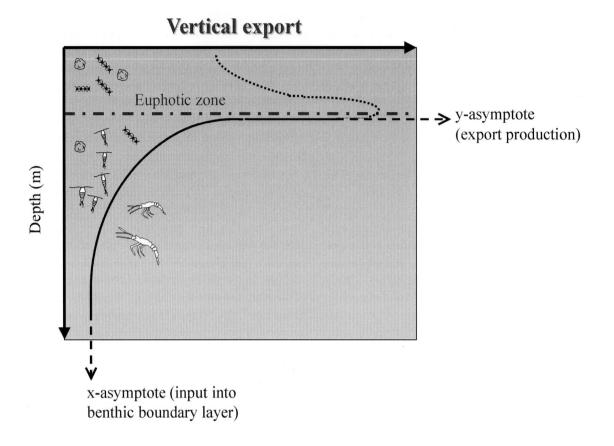

Fig. 1. Vertical carbon export under ideal conditions. See text for explanation.

The shape of the fitted power curve is determined by the parameters a and b (Equ. 1). The parameter a and b can be interpreted as the modelled vertical flux at 1 m below the asymtote and the relative attenuation of the vertical flux, respectively, at 1 m below the y-asymptote. The absolut attenuation of the vertical flux at any given depth is determined by the slope of the fitted power curve. Close to the y-asymptote the slope is steep which corresponds to high retention. The slope decreases at deeper depths corresponding to low retention.

Vertical export of phytoplankton, faecal pellets and detritus in the twilight zone: An example

The vertical export of biogenic matter in the marginal ice zone (MIZ) of the Barents Sea during spring was measured with small, cylindrical, gimballed sediment traps, equipped with a vane. They were deployed for 24 hours in the upper 200 m in a Lagrangian manner. No poisons were applied. The results revealed an export production of more than 1300 mg C m^{-2} d^{-1} and a strong attenuation in the 30-50 m depth interval (about 72 % decrease). Changes below 50 m were less prominent, but vertical export declines steadily. Phytoplankton carbon (PPC) represented a significant fraction and also faecal pellets carbon (FPC) contributed significantly from 40 m and downward. In contrast to the MIZ, the vertical export of POC in the open water of the central Barents Sea decreased less and the profile as well as the composition of the exported matter was more complex (Fig. 2). In the upper 60 m POC export was similar and increased slightly. An exponential decline

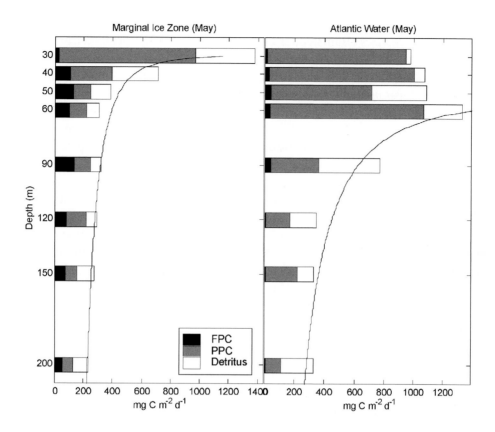

Fig. 2. Vertical flux of particulate organic carbon (POC; entire bar), phytoplankton carbon (PPC), faecal pellet carbon (FPC) and detritus (calculated by difference) in the upper 200 m of the marginal ice zone (left) and the open water (dominated by Atlantic Water, right) in the central Barents Sea in May 1998 (mg C m^{-2} d^{-1}). Unpublished results. Also shown is the exponential attenuation of POC sedimentation below the productive or mixed layer as a function of depth.

of the flux started first at 60 m depth. Between 60 and 120 m the relative decline of vertical POC flux was similar to that encountered in the marginal ice zone between 30 and 50 m (about 74 % decrease), but the depth interval was 3 times larger. The upper water column in the open Barents Sea was mixed and we suggest that the similar vertical export reflect vertical mixing. In the 30 – 60 m depth interval PPC clearly was the dominating fraction. From 50 to 200 m detritus also contributed to the vertical export while the portion of FPC was small.

These two Barents Sea stations characterise the vertical flux during a spring bloom in a productive environment. However, the Atlantic and MIZ stations differ considerably with respect to the

composition of suspended phytoplankton composition, composition of the settled material and the vertical mixing regime. In the MIZ the plankton in the strongly stratified upper 50 m layer was dominated by comparable shares of *Phaeocystis pouchetii* and diatoms while diatoms dominated the vertical flux in the upper traps with negligible contribution from *P. pouchetii*. At the Atlantic water station *P. pouchetii* dominated both the suspended biomass and the vertical flux. In both stations the vertical flux at the shallow depths was very high, > 1300 mg C m^{-2} d^{-1}, and decreased to 200 – 300 mg C m^{-2} d^{-1} at 200 m depth. However, the retention maximum at the MIZ station was just below the pycnocline (50 m) and largely explained by microzooplankton (Hansen et al. 1995) and meso-zooplankton grazing dominated by *Calanus fin-marchicus* and *C. glacialis* (Båmstedt et al. 1991). On the other hand, the Atlantic station was not stratified, the maximum retention occurred at depths 60 – 120 m and was probably more attributed to microbial decomposition than mesozooplankton grazing.

Although the attenuation of the vertical export of carbon can be calculated, that of PPC, FPC and detritus can not be estimated as easily. The reason for this is obviously the variability of different processes and diversity of involved key organisms. The vertical export of PPC depends mainly on the accumulation of phytoplankton and large cells in the upper layers, on their aggregation potential and on the grazing rates of zooplankton. The FPC fraction is determined by (a) the grazing rate of larger zooplankters, (b) the species composition determining the size and the sinking rate of the produced faecal pellets and (c) their ingestion (coprophagy) and destruction (coprorhexy) of faecal pellets. Thus the structure of the planktonic food web and food web function regulates the composition of the vertically exported matter. To understand and model the vertical export composition demands detailed information about the plankton food web and preferences and behaviour of key species. This is obviously demanding and at present this may only be possible in a few selected regions.

Conceptual model of vertical carbon export in the twilight zone

We suggest here a simple conceptual model of the vertical export of carbon throughout the twilight zone (Fig. 3). We distinguish 5 different horizontal layers: the euphotic zone, the upper aphotic zone, the mixed layer, the intermediate water and the deep water. The twilight zone comprises the layers between the aphotic zone to the bottom of the intermediate layer. Primary production in the euphotic zone gives rise to new production and suspended biomass. Over lengthy periods of time and assuming steady state, integrated new production represents the upper limit of export production, i.e. the carbon that can be exported vertically from the euphotic zone. Export production must be distin-

guished from the carbon flux at greater depths and is regulated by an assemblage of processes and factors (e.g. POC concentration in the euphotic zone, phyto- and zooplankton diversity, availability of mucilaginous matter, coagulation and aggregate formation, grazing and faecal pellet production). In the upper aphotic zone and under stratified conditions vertical carbon export is subjected to immediate and strong attenuation, caused by an entire range of heterotrophs that seek for elevated supply of high-quality food from the euphotic zone. After this zone of intensive attenuation vertical carbon exports declines more slowly with increasing depth. In the case that the mixed layer and the euphotic zone fall together, a in Equ. (1) approximates the vertical injection of carbon into the aphotic zone (i.e. the export production) and the y-asymptote approximates the vertical injection into the benthic boundary layer. b is the attenuation coefficient, i.e. a measure for the carbon retention efficiency in the water column. However, a and b are both model parameters is the case of an idealised model. In reality the variability of the data forces the y-asymptote to variable places which strongly influences the a and b value, as shown in figure 2.

Convective overturn increases the residence time of particles in the mixed layer (Kerr and Kuiper 1997). As a consequence, vertical export cannot be measured accurately with traps in the mixed layer or during mixed layer pumping caused by wind events (Gardner 2000). Vertical mixing is difficult to measure, but crucial for particle production and distribution (Huisman et al. 1999). Vertical mixing can reach deep down into the twilight zone, in particular at high latitudes. The variable depth of vertical mixing results in a dilution of suspended matter in the euphotic zone and a decline in the vertical carbon export attenuation in the upper layers (Fig. 3). As a consequence, vertical carbon flux decreases in the upper layer, but penetrates deeper into the twilight zone with increased rates. When carbon reaches the deep waters the differences are small. Equation (1) cannot be applied in the mixed layer, but first below this layer. In this case a in Equ. (1) is the vertical injection of carbon form the mixed layer while b is a measure for the carbon retention efficiency below this layer. The estimation of z is thus critical.

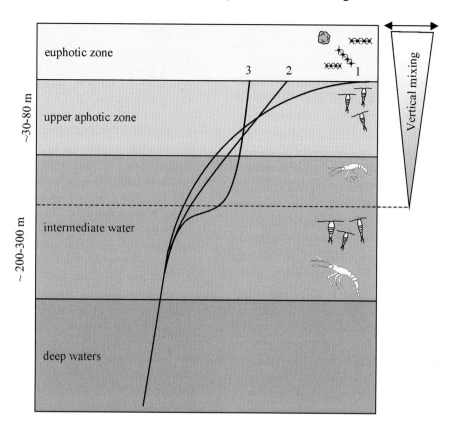

Fig. 3. Conceptual model of the vertical carbon export attenuation in the upper aphotic zone, intermediate and deep waters. Principle examples for vertical carbon export profiles during variable vertical mixing scenarios are shown. 1: stratified upper layer; 2 and 3: increasing vertical mixing. Increased sinking rates of bulk particles and increased grazing pressure in the upper layers decrease and increase the vertical carbon export attenuation, respectively. See text for details.

Where do we go from here?

Integrated research of plankton ecology and vertical flux has been rarely carried out in other than coastal and shelf regions. The emphasis on long-term measurements of vertical flux applying automated traps has not improved our understanding of the dependence of export production on total and new primary production or the regulation of vertical flux in the region where the greatest attenuation takes place, the upper twilight zone. We know what orders of magnitude we may expect and the basic seasonal patterns in deeper layers and the deep-sea. However, our knowledge regarding the functional biodiversity of the pelagic realm how it regulates the export and retention of biogenic mat-

ter is vague and immature. This implies that the conspicuous amounts of vertical flux data from all over the oceans must be interpreted on the basis of a rather vague and insufficient theory of vertical flux regulation. Modern oceanography cannot accept a situation where remote sensing-supported algorithms predict primary and new production, followed by a black box in the upper 500 m where approximately 90 % of the export production is retained, until data and understanding again become available.

The conceptual model and its interpretation may serve as a nucleus for further and more dedicated investigations. Vertical profiles of biogenic matter export and composition have to be investigated more closely and compared with each other. In particular the relationship between new production, suspended biomass and vertical carbon export needs attention, followed by investigations of the seasonal variation of the attenuation coefficient, retention efficiency below the euphotic or the mixed layer and the role of key organisms shaping the export scenarios. Lines of thought that explain biogeochemical cycling by bottom-up regulation have a dominant tradition in current oceanography. But the interpretation of vertical biogenic matter export and pelagic-benthic coupling is impossible without a balanced application of bottom-up and top-down regulation (Heiskanen 1998; Wassmann 1998). A quantification of top-down regulation is, however, only possible if the pelagic biodiversity and in particular the functional biodiversity of key organisms is adequately known. The analysis of functional biodiversity and key organisms regulating vertical export and pelagic-benthic coupling is a significant challenge, but the only mode to "shed light" into the obscurity of the twilight zone. It is thus not biodiversity as such, but the insufficient understanding of biogeochemical cycles in the pelagic realm that advocates for intensified research on planktonic key species and pelagic ecology.

For the over-all understanding of vertical flux regulation and the realism of carbon models the approach presented here has to be applied in ecologically disparate regions and throughout the productive period. A further step must be to include the algorithms into current carbon models addressing the oceans response to climate change, eutrophication and resource exploitation. Future investigations aiming at better carbon models have to focus on (a) concerted actions of system-ecological plankton investigations combined with short-term studies of vertical flux in the twilight zone and (b) sensitivity analyses of pelagic-benthic coupling applying mathematical models.

Acknowledgements

The research programme ALV (Norwegian Research Council), and OMEX (MAS3-CT96-0056) and CYCLOPS (EVK3-1999-00037) (European Union) supported this work.

References

Andreassen IJ, Wassmann P (1998) Vertical flux of phytoplankton and particulate biogenic matter in the marginal ice zone of the Barents Sea in May 1993. Mar Ecol Prog Ser 140:1-14

Andreassen I, Wassmann P, Ratkova T (1999) Seasonal variation of vertical flux of phytoplankton and biomass on the north Norwegian shelf break. Sarsia 84:227-238

Båmstedt U, Eilertsen HC, Tande K, Slagstad D, Skjoldal HR (1991) Copepod grazing and its potential impact on the phytoplankton development in the Barents Sea. In: Sakshaug E, Hopkins CCE, Øritsland NA (eds) Proceedings of the Pro Mare Symposium on Polar Marine Ecology, Trondheim, 12-16 May 1990. Polar Res 10:339-353

Banse K (1994) On the coupling of hydrography, phytoplankton, zooplankton, and settling organic particles offshore in the Arabian Sea. Proc Indian Acad Sci (Earth Planet Sci) 103:125-161

Berger WH, Smetacek VS, Wefer G (1989) Ocean productivity and paleoproductivity - An overview. In: Berger WH, Smetacek VS, Wefer G (eds) Productivity of the Ocean: Present and Past. J Wiley and Sons, Chichester pp 1-34

Betzer PR, Showers WJ, Laws EA, Winn CD, DiTullio GR, Kroonpnick PM (1984) Primary productivity and particle fluxes on a transect of the equator at 153°W in the Pacific Ocean. Deep-Sea Res 31:1-11

Burrell DC (1988) Carbon flow in fjords. Oceanogr Mar Biol Rev 26:143-226

Dagg M (1993) Sinking particles as a possible source of nutrition for the large calanoid copepod Neocalanus cristatus in the subarctic Pacific Ocean. Deep-Sea Res 40:1431-1445

Dam HG, Drapeau DT (1995) Coagulation efficiency, organic-matter glues and the dynamics of particles during a phytoplankton bloom in a mesocosm study. Deep-Sea Res II 42:111-123

Gardner WD (2000) Sediment trap sampling in surface water. In: Hanson RB, Ducklow HW, Field JG (eds) The Changing Ocean Carbon Cycle – Midterm Synthesis of the Joint Ocean Global Flux Study. Cambridge University Press pp 240-284

Gonzáles HE, Kurbjeweit F, Bathmann UV (1994) Occurrence of cyclopoid copepods and faecal material in the Halley Bay region, Antarctica, during January-February 1991. Polar Biol 14:331-342

Gonzáles HE, Ortiz VC, Sobarzo M (2000) The role of faecal material in the particulate organic carbon flux in the Northern HUmboldt Current, Chile (230 S), before and during the 1997-1998 El Nino. J Plankton

Res 22:499-529

González HE, Smetacek V (1994) The possible role of the cyclopoid copepod Oithona in retarding vertical flux of zooplankton faecal material. Mar Ecol Prog Ser 113:233-246

Hansen B, Christiansen S, Pedersen, G (1995). Plankton dynamics in the marginal ice zone of the central Barents Sea during spring: Carbon flow and structure of the grazer food chain. Polar Biol 16:115-128

Hargrave BT, Taguchi S (1978) Origin of deposited material in a marine bay. J Fish Res Bd Can 35:1604-1613

Heiskanen A-S (1998) Factors governing sedimentation and pelagic nutrient cycles in the Northern Baltic Sea. Monogr Boreal Env Res 8:7-80

Huisman J, Oostveen Pv, Weissing FJ (1999) Critical depth and critical turbulence: Two different mechanisms for the development of phytoplankton blooms. Limnol Oceanogr 44:1781-1787

Jackson G (1990) A model of the formation of marine algal flocs by physical coagulation processes. Deep-Sea Res 37:1197-1211

Kiørboe T, Andersen KP, Dam HG (1990) Coagulation efficiency and aggregate formation in marine phytoplankton. Mar Biol 107: 235-245

Kerr RC, Kuiper GS (1997) Particles settling through a diffusive-type staircase in the ocean. Deep-Sea Res 44:399-412

MacIntyre S, Alldredge AL, Gotschalk CC (1995) Accumulation of marine snow at density discontinuities in the water column. Mar Ecol Prog Ser 40:449-468

Noji TT, Rey F, Miller LA, Børsheim KY, Urban-Rich J (1999) Fate of biogenic carbon in the upper 200 m of the central Greenland Sea. Deep-Sea Res II 46:1497-1509

Olli K, Wexels Riser C, Wassmann P, Ratkova T, Arashkevich E (2001) Vertical export of biogenic matter, particulate nutrients and mesozooplankton faecal pellets off the NW coast of Galicia. Prog Oceanogr 51:443-466

Pace ML, Knauer GA, Karl DM, Martin JH (1987) Primary production, new production and vertical flux in the eastern Pacific Ocean. Nature 325:803-804

Peinert R (1986) Production, grazing and sedimentation in the Norwegian Coastal Current. In: Skreslet S (ed) The Role of Freshwater Outflow in Coastal Marine Ecosystems. Springer Verlag, Berlin Heidelberg pp 361-374

Reigstad M (2000) Plankton community and vertical flux of biogenic matter in north Nrwegian fjords: Regulating factors, temporal and spatial variations. Ph.D thesis, University of Tromsø

Reigstad M, Wassmann P, Ratkova T, Arashkevich E, Pasternak A, Øygarden S (2000) Comparative spring vertical export of biogenic matter in north Norwegian fjords. Mar Ecol Prog Ser 201:73-89

Smetacek V (1980) Annual cycle of sedimentation in relation to plankton ecology in western Kiel Bight. Ophelia, Suppl 1:65-76

Suess E (1980) Particulate organic carbon flux in the oceans: surface productivity and oxygen utilization. Nature 288:260-263

Svensen C, Kiørboe T (2000) Remote prey detection in Oithona similis: Hydromechanical vs chemical cues. J Plankton Res 22:1155-1166

Tande KS (1991) Calanus in North Norwegian fjords and in the Barents Sea. Pro Mare Symposium on Polar Marine Ecology, Trondheim, Polar Res p 389-407

Verity P, Smetacek V (1996) Organism life cycles, predation, and the structure of marine pelagic ecosystems. Mar Ecol Prog Ser

Viitasalo M, Rosenberg M, Heiskanen A-S, Koski M (1999) Sedimentation of copepod fecal material in the coastal northern Baltic Sea: Where did all the pellets go? Limnol Oceanogr 44:1388-1399

Wassmann P (1990) Relationship between primary and export production in the boreal, coastal zone of the North Atlantic. Limnol Oceanogr 35:464-471

Wassmann P (1991) Dynamics of primary production and sedimentation in shallow fjords and polls of western Norway. Oceanogr Mar Biol Annu Rev 29:87-154

Wassmann P (1998) Retention versus export food chains: Processes controlling sinking loss from marine pelagic systems. Hydrobiologia 363:29-57

Wassmann P, Hansen L, Andreassen IJ, Wexels Riser C, Urban-Rich J (1999) Distribution and sedimentation of faecal pellets on the Nordvestbanken shelf, northern Norway, in 1994. Sarsia 84:239-252

Wassmann P, Ypma JE, Tselepides A (2000) Vertical flux of faecal pellets and microplankton on the shelf of the oligotrohic Cretan Sea (NE Mediterranean Sea). Prog Oceanogr 46:241-248

Wexels Riser C, Wassmann P, Olli K, Arashkevich E (2001) Production, retention and export of zooplank-ton faecal pellets on and off the Iberian shelf, north-west Spain. Prog Oceanogr 51:423-441

Ecosystem Functioning and Biodiversity

C. Heip[1*], A. Brandt[2], J.-P. Gattuso[3], A. Antia[4], W.H. Berger[5],
J. Boissonnas[6], P. Burkill[7], L. d'Ozouville[8], G. Graf[9], G.J. Herndl[10],
J. Patching[11], K. Reise[12], G. Riou[13], R. Simó[14], V. Smetacek[15]
and P. Wassmann[16]

[1]NIOO - CEMO, P.O.Box 140, 4400 AC Yerseke, The Netherlands
[2]Zoologisches Institut und Museum, Martin-Luther-King Platz 3, 20146 Hamburg, Germany
[3]Laboratoire d'Océanographie, BP 28, 06234 Villefranche-sur-mer Cedex, France
[4]Institut für Meereskunde, Düsternbrooker Weg 20, 24105 Kiel, Germany
[5]Scripps Institution of Oceanography, UCLA, San Diego, La Jolla, CA 92093-0215, USA
[6]49 Avenue de l'Escrime, 1150 Brussels, Belgium
[7]Individual Merit Scientist, Plymouth Marine Laboratory, Prospect Place,
Plymouth PL1 3DH, UK
[8]ESF Marine Board, 1 quai de Lezay-Mernésia, 67080 Strasbourg Cedex, France
[9]Universität Rostock, Institut für Aquatische Ökologie, Freiligrathstr. 7/8,
18055 Rostock, Germany
[10]Dept. Biological Oceanography, NIOZ, P.O.Box 59, 1790 AB Den Burg, The Netherlands
[11]National University of Ireland, Marine Ryan Institute, Galway, Ireland
[12]AWI, Wattenmeerstation Sylt, 25992 List, Germany
[13]FREMER, Direction de la Technologie Marine et des Systèmes d'information, BP 70,
29280 Plouzané, France
[14]Institut de Ciències del Mar, Pg. Joan de Borbó s/n, 08039 Barcelona, Spain
[15]Alfred - Wegener - Institut für Polar- und Meeresforschung, Postfach 12 01 61,
27515 Bremerhaven, Germany
[16]Norwegian College of Fishery Science, University of Tromsø, 9037 Tromsø, Norway
* corresponding author (e-mail): heip@cemo.nioo.knaw.nl

Abstract: The task of working group 4 was to examine the relationship between ecosystem functioning and biodiversity in marine systems. The definition of biodiversity used by the working group is the biological variability in ecosystems at the genetic, species and habitat level. The inventory of marine life is much closer to completion in Europe than in many other areas but some geographic areas, such as the Mediterranean Sea, and specific taxa, mostly in the small size range (viruses, bacteria and protists), require more exploration. Also required is a European synthesis, including studies of both horizontal and vertical gradients of biodiversity, as well as the relationship between diversity and environmental data. The most efficient manner to investigate how species impact ecosystem functioning is to understand the role of the relatively few key organisms. These species should be identified and preferentially attract the attention of ecologists and biological oceanographers. Their study would provide a functional understanding that could be used to model and predict the response of marine ecosystems to global environmental change. In addition to global issues, working group 4 also examined a large range of regional and local issues that also

From WEFER G, LAMY F, MANTOURA F (eds), 2003, *Marine Science Frontiers for Europe.* Springer-Verlag Berlin Heidelberg New York Tokyo, pp 289-302

require attention because it is at these scales that most burning societal questions occur. The working group identified a need for additional support to strengthen existing large-scale research infrastructures and establish new ones in the Mediterranean and on the Atlantic coast. These are invaluable tools to investigate the interactions between biodiversity and ecosystem functioning. Finally, the loss of taxonomic expertise due to the declining number of marine systematists should be a matter of great concern to Europe, which must be dealt with urgently.

Description of the scientific domain

Marine ecosystems provide a series of goods and services that are of great importance to mankind (Costanza et al. 1997). These include food, from fisheries and aquaculture, for man and domestic animals; a number of mineral resources including oil, gas, sand and gravel; and ingredients for biotechnology (bioactive chemicals and medical products). Moreover the seas provide a free transport system, a buffering system for climate change, a treatment system for human and animal waste and a sink for pollutants from air and land. People not only exploit the seas but also make use of it for educational and recreational purposes, e.g. water sports, sport fishing, wildlife observation and tourism in general.

Biodiversity is the biological variability in ecosystems at the genetic, species and habitat level. Many marine products are species specific and their preservation is therefore directly based on preserving biodiversity. Sustaining the present functioning of marine ecosystems as well will depend to a large extent on preserving biodiversity. Marine organisms play crucial roles in many biogeochemi-cal processes that sustain the biosphere. The rate and efficiency of any of the processes that marine organisms mediate, as well as the range of goods and services that they provide, are determined by interactions between organisms, and between organisms and their environment, and therefore by biodiversity (Heip et al. 1998).

The effect of biodiversity on ecosystem functioning has become a major focus in ecology (Naeem et al. 1994; Boucher 1997; Naeem and Shibing 1997; Huston et al. 2000; Naeem et al. 2000). The insurance hypothesis (e. g. Yachi and Loreau 1999) is a fundamental principle for understanding the long-term effects of biodiversity on ecosystem processes. High biodiversity insures ecosystems against decline in functioning because the more species an ecosystem carries, the greater the guarantee that some species will maintain functioning in the absence of others. The strength of this insurance, and the relationship between biodiversity and ecosystem functioning in general have not been quantified in the marine realm yet (Boucher 1997). We are at present unable to predict the consequences of changing ecosystem functioning as well as the loss of biodiversity resulting from environmental change in ecological, economic or societal terms.

Most ecological theories are based on experience from terrestrial ecosystems (Tilman et al. 1996, 1997) but marine biodiversity does not necessarily comply with terrestrial paradigms. Our understanding of the role and regulation of marine biodiversity lags far behind that of terrestrial biodiversity, to such an extent that we do not have enough scientific information to underpin management issues such as conservation and the sustainable use of marine resources.

The common perception that marine biodiversity is not threatened or far less threatened than terrestrial biodiversity is unfounded. In coastal areas there are many threats of loss and degradation of bio-diversity, but also in deeper water exploitation, mainly by deep-water trawling, is increasing. Direct threats include overexploitation of species, introduction of exotics including toxic species, fragmentation and loss of natural habitats, the impact of coastal aqua-culture, pollution, and destruction of the sedimentary systems through fishing, mining, dredging and dumping. Indirect threats include the development of rivers and the coastline for industrial development, tourism and residential purposes and the many disturbances linked to leisure activities. Socio-economic factors are the difficult economic situation in many countries,.the weakness of legal systems and institutions, the absence of adequate scientific knowledge and ineffective dissemination of what information exists to the general public.

Economic sectors in Europe that are directly concerned with marine exploitation are the oil and gas industry, shipping, fisheries and aquaculture, some pharmaceutical industries and, perhaps economically the most important, tourism. Tourism also directly depends on a healthy biodiverse marine system.

Key questions 1: What is the biodiversity of European Seas?

A number of recent initiatives (European Register of Marine Species, Natura 2000, the Census of Marine Life) have produced, or will produce in the near future, lists of the known species in Europe. This information will over the coming years become accessible in data bases that also contain additional information, e.g. on environmental variables (OBIS, GBIF). For most European marine areas, the inventory of marine life is much closer to completion than for any other area in the world, although some geographic areas such as the Eastern Mediterranean have been less well explored and still need some basic inventory. What is still lacking is the synthesis of such inventories and the link with environmental data to describe biogeographical patterns and their change on the European scale.

The inventories are also more complete for higher than for lower marine phyla, but gaps exist also for higher phyla. As an example, during the 1986 North Sea Benthos Survey of the ICES, about 40 % of all benthic copepod species found were new to science (Huys et al. 1992). The numerous protist phyla remain poorly known almost everywhere, also in Europe. A complete inventory of microbes and viruses using molecular techniques does not seem to be within reach in the next five years because of the enormous number of species or phylotypes present. The genetic information of new microbial species and phylotypes are, however, added to the existing gene banks at a rapidly growing rate. Also special habitats continue to yield new discoveries: examples are the marine cave fauna in the Mediterranean, the shallow hydrothermal seeps in Greece, and gas seeps in the North Sea.

Taxonomy is essential in biodiversity studies and species inventories are basic tools in applied areas such as fisheries, nature conservation and environmental impact studies. Inventories of species and especially communities are also needed for adequate assessment of the changes in biodiversity following different scenarios of global change (changes in surface currents, surface and deep-water temperature, productivity etc.). There are at present no adequate assessments comparable to that of terrestrial environments (Sala et al. 2000). Accurate identification and recognition of species remains a fundamental underpinning of biodiversity research, both basic and applied. For the small-sized taxa new efforts are required. For the larger-sized taxa (meiofauna to megafauna), the taxonomic keys and identification literature are mostly old. They must be improved, updated, standardised and biogeographical information must be included. On the basis of these data sets, atlases should be established and all data should entered in electronic databases for future updates and for a better co-ordination of the national efforts in surveying and monitoring the marine environment.

There is a long and strong tradition of research in systematics in Europe, and this field of science is undergoing a rapid and revolutionary change with the advent of molecular techniques. Many of the systematic relationships in the animal and plant kingdom are being re-examined and in this exercise there is a special interest in the oceans since the diversity of living organisms at higher taxonomic levels is much higher in the oceans than on land. Molecular techniques have not only revolutionised systematics they have also opened new possibilities for identifying genetic variability and distinguishing species (e.g. cryptic or sibling species). We are now in need of molecular markers and probes for real-time monitoring of biogeochemical functions, such as the nif gene, which can be identified in order to determine the functional role of biodiversity of these ecosystem components.

Whereas the flora and fauna of Europe are reasonably well known, large scale patterns in the distribution of plant and animal species and ecological communities are far less well known. This is because most effort in the hundreds of years of classical biological exploration and research has been local and restricted in time. Gradients in space (longitudinally, latitudinally, and vertically) and time (from geological to more recent time scales) influence biodiversity considerably. Studies of gradients should be supported by putting together data from long-term local and regional monitoring programs.

The identification of key species (see below) at large geographical scales is required for the monitoring of the biodiversity of the European Seas, especially for higher taxonomic categories. Identification of biodiversity hotspots, considering different ecotones and climatic regions, will permit a rational approach to maximising conservation at minimum expense. A typical hotspot example is found in frontal areas both for pelagic and shallow-water benthic organism and in areas where biogeographical provinces meet. Areas of high genetic and species diversity should be identified because they are the prime targets for conservation.

Key questions 2: What is the functional role of biodiversity?

Key organisms and their functional role

The overall functioning of an ecosystem is often impacted by a few key species that exert a major impact on overall metabolic processes such as primary and secondary production, remineralisation, vertical export and bioturbation. This impact can be disproportionate to their abundance or biomass.

Key species exist in the pelagic and benthic realms. The characteristics of these habitats differ fundamentally. The pelagic habitat is renewed continuously by advection and mixing whereas the benthic habitat is more heterogeneous but stable for longer periods. Hence dominant organisms in the benthos tend to be competitive space-holders whereas dominant planktonic species either have high growth rates, low mortality rates or life cycle strategies that enable them to maintain populations in a given locality. Consequently, the attributes of key species in both habitats are very different.

The various successional stages in the plankton tend to be dominated by a few species or genera: the diatoms *Skeletonema*, *Chaetoceros*, *Thalassiosira* and the haptophytes *Phaeocystis* and

Emiliania in the early or late spring bloom and the dinoflagellates *Prorocentrum* and *Ceratium* in summer and fall. Dedicated studies on growth rates and life cycles of *Phaeocystis* (because of its nuisance status) and *Emiliania* (Fig.1) (because it is a major calcifier) have yielded much interesting information on pelagic ecosystem functioning. This type of autecological study is recommended for other dominant species as well. This also applies to zoo-plankton species such as *Calanus helgolandicus*, *C. finmarchicus* and *C. glacialis* that shape the flow patterns of biogenic matter along the European margin by extensive grazing on larger phyto-plankton cells or larger protozooplankton, faecal pellet production and fuel pelagic fisheries. The link between benthic and pelagic systems is often strongly determined by filter feeders in the benthos. In the benthos, organisms such as *Mytilus gallo-provincialis*, *M. edulis* or scallops such as *Pecten maximus* or *Chlamys islandica* biodeposit massive amounts of biogenic matter and dominate the transfor of carbon and therefore the carbon flow in benthic systems. On rocky bottoms a wide diversity of filter feeders (sponges, gorgonians, bryozoans, crustaceans and others) exists as well.

Fig. 1. Coccolithophorid blooms in the North Atlantic and along the coasts of UK and Ireland (18 May 1998). Photo credit: Provided by the SeaWiFS Project, NASA/ Goddard Space Flight Center and ORBIMAGE.

The diversity of soft-bottom communities is strongly influenced by ecosystem engineers who manipulate the sea floor and create microhabitats for meiofauna, Protozoa and bacteria. Such ecosystem engineers may be either plants (e.g. seagrasses) or animals. Because of this more indirect interaction, food web links are less well understood than in the pelagic environment. Macroalgae and seagrasses shape a microenvironment around their rhizoids or roots, and large animals determine the thickness of the bioturbated sediment layer in which the major decomposition of organic matter occurs. Although animals may respire an important part of the organic matter in sediments (Heip et al. in press), the bulk remineralisation is carried out by microorganisms. The functional role of animals is to increase fluxes and to mix or stabilise the sediments. Thus, key species of the benthic system that have to be studied are those which provide structure (reef builders, plants) and those that are the major bioturbators (e.g. builders of permanent burrows) and provide biodeposition and resuspension, bio-irrigation and sediment mixing. Since the diversity of soft bottom communities is much higher than the diversity of pelagic communities, and varies significantly along the European coasts, it is not possible to name a few species. Among the seagrasses, *Zostera* and *Posidonia* are the most obvious key organisms. Among the animals in many coastal waters large crustaceans such as Thalassionoidea (Fig. 2, *Callianassa subterranea*), or large poly-chaetes (*Arenicola marina*, *Sabella pavonina* and *Lanice conchilega*) and echiuroids will be major bioturbators, i.e. key species.

Hard bottom communities, which exist mainly in the littoral zone, are more affected by competition for space and by keystone predators and in this way perhaps resemble more terrestrial systems than other marine systems. Sedentary organisms such as the various sea weeds (greens, browns and reds), sponges, gorgonians, corals, bryozoans, bivalves and so on form what must be the most highly diverse communities in Europe. Moreover they are highly attractive to humans, at least during summer time, when they are visited by countless numbers of snorkelers, spearhunters and divers. Some of the basic ecological theory on species interactions has been developed from studies

Fig. 2. *Callianassa subterranea* in its burrow. Photo credit: G. Graf, University Rostock, Germany

in the intertidal of rocky shores and the good knowledge that exists on these communities, both in the Atlantic and the Mediterranean, should be exploited further to strenghten the theoretical base of the regulation of community composition.

Marine vertebrates are often ignored by general oceanographers and ecosystem ecologists focusing on ecosystem processes, as they usually contribute very little to the overall energy flow. But they as well are often key-species that may structure marine food webs. This function is not well known. Marine fish, birds and mammals are studied by fisheries biologists, ornithologists or marine conservation scientists, and a lot of information exists on many of the perhaps several hundred species of importance in Europe.

In summary, the essential characteristics of a large number of ecosystems are shaped by a relatively small number of key species that preferentially should attract the attention of ecologists and biological oceanographers. A concerted European action to understand the role of these relatively few key organisms rather than local investigations of the plethora of organisms that are present, is the most efficient manner to study how species impact ecosystem functioning. Such information will provide a functional understanding of biodiversity and species composition that can be used to model and predict the response of biogeochemical processes to global environmental change (see Antia et al. this volume).

What is the role of redundancy and how is it regulated?

Diversity is related to stability, i.e. the persistence of an ecosystem. Whereas genetic diversity may stabilise a population, additional functional diversity stabilises the ecosystem. The main question in this context is the role of redundant species, which take over certain functions in the case of species loss due to reduction or succession. Are weak links in the food web the major reason for stability and what is the importance of co-dependency? The overall question is how many species and functional groups are needed to sustain a stable ecosystem.

Fig. 3. Controlled laboratory experiment with a chemostat culture of the coccolithphorid *Emiliania huxleyi.* Photo credit: J.-P. Gattuso.

Microbes and viruses

Molecular techniques now allow the characterisation of non-cultivatable, morphologically indistinguishable microbes (viruses, bacteria, archaea and protists) (Fig.3) (Caron et al. 1999; Massana et al. 1997; Moeseneder et al. 2001). This molecular approach has already revealed the enormous richness of bacteria, the ubiquitous presence of archaea and the phylogenetic heterogeneity of pico-eukaryotes in the ocean (Giovannoni and Rappé 2000).

The major scientific challenges for the near future are:
i- to determine the key "species" among the bacteria and archaea focusing on the abundance of these key species and their role in biogeochemical cycling. Thus, phylogenetic probes should be coupled with functional probes targeted towards key metabolic activities. Of particular importance might be species consortia, i.e. species meta-

bolically interacting and thereby driving biogeochemical processes.
ii- The microbial environment needs to be investigated at the appropriate microbial scale. There is evidence that microbes are structuring their microenvironment leading to a larger spatial heterogeneity than we anticipated in the past. While there is considerable information already available for sediments, there is only circumstantial evidence that this might also be true for the pelagic realm (Azam 1998; Blackburn et al. 1998).
iii- The role of microbial-derived substances controlling intra- and interspecific interaction with other organisms might be a research area with a high potential for biotechnological applications.

The critical lack of information on the role of viruses (Fig.4) in regulating microbial populations needs to be addressed and virus-host systems established for the key species. This has the poten-

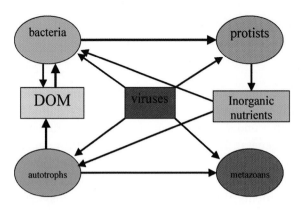

Fig.4. The interactions between the different compartments of the microbial food web (phytoplankton, bacteira, protists, phytoplankton, viruses) and the non-living organic (DOM-dissolved organic matter) and inorganic matter (inorganic nutrients).

tial to reconstruct the occurrence of specific organisms in the past if sediment cores are analysed for the presence of the specific virus. Together with sediment dating this allows retrospective analysis of events of specific interest for humans such as harmful blooms, even if the particular species does not leave any marker in the sediment.

These research directions should result in a better understanding of the temporal and spatial variability of microbial population dynamics and on the regulatory mechanisms of microbial species distribution.

Key questions 3: How will global and local environmental changes affect species composition and its role in ecosystem function and services?

The response of marine ecosystems to global and local environmental changes and the consequences of changes in marine ecosystems on climate and human societies along the coastlines of the world are major issues faced by marine science. Numerous biotic and abiotic factors control, often interacting, the structure and function of marine ecosystems. (Gattuso and Buddemeier in press). Natural and human forcing affect most of them and there is no doubt that the properties of marine eco-

systems will be altered, with consequences through feedback processes on climate patterns and human societies. These alterations are poorly understood; yet management and policy initiatives are critically required either to prevent or minimize them, or to enable adaptation of human society if the expected consequences appear inevitable.

Global changes

Shifts in biogeographical regions have been observed both horizontally and vertically (physical settings such as temperature, light and ocean circulation). Shifts in dominant species such as blooms, for example of toxic algae, and invasions may occur. Recent data demonstrate that community composition has already changed in various geographical areas (e.g. the Californian coast and the North Atlantic) in response to changes in physical features such as temperature. It is likely that this will continue and perhaps be amplified in the foreseeable future. It is essential to continue long-term monitoring because as long as seasonal, annual and decadal variability remain unknown, it is extremely difficult to assess anthropogenic changes. The key organisms identified in a previous section should be added to the monitoring programs when needed. It must also be determined how these changes in community composition translate to changes in the food-web structure, on species of commercial interest, on the flux of matter to the deep ocean and on the size of the reservoirs of particulate and dissolved organic matter.

In addition to changing the community composition, global environmental changes can affect physiological performances. Parameters other than physical ones must also be investigated. For example, the considerable change in the seawater carbonate chemistry due to elevated pCO_2 is known to be detrimental to calcium carbonate production. It is therefore possible that coccolithophorids and other calcifying species will be partly replaced by non-calcifying species. This may have severe consequences on the carbon cycle because the ratio of inorganic:organic downward flux of matter (rain ratio) is one the factor controlling atmospheric pCO_2. It seems that primary production is generally not CO_2-limited because primary producers can use the very large reservoir of bicarbonate to accomplish inorganic carbon fixation.

Finally, the data collected must be used to establish and refine models to predict the response of community composition and of the biogeochemi-cal cycling of elements. Prediction of the response of ocean biology to climate and environmental changes and its feedback on future climate is hampered by serious limitations related to the lack of mechanistic understanding of many interactions and by the very modest database upon which models and predictions are based. One of these limitations relates to the response of marine organisms to environmental changes. This has mostly been investigated during short-term experiments in the laboratory. It remains to be demonstrated that acclimation and adaptation will not mitigate acute physiological responses measured in the short-term. Also, physiological responses have often been investigated by manipulating one parameter at a time. There is a pressing need to manipulate simultaneously the various parameters of interest in order to determine their interactive effects.

The biogeochemical significance of physiological response is also an issue that has received relatively little attention (see Antia et al. this volume). Preliminary results suggest that reduced nutrient supply and increased light efficiency that result from increased stratification will lead to a small global decrease (-6%) in the magnitude of the biological pump (Bopp et al. in press). Significant regional differences are predicted with a large decrease of export production in the tropics (-15%) due to reduced strength of upwellings, and an increase in the high latitudes, especially the Southern Ocean (+10%), due to a longer growing season. It must be emphasized that the model used to make these predictions does not include possible changes in community composition, strength of the carbonate pump, nor physiological response to changes other than light and phosphorus availability. Both long-term monitoring and experimental approaches would be instrumental to refine models and get more reliable predictions (Antia et al. this volume).

Key questions 4: Regional, local and specific issues of European interest

The study of the interaction between species (biodiversity) and ecosystem functioning is required to better understand how marine ecosystem function in general. It is recognized that this still requires basic research. However, besides lack in fundamental, general knowledge, there are a number of local problems that involve species and ecosystem functioning and for which specific European actions are required. We cannot provide a detailed discussion on such topics in this document, we limit ourselves therefore to a survey of the main issues.

Effects of aquaculture

Although the problems associated with intensive aquaculture such as the salmon industry in Norway and the British Isles or the shellfish industry in the Netherlands and Spain are increasingly being recognized, the ecosystem consequences have not been given proper attention. Most of these industries are in semi-enclosed coastal systems and their operation can have profound impacts on the functioning of those systems. A better scientific evaluation of such ecosystem consequences is required for the sustainable development of aquaculture in Europe.

Effects of fisheries

Fisheries have arguably the largest effect on marine biodiversity and ecosystem functioning besides global climate change. It is now well documented that the bottom trawling in the North Sea and elsewhere has fundamentally changed the characteristics of the sediment ecosystem by causing the disappearance of long-lived species and the increased disturbance of the sediments. Many prized species are top predators, the best example being the tuna species, and their disappearance changes the entire food chain through a process known as a trophic cascade. The short and especially the long-term consequences of changing food chains in the benthic and pelagic domain are not yet properly understood.

Effects of tourism

Beaches as well as rocky shores have great appeal for humans. Beaches and the associated dunes are important for coastal defence, and this function is often in conflict with their use in summer as areas for swimming, walking and sunbathing. The pressure on living resources is mainly indirect through trampling and compacting of the sediments, although some direct exploitation exists as well. Rocky shores are visited by sport fishermen, snorkelers and divers and are directly exploited for edible species,

Coastal eutrophication

The major effects of eutrophication in coastal environments are well known: the appearance of nuisance or toxic algae, changes in the pelagic food web, increased sedimentation resulting eventually in widespread anoxia with massive mortality in the benthos. Such events have been described for coastal lagoons all over Europe where they occur regularly but even in more open areas such as the German Bight of the North Sea or the Skagerrak. The economic consequences are very important as they affect fisheries, aquaculture and tourism.

The causes of eutrophication were the subject of intense political attention in the 1980s and have resulted in European legislation. One of the major objectives of this legislation, the reduction in nutrient inputs by half (for the North Sea), has been achieved for phosphorus but not for nitrogen. The consequent distortion of the nutrient ratios in river water has had negative effects on the phytoplankton. What the long-term consequences of these changes and how they pervade the entire marine food web is not well known.

Waste and CO_2 disposal

The possibility that urban waste and CO_2 will be deposited in future in the deep-sea requires an extensive research programme aimed at evaluation of the consequences of such disposal for deep-sea biodiversity and ecosystem functioning. The deep-sea is a high diversity environment with many long-lived, slow growing species and its resilience and stability has to be charted before any large impacts can be delivered.

Introduced and invading species

Marine alien species are now introduced in Europe at a rate of about one species each week in some areas. A number of recent cases of successful introductions or escapes in Europe has lead to widespread speculation about the effects of such introductions on Europe's marine ecosystems. Yet, there is few scientific evaluation. Many invasions seem not to lead to disappearance of local species, perhaps because biodiversity in Europe is still recovering from the ice ages and many niches are still left unoccupied. A careful evaluation of case studies on important invaders (e.g. the polychaete *Marenzelleria*, the cnidarian *Mnemiopsis* and the green alga *Caulerpa taxifolia*) would help in charting the factors that determine success or failure of invasions.

Pathogens and parasites

There is little known on pathogens and parasites of marine plants and animals other than for species with commercial importance. With the recent outbreaks of disease in cattle in Europe and elsewhere the attention to the rapid spread of pathogens and parasites through human vectors has returned in full force. The problem also exists in the marine environment. Some of the exotics that are transported by man mentioned in the previous paragraph may be pathogens or parasites. Some species may extend their distribution with changing temperature and current patterns. The recent attention to the spread of *Vibrio cholerae* in estuarine waters through the ballast water of ships and the spread of disease associated with aquaculture require new methodologies of rapid detection of pathogens and parasites.

Reintroduction of lost species

From historical reconstructions it has become evident that a number of 'charismatic', large species that were once present along Europe's coasts have become extinct in certain areas or have disappeared altogether. Spectacular examples are the findings of remains of grey whales and pelicans in the Wadden Sea. The near disappearance of monk seals and sea turtles in the Mediterranean and the sturgeon in Western European rivers are other examples. Programmes of reintroduction have sometimes been started but a historical reconstruction of the vertebrate fauna of Europe on which such reintroductions should be based has not been attempted yet.

Bio-remediation and eco-engineering

The recovery of marine ecosystems from impacts such as oil spills and the possibility of creating new habitats and increasing biodiversity through carefully planned construction works require a good understanding of the basic processes that generate and maintain marine biodiversity. Examples of such constructions are artificial islands and reefs, windmill parks, oil rigs, dams and jetties, marinas etc.

Requirements for implementing the scientific challenge

All systems are now responding in some way to environmental change induced by human activities. Marine systems are characterised by processes occurring over a wide range of scales in space and time and all these processes are now being modified, possibly even the global circulation patterns. The major challenges in topic 4 are to understand how the performance of marine species and communities, food webs, their productivity and biodiversity, will be modified by the globally changing climate on which a multitude of local anthropogenic impacts are superimposed, and, vice versa, what the consequences of expected changes in biodi-versity are for the functioning of ecosystems in the oceans. For this we need to better exploit existing data and theory, for instance in assessment of scenarios of future change, as well as new knowledge especially at scales that are not normally covered within present research strategies.

How can we capture past and present changes and predict future ones? The key to track and understand changes is carefully planned observations and experiments. Results from observations and experiments must fuel the models we need for prediction. Implementation of theme 4 thus involves a combination of long-term monitoring, analysis of large scale patterns, manipulative field experiments and adequate laboratory facilities, and a major effort in modelling and data assimilation.

All this will not be achieved without concerted efforts involving agreement among scientists on the priorities of research in the coming years, willingness of scientists to participate in multidisplinary research, willingness of funding agencies to support long-term and therefore costly projects, cooperation between academic institutes and governmental and international agencies, involvement of private industry engaged in the exploitation of marine resources etc. The role of the different marine platforms in Europe for promoting and guiding marine research in general will be essential if these goals are to be attained.

Research infrastructures

Inventorying and explaining changes in biodiversity must be linked to ecosystem properties. Investigating the relationships between biodiversity and ecosystem functioning as well as their response to environmental change requires both field and controlled experiments in the laboratory (Fig.5) and in mesocosms. Field experiments need the adequate support of research vessels. There is a considerable need to expand the available laboratory and mesocosm facilities. As pointed out above, the interaction between the numerous changing environmental parameters need to be taken into account. This requires sophisticated controlling systems as well as a large number of tanks and mesocosms in order to adequately replicate the experimental conditions.

The facility required is similar to the Ecotrons developed for investigating terrestrial populations and ecosystems (Lawton et al. 1993). It could be a European Large Scale Facility dedicated to the study of the response of marine organisms and communities to environmental changes. It should have an Atlantic branch as well as a Mediterranean branch.

The infrastructure required to study long-term and large-scale patterns in European seas can be

Fig. 5. Bacteria and viruses stained with SYBRGreen. The sample was taken from the Bay of Villefranche-sur-mer. Large, bright dots are bacteria and smaller, dimmer dots are viruses. Photo credit: M. Weinbauer.

Long-term monitoring of ecosystem functioning and biodiversity in the European seas on an adequate spatial scale is a challenge not yet properly solved. It requires the identification of adequate indicators, determination of sampling frequencies, standardization of methodologies etc. These methodologies should be cost effective and will involve the deployment of automatic measurement devices on permanent platforms (buoys, oil rigs, dedicated research vessels or ships of opportunity), remote sensing from planes and satellites. It is a major challenge for the scientific community to select the indicators, to define the needs for instrumentation, and to support the development of the technology required.

Human resources and interdisciplinarity

Research on biodiversity requires the involvement of systematists. Systematics is a science that is practiced at many universities and a number of large museums in Europe. The communication between systematics and ecology should be encouraged and facilitated. Recent workshops (MARS-Museums, Plymouth 1992; CoML Crete 2000) have discussed this problem in detail and their recommendations should be considered seriously and implemented where feasible.

Europe has a strong tradition in taxonomy but considerable expertise will be lost soon. Therefore, we face a pressing need to preserve and strengthen this expertise by education as well as by building bridges with the larger number of taxonomic experts available in Eastern Europe. Interdisciplinarity is of course essential; the research programmes must involve taxonomists, physiologists, biogeo-chemists and modelers.

An important biodiversity and food web issue concerns fisheries. Most fisheries research is performed at government laboratories that are somewhat isolated from mainstream academic research. Fisheries as well as conservation biology are species based and concerned with applied aspects. Initiatives to promote the interaction between systems ecologists and biologists studying the ecology and evolutionary biology of invertebrate and vertebrate species will almost certainly improve understanding of the interactions between

provided by the network of marine stations and governmental institutes that covers large parts of Europe's coasts. Many of the marine stations are now linked in the Network of European Marine Research Stations MARS. These stations again have links with most university laboratories engaged in marine education and research in Europe.

Forward looking technology

Technological breakthroughs are required on the very small and the very large scale. Several methodology and technological developments need to be used to advance our understanding of microbial diversity and functioning. For micro-organisms, fluorescent *in situ* hybridisation (FISH) linked to auto-radiography has been recently applied and should be more widely used to link biodiversity (richness and evenness) with function. These single cell approaches should be combined with already widely applied fingerprinting methods such as DGGE and T-RFLP and sequencing. Automated "species" identification and the use of *in situ* detection systems are currently being planned and it is expected that these efforts will contribute to a higher spatial and temporal resolution of microbial population dynamics and function. Microsensor technology also advances rapidly which enables us to better characterise the microenvironment of the microbes.

biodiversity and ecosystem functioning as well as of the important problem of predicting adaptation of marine species to their changing environment.

The main key species in the coastal marine environment is man. One cannot escape from the rationale that understanding changes in the biodiversity and functioning of coastal ecosystems will at some stage have to include human sociology, economics and behaviour. The tendency in the natural sciences to consider human activities as forcing functions that are external to the system studied, and the tendency in sociology to more or less ignore environmental issues and species other than man as being irrelevant to human behaviour, are both things from a monodisciplinary past. To resolve the difficulties of dialogue, case studies may be the way foward.

Predictive models must be developed to describe the future changes in, for example, seawater temperature, CO_2 partial pressure, water circulation and stratification. To these prediction of ecological components should be coupled. This requires a platform where physical oceanographers, biogeo-chemists and ecologists could interact.

Besides making better use of the expertise available in Europe there is also the challenge to recruit good students to the different fields of marine research. In many countries in Europe marine science is suffering from diminishing funds. Marine biodiversity is not a subject of serious attention in many science programmes yet, but the tide is turning somewhat.

Financial resources and long-term commitment

The major obstacles for implementing research at adequate temporal and spatial scales and the actions required have been discussed by the Marine Board of ESF (Heip and Hummel 2000). It was recognized that in the present funding and institutional situation in Europe large-scale and/or long-term marine biodiversity research is not possible.

The need for long-term and large-scale research to solve some of the major problems in inventorying, explaining and modelling marine biodiversity has been well argued in a series of major scientific discussions in Europe summarized by the Marine Board of ESF. The coordination and networking of the scientific potential in this field in Europe will be attempted in the coming years through a series of actions explained in Heip and Hummel (2000). The core of the effort, as it has been supported through a EC Concerted Action BIOMARE (http://www.biomareweb.org) , will be an agreement on a series of European Marine Biodiversity Research Sites, spread along Europe's coasts, that will be under the responsibility of a dedicated research institute. Two types of research sites are proposed, reference sites which are close to unimpacted and focal sites which most often are subject to clear anthropogenic impact. At some reference sites a complete inventory of marine biodiversity will be attempted whereas others will be based on the existence of long-term time series, often from local monitoring programmes. At focal sites a selection of indicators will be surveyed for long periods of time using a common methodology.

The BIOMARE initiative will be finished in October 2002 and the further implementation and responsibility will be handed over to the European Marine Research Stations Network MARS (http://www.marsnet.org) under the provisional name of the European Marine Biodiversity Initiative (EMBI), which is open to any interested scientist or Institution.

European and societal dimensions

Europe's seas are a genuine European area, both geographically and in terms of science and scientists. Europe has the longest coast relative to its surface of all continents and the coast is an important factor in generating wealth and well-being of Europe's citizens. Europe not only depends on the adequate functioning of its marine ecosystems but also has an obligation to preserve them and the species they contain. Most European countries and the EU are signatories to a number of treaties aiming at protecting the marine environment and its species. However, most efforts of surveying and monitoring the marine environment are based on national programmes, most marine research is still being paid out of national agencies and national interests still dominate policies of exploitation of living and non-living marine resources. All this does not allow for an adequate strategy to deal with the problems of the seas to be developed.

European coordination is required on the following issues:
• Scientific support for the implementation of legally binding conventions (Rio, Barcelona, Ospar) and EU directives (Habitat, Water).
• Coordination of research on large-scale patterns and long-term trends in biodiversity and ecosystem functioning
• Scientific assessment of future changes in biodiversity and ecosystem functioning
• Standardization of methodology
• Support of observation systems (GOOS, Euro Goos, MedGoos)
• Development of (rapid) assessment technology
• Access to data and compilation of data and inventories
• Updating of taxonomic literature, conversion to electronic carriers of taxonomic guides
• Maintaining and improving the scientific potential in marine sciences and its European orientation.
• Promoting public awareness and understanding

Socio-economics

One of the major research goals outlined in this document is to predict how marine biodiversity and ecosystem function will change in the next 100 years. The ultimate objective is to provide the scientific basis to establish policies and directives aimed at either counteracting the predicted changes, or at adapting the European society to them. The use of the scientific basis for policy-making requires several bridges, such as cost-benefits analysis.

Science, Society and Citizens (ethics, public awareness, cooperation with developing countries)

The open oceans have no owner and exploitation of marine resources is not based on any consideration of sustainability or responsible ownership. Many sad examples of poor exploitation practices of marine species exist and continue until this day. Often science has not been able to support the

regulation of marine exploitation well enough to change these practices. This is partly a consequence of the relatively small research community that is engaged in marine research and of its fragmentation (e.g. the gap between fisheries scientists and marine ecologists).

Public awareness of the fragility of marine life and the problems of human impact on the oceans exists and the numerous and often very professional documentaries on marine life that are shown regularly on television do a lot to support this attitude. Biodiversity is a key element in this awareness as the state of the marine environment is often translated in the common perception that 'charismatic' species (whales, dolphins, even sharks) are threatened. People in the affluent western countries enjoy the seas during holidays when especially clear blue waters and spectacular biological habitats (coral reefs, rocky shores) are well appreciated.

This in general positive and informed attitude of the general public in Europe is not reflected in the interest from politicians in most European countries. Human perceptions, attitudes and behaviour directly or indirectly affect the conservation and the sustainable use of biological diversity. Changing these factors requires long-term concerted efforts in education and public awareness in general, and some specifically targeted at sectors like sea professions, coastal city and coastal recreation area managers, water management, both at national and European level.

References

Antia AN, Burkill PH, Balzer W, de Baar HJW, Mantoura RFC, Simó R, Wallace D (2003) Coupled biogeochemical cycling and controlling factors. In: Wefer G, Lamy F, Mantoura F (eds) Marine Science Frontiers for Europe. Springer Berlin pp 147-162

Azam F (1998) Microbial control of oceanic carbon flux: The plot thickens. Science 280:694-696

Blackburn N, Fenchel T, Mitchell J (1998) Microscale nutrient patches in planktonic habitats shown by chemotactic bacteria. Science 282:2254-2256

Bopp L, Monfray P, Aumont O, Dufresne J-L, Le Treut H, Madec G, Terray L and Orr JC (in press) Potential impact of climate change on marine export production. Glob Biogeochem Cycl

Boucher G (1997) Species diversity and ecosystem function: A review of hypothesis and research perpectives in marine ecology. Vie et Milieu 47(4):307-316

Caron DA, Gast RJ, Lim EL, Dennett MR (1999) Protistan community structure: molecular approaches for answering ecological questions. Hydrobiologia 401: 215-227

Costanza R, d'Arge R, de Groot R, Farber S, Grasso M, Hannon B, Limburg K, Naeem S, O'Neill RV, Paruelo J, Raskin RG, Sutton P and van den Belt M (1997) The value of the world's ecosystem services and natural capital. Nature 387:253-259

Gattuso J-P and Buddemeier RW (in press) Coral reefs: An ecosystem subject to multiple environmental threats. In: Mooney HA and Canadell J (eds) The Earth System: Biological and Ecological Dimensions of Global Environmental Change. New York Wiley

Giovannoni S, Rappé M (2000) Evolution, diversity, and molecular ecology of marine prokaryotes. In: Kirch-man DL (ed) Microbial Ecology of the Oceans. Wiley-Liss, New York pp 47-84

Heip C, Warwick R and d'Ozouville L (1998) A European Science Plan on Marine Biodiversity. European Marine and Polar Science Board Position Paper 2

Heip C and Hummel H (2000) Establishing a framework for the implementation of marine biodiversity research in Europe. EMAPS Position Paper. 48 pp

Heip C, Duineveld G, Flach E, Graf G, Helder W, Herman PMJ, Middelburg JJ, Lavaleye M, Pfannkuche O, Soetaert K, Soltwedel T, de Stigter H, Thomsen L, Vanaverbeke J and de Wilde P (in press) The role of the benthic biota in sedimentary metabolism and sediment-water exchange processes in the Goban Spur area (N.E. Atlantic). Deep-Sea Res

Huys R, Herman PMJ, Heip CHR and Soetaert K (1992) The meiobenthos of the North Sea: Density, biomass trends and distribution of copepod communities. ICES J Mar Sci 49:23-44

Huston MA, Aarssen LW, Austin MP, Cade BS, Fridley JD, Garnier E, Grime JP, Hodgson J, Lauenroth WK, Thompson K, Vandermeer JH and Wardle DA (2000) No consistent effect of plant diversity on productivity. Science 289(5483):1255

Lawton JH, Naeem S, Woodfin RM, Brown VK, Gange A, Godfray HJC, Heads PA, Lawler S, Magda D, Thompson LJ and Young S (1993) The Ecotron: A controlled environmental facility for the investigation of population and ecosystem processes. Philosophical Transactions of the Royal Society of London Series B 341:81-194

Massana R, Murray AE, Preston CM, DeLong EF (1997) Vertical distribution and phylogenetic characterization of marine planktonic Archaea in the Santa Barbara Channel. Appl Environ Microbiol 63:50-56

Moeseneder MM, Winter C, Herndl GJ (2001) Horizontal and vertical complexity of attached and free-living bacteria of the eastern Mediterranean Sea determined by 16S rDNA and 16S rRNA fingerprints. Limnol Oceanogr 46:95-107

Naeem S, Hahn DR und Schuurman G (2000) Producer-decomposer co-dependency influences biodiversity effects. Nature 403(6771):762-764

Naeem S and Shibin LI (1997) Biodiversity enhances ecosystem reliability. Nature 390:507-509

Naeem S, Thompson LJ, Lawler SP, Lawton JH and Woodfin RM (1994) Declining biodiversity can alter the performance of ecosystems. Nature 368:734-737

Sala OE, Chapin FSI, Armesto JJ, Berlow E, Bloomfield J, Dirzo R, Huber-Sanwald E, Huenneke LF, Jackson RB, Kinzig A, Leemans R, Lodge DM, Mooney HA, Oesterheld M, Poff NL, Sykes MT, Walker BH, Walker M and Wall DH (2000) Global biodiversity scenarios for the year 2100. Science 287(5459):1770-1774

Tilman D, Wedin D, Knops J (1996) Productivity and sustainability influenced by biodiversity in grassland ecosystems. Nature 379:718-720

Tilman D, Lehman CL, Thomson KT (1997) Plant diversity and ecosystem productivity: Theoretical considerations. Proc Natl Acad Sci 94:1857-1861

Yachi S and Loreau M (1999) Biodiversity and ecosystem productivity in a fluctuating environment: The insurance hypothesis. Proceedings of the National Academy of Science USA 96(4):1463-1468